Dietrich Sahrhage
Johannes Lundbeck

A HISTORY OF *Fishing*

With 133 Figures

Springer-Verlag
Berlin Heidelberg New York
London Paris Tokyo
Hong Kong Barcelona
Budapest

Prof. Dr. Dietrich Sahrhage
Prof. Dr. Johannes Lundbeck (Deceased 1974)

Both formerly Directors
Institut für Seefischerei
Bundesforschungsanstalt für Fischerei
Palmaille 9
W-2000 Hamburg 50, FRG

Present address

Prof. Dr. Dietrich Sahrhage
Waitzstraße 48
W-2000 Hamburg 52, FRG

ISBN-13: 978-3-642-77413-3 e-ISBN-13: 978-3-642-77411-9
DOI: 10.1007/978-3-642-77411-9

Softcover reprint of the hardcover 1st edition 1992

31/3145-5 4 3 2 1 0 – Printed on acid-free paper

Preface

From 1972 to 1979 Johannes Lundbeck published an extensive manuscript in German, entitled *Die Fischerei von den Naturvölkern bis zur modernen Technik und Wirtschaft* (The fishery from primitive people until modern techniques and economies) in the *Mitteilungen aus dem Institut für Seefischerei*, Hamburg. Of the seven volumes with almost 950 pages, four volumes were issued after Lundbeck died in 1974. Unfortunately, the publication could be produced only in mimeographed form. This, and the presentation in German only, limited its international distribution.

Lundbeck's inspiring work provided a strong impetus to the preparation of this book. Parts of his publication, concerned with historical aspects, were used as a basis for the English text. Furthermore, his contribution was substantially amended and expanded to cover in more detail the development of fisheries until the present time and to evaluate the perspectives for the future.

Other parts of Lundbeck's publication in the fields of biology, ecology, fishing gear and methods and fishing vessels are not considered here, since they deal with aspects which have meanwhile been presented by other authors in extensive reviews.

It is my pleasure to acknowledge gratefully the contribution of the late Johannes Lundbeck to this book. I thank Mrs. Elsbeth Lundbeck, his daughter and assigner of the related copyright, for her kind permission to make use of the manuscript.

Hamburg, September 1991 D. Sahrhage

Acknowledgements

The authors are indebted to many persons for useful comments, encouraging discussions and other assistance. Special thanks are due to P.Dr.P.Becker (Trier), Prof.A.Ben-Tuvia (Jerusalem), Dr.P.Breunig (Frankfurt/M), Prof.R.J.H.Beverton (Langstone), Dr.W.Fischer (Rome), Dr.H.-J.Heinrich (Hildesheim), J.Krohn (Hamburg), Dr.M.Riedel (Köln) and P.B. Sandner (Maria Laach). The Office for Information and Documentation of the Federal Research Centre for Fisheries in Hamburg and its library was very helpful during the preparation of the manuscript. In particular, the kind assistance of Dr.W.P.Kirchner, the leader of this office, is gratefully acknowledged. Many institutions, publishers and authors kindly granted permission to reproduce figures from earlier publications. The authors wish to express their sincere appreciation for all help received.

Contents

Introduction

Books on historical events are mostly concerned with politics, dynasties and wars. Less often they describe the development of such basic sectors like food production, related industries and the role of people in this regard. Yet, these developments also played a decisive part in human society and culture. They influenced the daily life and the well-being of many generations, as well as misery and famine suffered by them. Conflicts and, all too often, bloody wars arose from competition for food resources and their economic utilization. This holds true not only for agriculture but also for fishing.

Under these premises it appeared a worthwhile task to write a concise overall account of the history of fishing from earliest times until the present, addressed not only to specialists in fishing but also to a wider circle of readers interested in this subject. The authors realize that it would be too ambitious to try to cover fully all aspects and all countries. But an attempt was made to describe the general trends in fisheries development in relation to the basic natural conditions in various parts of the world, the techniques applied, and the economic and social problems involved.

The authors could base their work on numerous publications containing relevant information. In the interest of better readability no bibliographic references are included in the text; however, a selected list for each Chapter is given in the bibliography at the end of this book. Historical aspects of fisheries development are usually rather scattered through the literature and are not always easy to find. This book will help to facilitate access to publications in German.

Reference is made to some publications of special interest as they are closely related to the scope of this book and provide much supplementary information: a book by Cutting (1955) concentrates mainly on the history of fish processing but also contains information on the development of certain fisheries. Kreuzer (1974) considered the place of fish in culture with special reference to the role of fish in myths, legends and lore, art, religion, and the diet of man. A review of fish catching methods of the world was published by von Brandt (1984). The recent book by Cushing (1988) describes the history of a number of fisheries and whaling activities and evaluates the development of fisheries science and related institutions particularly with regard to resources management.

Since the early days of mankind fish and other aquatic organisms have played a most important role as food and as a source of materials for housing, clothing, tools and other technical purposes. In contrast to hunting, fishing has, since its inception in prehistoric times, always retained its importance in food production. Starting from very low levels, this production has today reached 100 million metric tons (t) per year. Fish and fishery products currently provide around 16% of the total supply of animal protein in the world. If the indirect contribution from fish meal, used as animal feed, is included, fisheries produce about 24% of the animal protein consumed by man. Today, about 15 million people in all parts of the world are directly engaged in professional fishing, and the number of people indirectly dependant on this sector is many times larger.

It is amazing that this enormous development has taken place although the methods of fishing have not, in principle, changed much over thousands of years. Most gear, like hook and lines, fish fences, traps and baskets, and various types of fixed and towed nets, are based on primeval types of fishing instruments, but the efficiency has been improved over long periods with increasing experience. It is also fascinating to see that similar fishing gear, fishing methods and, also, artistic portrayal of fish have developed almost in parallel in quite different parts of the world. Certainly much of the knowledge was transferred from one area to another with increasing communication between cultures and people. However, there is no doubt that during early times many analogous evolutions took place quite independently, for instance in Europe and South America. It must be the common way of human thinking and inspiration, as well as the kinds of material usually available for the production of fishing tools and artistic objects, which is responsible for such coincidence.

A major contribution was the construction of seaworthy ships and their adaptation to fishing purposes. This development allowed the gradual expansion of fishing from rivers and nearshore waters to the open seas. The most revolutionary step was the replacement of sailing vessels by ships with steam engines and, just a little later, ships with motors. Mechanization also led to improvements in the treatment of catches and to progress in product development. At the same time the building of central fishing ports and railroads resulted in better marketing and distribution systems. New fishing vessels became bigger and more powerful to tow larger nets which were technically improved to increase their efficiency. Production rose further with the introduction of the most effective mid-water trawls and purse seines, and with the application of echosounding techniques for the location of fish and other animals.

Good catches and high economic returns, based on rapidly growing demand for food and for fish meal and oil, provided great incentives to build more ships. Thus, fishing fleets increased until they reached a present size of more than 20 000 ships larger than 100 GRT (gross register tons) with a total of over 9 million GRT.

It soon became apparent that the living resources were by no means sufficient to bear the heavy overexploitation caused by the fleets. In many parts of the world one commercial fish stock after another was depleted. The situation became particularly grave in the North Atlantic; with the depletion of stocks the yields of fish declined in spite of increased efforts and this resulted in economic losses and in social problems. Economic difficulties, rather than scientific advice, eventually forced governments to agree on more rigid measures for the protection of stocks and management of resources. This was even more so in the fields of commercial whaling and sealing. In many respects the measures that have been taken so far are not yet adequate to remedy the situation, but fortunately there is now growing awareness that in the interests of future mankind the living resources and their environment must be treated in a much better way than they were in the past. The widespread fundamental change in public opinion towards a new feeling of responsibility for the wealth of our world gives hope for the future. The authors hope that this book will contribute to positive development by learning from history.

1 Early Times

1.1 Stone, Bronze and Iron Ages

1.1.1 Europe

Fishing is one of the oldest professions of man and its origins lie in the days of yore. It developed gradually when man moved from the unselective and unplanned collection of things found in nature to the first systematic utilization of food, applying experience and newly invented simple techniques to the preparation of artefacts. During the earliest stages of civilization fishing was done with the same tools as hunting. It is an idle question as to whether fishing or hunting has a longer tradition. There is, however, a marked difference between hunting and fishing. The former lost its great importance in most parts of the world at a fairly early stage during development and became a subsidiary element as man settled and changed to agriculture and stock-farming. Also, hunting increased as a noble activity. On the other hand, fishing has retained its importance up to the present time but is often considered to be the profession of poor people of a socially lower standing, with the exception of sport fishing.

Information on fishing during the Palaeolithic Age is rather sparse and not very reliable. It seems that fishing was already practised by the Prehominids. In the Mousterian (Third Interglacial), about 100 000 years ago, Neanderthal man had a certain preference for waters where trout and other fish could be caught by hand, perhaps with the help of stone barriers in rivers. There are indications that near the coasts salmon was caught but obviously no real sea fish.

Only a few indications of fishing activity also remained from the Late Palaeolithic period when *Homo sapiens* appeared in Europe about 50 000 years ago. They show a development from stone-made tools to the application of gear made of wood, bone, ivory and horn. The Aurignac people were the first Europeans to make regular use of the fish resources in rivers. They produced pointed pieces of bone and horn, indented on one or

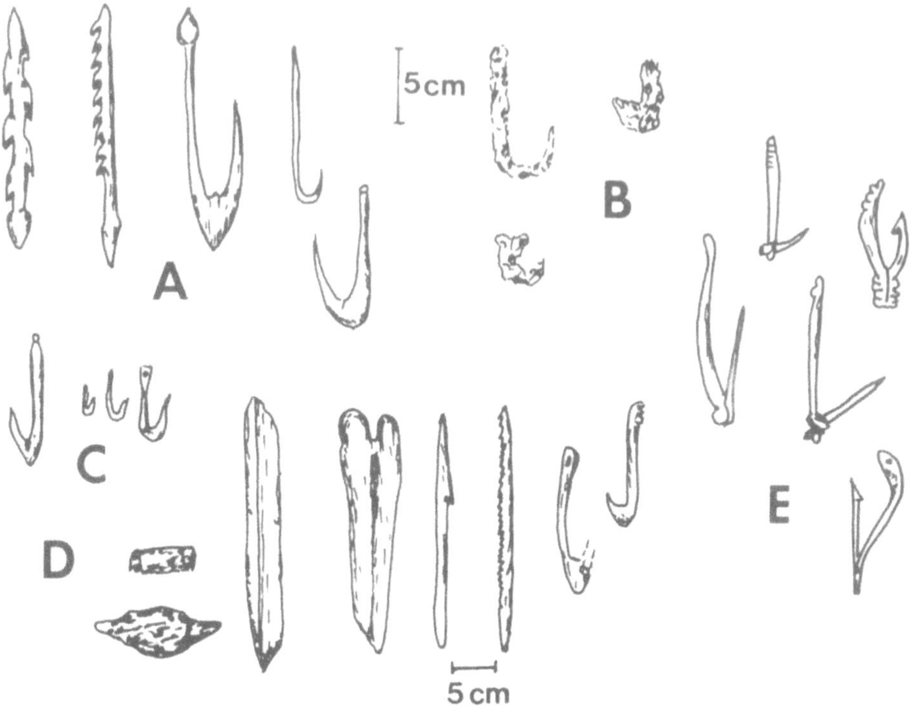

Fig. 1A-E. Fishing tools from the Stone Age. **A** Mesolithic harpoons and hooks of bone from northern Europe; **B** hooks of stone from the same period and region; **C** hooks of bone from the Mesolithic period in North America; **D** Neolithic tools from northeast Europe: ice chisel with cutting edge sloped at 45°, harpoons and hooks of bone, float of fir bark and sinker made of pebble rolled in birch bark; **E** prehistoric compound hooks of bone and horn from northern Eurasia (Finmark to Angara and Japan)

both sides, which were used on lances both for hunting large game and for fishing (Fig. 1). Such barbed tools greatly facilitated the capture of animals.

It is not certain whether real harpoons, i.e. detachable tools with a line to a shaft, were already being used, but such instruments are exemplified for the Middle Neolithic. For many tools it is not possible to prove that they were used for fishing rather than for hunting mammals, birds and other animals. There are indications that fish lances and arrows shot from bows were already being used before the end of the Palaeolithic Age. However, it is doubtful that fishing with lines and hooks was known during this early period. Perhaps small bony objects, gorges pointed at both

Fig. 2. Fish images in Palaeolithic art. *Above*, deer and salmon engraved on reindeer horn (23 cm), from cave of Lorthet (Pyrenees), Magdalénian; *below*, flatfish, carving from Lespugue (Haute-Garonne, France), length 4.4 cm. (Musée des Antiquités Nationales, St. Germain-en-Laye)

ends and with a ring furrow in the middle, found close to the refuse from fish meals (mainly salmon) were used for this purpose. Significant progress was the control of fire which allowed food to be cooked or dried for preservation by heat and smoke.

The Magdalenian (30 000 - 10 000 B.C.) was a period of advanced culture, as documented by many discoveries, particularly in the region between northern Spain, France, southern Germany and Bohemia. Most admirable are the cave paintings of Franco-Cantabrian art in Spain and France (e.g. Altamira, Lascaux), interpreted as magic figures for successful hunting. Besides bison, horses, reindeer and many other animals, these paintings occasionally show effigies of fish. Such pictures, mostly of salmon and trout, were found engraved in reindeer horn (Fig. 2) and also in clay as, for instance, at the bottom of the cave at Niaux in southern France. Fish and dolphins were painted in black and red on stalactites in the cave of Nerja near Motril in southern Spain. Bones of salmon, trout, pike, bream, eel and other fish were frequently discovered in refuse heaps made by the cave dwellers in the river valleys of Dordogne and Vézère in France. During the Magdalenian the replacement of stone tools by instruments of bone, ivory and horn advanced further.

With the end of the last ice age (Würm Glacial) and the warming of the climate about 10 000 years ago, this culture perished. Possibly the conditions for hunting deteriorated in the south due to the advance of the arid zone. In contrast, conditions improved further north and people followed the retreating ice edge northwards.

During the Mesolithic Age (10 000 - 6 000 B.C.) there was a transition from large game and reindeer hunting to the earliest farming cultures. With improving climatic conditions forest areas expanded, hunting became more difficult, and this deterioration in the food basis created an incentive for cultivation of plants and breeding of domestic animals. Fishing greatly increased in importance, particularly in swamp and tundra areas that were previously covered by ice, from Britain and the region that is now North Sea (land south of Dogger Bank), Denmark and along the Baltic Ancylus Lake to Russia and beyond. People settled, first only during the summer, then permanently, along the banks of rivers and lakes, and later also by seashores, so that they could fish.

The earliest stage of such a culture is named after the site of Maglemose in Denmark and is also known from excavations in Germany (Schleswig-Holstein and the Havel area) where many tools like harpoons and fishing hooks of bone and horn were found. With further improvement in living conditions during the warm Litorina phase, when the Baltic turned into a salty water, people settled permanently along the shores as fishermen and oyster eaters. They left large piles of mussel shells which are known in Denmark as "kökkenmöddinger" (kitchen middens) and are significant of the Ertebölle culture. Corresponding sites can be found in Germany at the oldest remains of settlements near Ellerbek (Kiel) and the Federsee in southwestern Wurttemberg as well as in Switzerland and in the east in the Baltic provinces and Poland (Kunda culture).

The mounds of mollusc shells remaining from the meals of these ancient people are found from the mouth of the Tajo River along the coasts of the Atlantic northwards. With lengths of several hundred metres, widths of 50 m and heights of 1-3 m these mounds are largest and most numerous in Denmark. Fireplaces and tools of flint, bone and horn certify the presence of man. The shells indicate that the mollusc species were fairly much the same as those found today in coastal areas. The bulk are oyster shells, followed by, in order of importance, shells of blue mussels *(Mytilus edulis)* and snails. Also fish bones from flounder, herring, eel, cod and haddock can be found. In the west of Scotland (Oronsay) bones of conger eel, sea bream *(Pagellus centrodontus)*, thorn-back rays, sharks and small cetaceans were also found. Thus, it is likely that fishing from boats at some distance from the shore already existed at that time.

Typical fish hooks appeared in the Maglemose period. They were made of bone, horn or wood, were in one piece and were initially without barbs but with a button-like widening at the end for fixing a line to. It is uncertain whether the straight pointed gorge is older than the curved hook. Most likely both types were developed independently, the former from straight pieces of material and the latter from crooked thorns or branches of trees. Comparative studies showed that possibly the true artificially prepared hook was invented in northern Eurasia by the old subarctic hunter culture. From there the hook, in its typical U-form, reached northern Europe during the Mesolithic Age. It was not found from that period in western Europe, where only the gorge was known since the Palaeolithic. To the east, the curved hook was originally distributed to China and Japan but not further south. Later, this type of hook was improved to perfection in Japan, Polynesia and along the shores of northwest America. It seems that, like the simple curved hook, a little later the compound hook (Fig. 1) also appeared first in northern Asia. The transition from the simple hook to the compound hook can be seen, for instance, in the Angara culture on Lake Baikal. In the area around this lake, settlements with fish pits and effigies of fish in stone already existed at around 4000 B.C. and judging from the large number of fish hooks found, people still lived mainly by fishing 2000 years later.

Certain pointed bone tools have been interpreted as fish-scale knives and ice chisels. Pieces of baskets made of wickerwork from about 8000 B.C., widely distributed in Scotland, Denmark and Estonia, netting from willow and lime-tree bast with stone sinkers (from 5000 B.C.), floats of pine wood or birch bark and harpoons with lines of animal fibres or leather straps testify to the progress made in fishing.

Fig. 3. Oldest European dugout, about 8300 years old and 3 m long, excavated near Pesse, the Netherlands. (After Schwabedissen in Ellmers 1976)

Most likely, boats of wood or skin had been used for a long time but the authentic proof dates from the Mesolithic period. The oldest remains are paddles that were excavated in England, Denmark, Germany and Sweden. A little younger are the remains of pine dugout canoes found near Perth (Firth of Forth) and near Pesse in the Netherlands (Fig. 3).

On several sites in southwestern Germany remains of various fish species were found in caves and under rock shelves in layers that date from the Aurignacien until the Neolithic period (36 000 - 3900 years old). In two places remains of harpoons documented fishing activities. The fish remains are mostly from grayling *(Thymallus thymallus)*, river charr *(Hucho hucho)*, burbot *(Lota lota)*, chub *(Leuciscus cephalus)*, roach *(Rutilus rutilus)*, makrel *(Chondrostoma nasus)*, perch *(Perca fluviatilis)*, bream *(Abramis brama)* and pike *(Esox lucius)*.

The early importance of fishing in Scandinavia is documented by numerous Mesolithic and Neolithic pictures carved into rocks which were found in many places along the Norwegian coast as far as the North Cape and, further inland, also in Sweden. Like the Palaeolithic paintings in France and Spain, they are considered to be magic figures for successful hunting and fishing but they are located in open air sites and not in caves. Most of these rock carvings display ships, elks and reindeer; other pictures present fish, often interpreted as halibut, whales, seals and porpoises (Fig. 4). Often the images also show the inner organs of the animals, a style similar to that found far away in rock paintings by the Australian aborigines.

Fig. 4. Rock carvings from Scandinavia. *Above*, elks and fish, possibly halibut, caught with line from a boat. Forselv, Skjomen, Norway; *centre*, elks, fish and a whale from Askollen, Drammen near Oslo. Length of elk in upper part 1.81 cm; inner organs shown; *below*, fish (2 m), later covered with boats. Meling, Rogaland, Norway. (Kühn 1952)

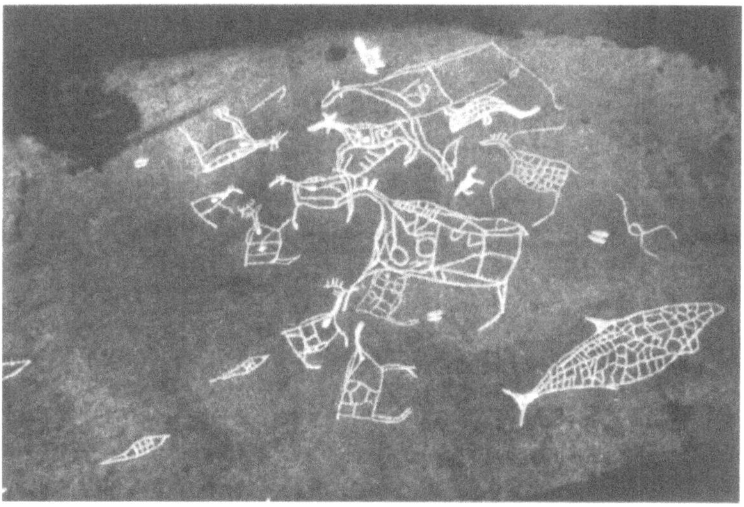

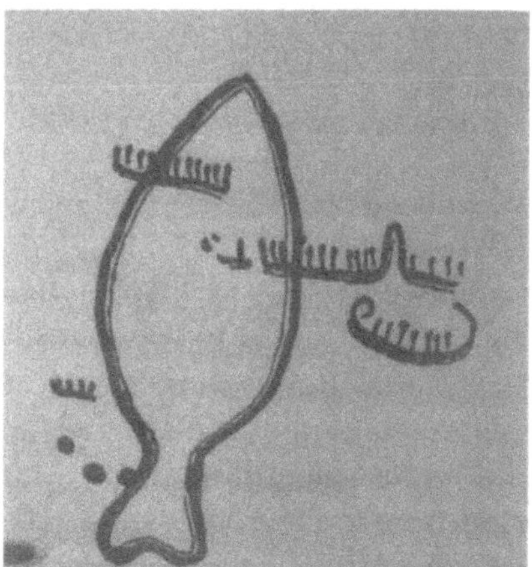

During the Neolithic Age (about 4000 - 1700 B.C.) the peasantry, of more or less permanent settlements, developed and populations grew more rapidly. Fishing developed further but generally its importance decreased in comparison to agriculture and stock-farming. Groups which relied more than others on fishing were the people living in the region from northern Germany to Scandinavia and people of the culture with Megalithic burial places from Portugal to Ireland and Scotland. Since wide-ranging trade connections have been proved for the region between the Mediterranean, the British Isles and the Baltic, it can be assumed that sea-going vessels existed which may have permitted fishing at some distance from the coasts. Very little is known about this fishery. It is certain, however, that from Norway there was already fishing not only for cod but also for ling, tusk, redfish and halibut.

A more comprehensive impression of the living conditions at that time is available from the lake-dweller settlements which existed in large numbers mainly in the region of the Alps in France, Austria, Germany, Switzerland and Italy. Remains were well preserved in the swamps. There is no doubt that the inhabitants were farmers, but fishing was also an important sector of food production since the settlements were built at lakesides. Romantic ideas of settlements over the open waters of lakes were possibly influenced by antique descriptions given by Homer, Hippocrates and Herodotus of the conditions at Lake Prasias (Thrace), where buildings were erected on scaffolding above the water and fish caught with baskets lowered through holes in the floor of the huts. But vertical beams are often missing in the alpine lake-dweller settlements. Such settlements, up to 500 m long and 100 m wide, are today located sometimes up to 70 m beyond the present lakeside. Investigations showed that it is most likely that the settlements were originally located on dry land along the ancient lakeside. The difference can be explained by the lower water level resulting from the greater aridity during the Stone Age and Bronze Age (about 1700 - 800 B.C.).

The well-preserved fishing tools from such settlements indicate further technical improvements (Fig. 5). Besides pointed gorges, curved fish hooks, mostly barbed, appeared in increasing numbers. They were made of bone or bronze. Holes were drilled into the bones for the preparation of barbed hooks. The use of metal, at first bronze and later iron, allowed the construction of finer, more pointed and more durable hooks than before. Old materials like flint were also still used for a long time in the same conservative manner as later bronze hooks (7 - 10 cm long) were used until the 11th century. Fish lances and harpoons were in use and metal spears with two to three or more barbed points appeared for the first

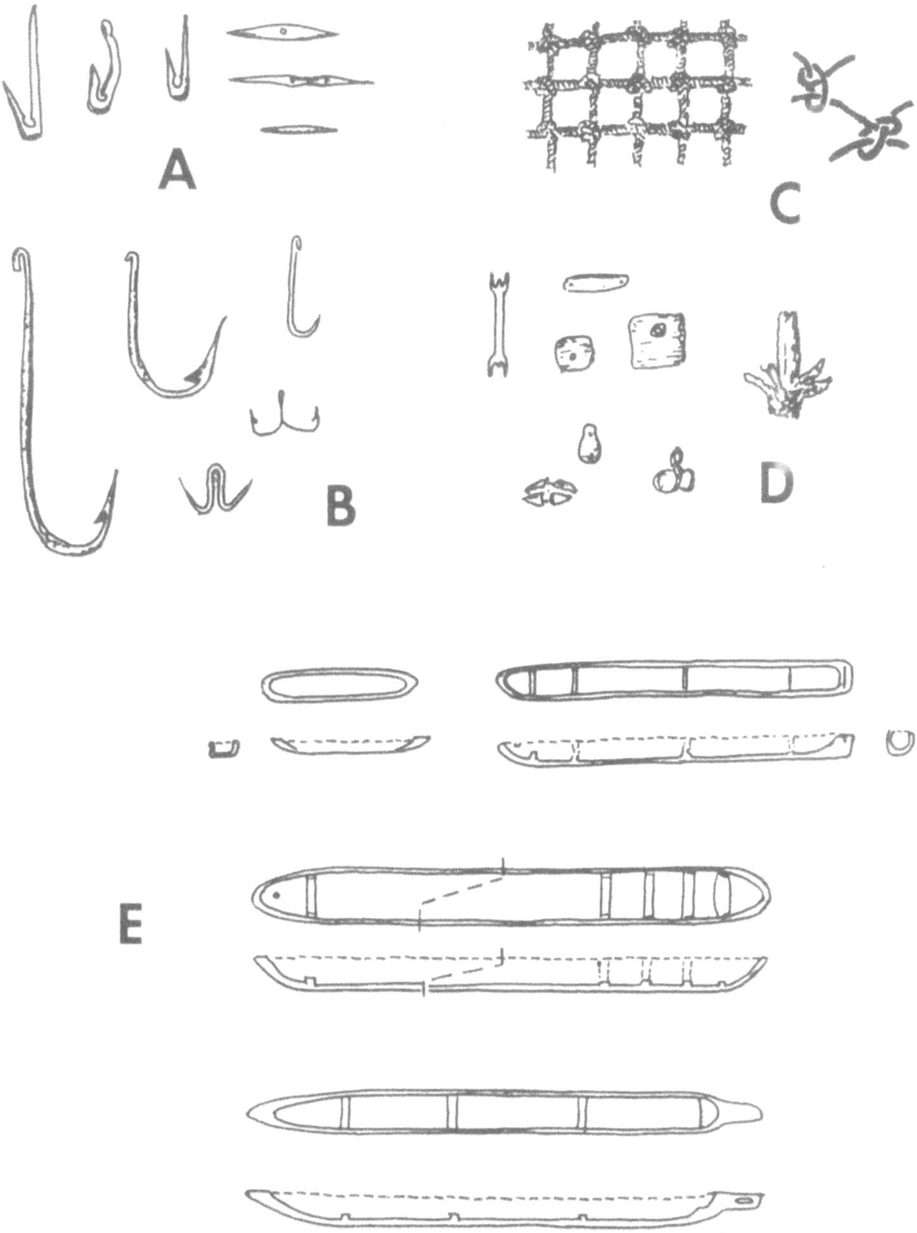

Fig. 5A-E. Fishing equipment from lake-dweller settlements in the Alps region (Neolithic and Bronze Ages). **A** Hooks and gorges of bone; **B** hooks of bronze; **C** remains of netting knitted with flax twine and of mesh-size 3 cm from knot to knot. On the *right* lake-dweller knots according to von Brandt (1970); **D** net needle (bronze), floats (wood or bark), sinkers (stone) and wooden knot for hauling nets or lines; **E** dugout canoe

time. Remains of netting show that nets were made of hemp *(Cannabis sativa)* and flax *(Linum angustifolium)*. They had mesh-sizes of 5 - 45 mm from knot to knot. It is amazing that the same type of knots were not only found in these lake-dweller settlements but also in various other places in northern Europe, Africa, Peru and even Oceania. Accessory tools were sinkers of stone or earthenware, floats of wood or bark, distaff fly-wheels and net needles. It is uncertain how the nets were used. Fish baskets were wickerwork or built from bark or reed. There are also indications of the existence of fish-hedges.

From the hundreds of dugout canoes excavated some can be dated precisely; some originate from the lake-dweller settlements. These old dugouts are 3 - 14 m long, mostly around 5 m long, and their width is 40 - 100 cm, usually 60 - 70 cm. Their height was measured as between 30 and more than 100 cm. The limitation for such vessels is certainly related to the diameter of the available tree trunks. These boats were mostly cut from oak and beech trees but also from pine, silver fir, poplar and other wood (Fig. 5). Dugouts made from hollowed oak were also unearthed in Scotland. Stability was achieved by ribs or bulkheads left in the wood during the cutting and later, by also inserting bowed pieces of wood. It seems that these dugouts were moved by stakes over shallow waters or by paddles.

Layers of several centimetres depth of fish bones and scales are found in areas of such settlements, indicating the mass of fish consumed by the people. Fish species were the same as those abundant at present in small boggy waters or large deep lakes. Salmon appears to have been a major food fish besides carp, pike, perch and other species. Gwyniad *(Coregonus wartmanni)* was eaten from such waters as Lake Constance. Deterioration of the climate and warlike disturbances during Roman times and a period of mass migration put an end to this relatively peaceful era.

It is likely that the people characterized by the lake-dweller settlements were widely distributed in Europe. They were farmers who also went hunting and fishing. In the northern zone populations that had hunted and fished since the Mesolithic Age remained for a long time at the same cultural level although, little by little, they took over innovations from their neighbours. This was the case in the east, where similar conditions to those in northwestern Europe had already developed further south. In southern Russia the importance of fishing increased with the introduction of bow and arrow and, later, boats, fishing tackle and nets. It is assumed that fishing lost some of its significance as a result of the development of agriculture and stock-farming during the Neolithic era.

Fig. 6. Golden fish from the Scythic treasure of Vettersfelde, province of Brandenburg, sixth century B.C.; length 41 cm. (Photo: I. Luckert, Antikenmuseum Berlin, Staatliche Museen Preußischer Kulturbesitz)

Advances to the northern coasts were delayed and the White Sea culture dates from about 2000 B.C. to 1000 B.C.. Obviously the coasts were visited only sporadically and winter settlements were built near inland waters. In these settlements in northern Russia simple fish hooks of bone without barbs were found from this period as in northwestern Europe at an earlier period, during the Mesolithic. Obliquely pointed ice chisels are significant. Here, as in northern Asia, the simple fish hooks were later gradually replaced by compound hooks originating from regions further south.

On the eastern side of Lake Onega in northern Russia and on the west coast of the White Sea, Neolithic rock carvings can be seen which are rather similar to those in Scandinavia (Fig. 4), and some of them depict fish images. For instance, the carving at Vyg-Ostrov shows the harpooning of a fish from a boat. It was estimated that these carvings were engraved around 500 B.C., whereas those in Scandinavia were originated between 8000 and 1000 B.C. and the cave paintings in France and Spain date back even as far as 40 000 - 60 000 years ago. This indicates the advance of the Stone Age culture from west to east.

A non-indigenous fish image, imported to central Europe from the southeast, was discovered in Germany at Vettersfelde in the province of Brandenburg (Fig. 6). With other objects of treasure, mostly of gold, it documents the influence, and possibly the invasion, of the Scythians, nomadic horsemen from the steppes north of the Black Sea, around 500 B.C.. The fish is covered with figures of sacred animals and ornaments. It was found on top of a horse skull and served apparently as an amulet. Originally, Anahita, the horse deity of the Scythians, was believed to be a fertility demon which was surrounded by water symbols and fish. The style of the gold work is Greek.

1.1.2 Africa

Similar progress in the production of Palaeolithic tools for fishing was observed in Africa and many stone artefacts from this period have been found. The Sahara region was fairly densely populated as indicated by rock paintings. Game hunting and fishing supplied the food. After an arid phase with a decrease in the population the number of people rose again during Neolithic times. Old fishing villages with remains of fish bones

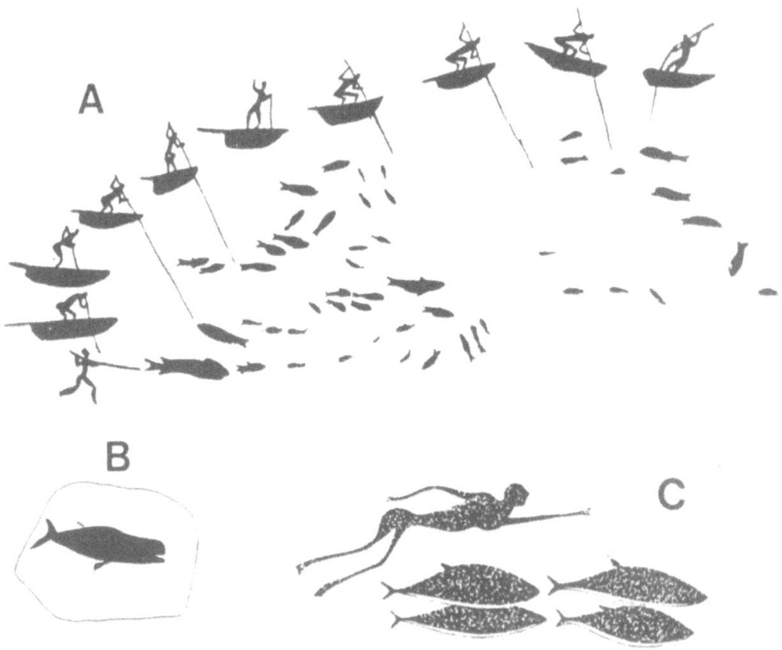

Fig. 7A-C. Prehistoric rock paintings from South Africa. **A** Spear fishing by a group of Bushmen. Tsielike River; **B** whale on a burial stone from the south Cape coast; **C** man swimming with four dolphins. Carving on pebble from Klasies River mouth, 2300 years old, securely dated (Pager 1975)

and mollusc shells were discovered south of the Hoggar mountains, in Tanezrouft and Ténéré. Fishing was carried out in rivers and lakes, jointly with hunting for hippopotamus.

Also, at the southern end of the continent, a centre of prehistoric human activity developed. Thousands of cave paintings and rock carvings in South Africa and Namibia, the oldest of which date back about 25 000 years, show a striking similarity to those in France and Spain. Several of them demonstrate fishing activities (Fig. 7). Fish were speared from boats and floats or caught in nets, baskets and stone traps. Lines were fitted with gorges but curved fish hooks were not found in South Africa. From painted burial stones in shell middens and paintings in rock shelters with shell deposits it is known that people lived also on shellfish, probably collected by the women, and on stranded whales and seals; freshwater dolphins were also depicted. From the fish paintings mudfish *(Labeo capensis)*, barbels, mormyrid and cyprinid species have been identified. Numerous paintings of fish are associated with rain animals, figures from Bushman mythology concerned with water and thunder.

Taking into account the remarkable abilities of African people to invent and apply fishing methods and tactics, it can be assumed that fishing also has a long tradition in the tropical zone. However, due to the climatic conditions there are hardly any remains from the old times.

1.1.3 America

The originally uninhabited continent of the Americas was populated from Asia via the Bering Strait. Most archaeologists assume that this migration began around the end of the latest (Wisconsin) ice age, about 13 000 - 15 000 years ago and proceeded in several waves with people of different cultural levels. It is also possible that the immigration started earlier, perhaps 30 000 years ago. Indications of human activity in the southern part of the double-continent, like rock paintings in Patagonia, date back to 8500 - 9000 B.C.. Archaeological findings show that the early immigrants lived by hunting, fishing and collecting their food. As in Europe, a transition from large game hunters and plant collectors to agriculture and stock-farming can be observed, and some peoples remained also as fishing tribes. Many of the ancient tools provide evidence of their Asian origin. It is much disputed, though, to what extent the old American cultures are based on diffusion from Asia and to what extent they are endemic developments in America. Old types of tools, made of stone, bone, horn, ivory and, later, iron, were used widely until the European immigration after the end of the 15th century introduced many new elements. In some remote areas, like the arctic region, stone age types of equipment are still used up to the present time.

Archaeological discoveries show that in the arctic region of America, and from there also in Greenland, the immigration of the Eskimos (now called Inuit) proceeded gradually, and sometimes overlapping, with at least four waves of people at various cultural levels. The first can be dated back to 2000 B.C.. Since these early times the poverty of resources on land forced the utilization of marine animals such as seals, including walruses, in the ice-covered waters and fish and whales in the more open waters of the subarctic. The catches served not only as food but also as basic material for most of the equipment. Seal skins were used for clothing, for summer tents and as covers for boats. Animal oils were burnt in lamps for light. Fishing and sealing was done from the "kayak", a one-man light canoe with a maximum length of 5 m, constructed with a wooden scaffold covered with seal skin (Fig. 8). In Canada kayaks for two to three persons were also occasionally used. These very mobile boats were moved with paddles of wood, often strengthened by bone. Such boats are very seaworthy since they are almost completely closed with only a small opening for the man's body; the hole is made watertight by his sealskin dress. It seems that the kayak was first used by the Thule Eskimos perhaps 1000 years ago.

The main gear for sealing was the harpoon, mostly made of bone, which was connected to a line. The spear was thrown with the help of a sling. One or several air bags of intestine or skin were carried to keep the prey afloat. During seal hunting a white sail served as camouflage. Animals hunted were mainly ringed seals but harp and hooded seals were also caught. Walruses, particularly valuable for their long teeth of ivory, were also captured with large gorges.

While sealing was usually an individual activity, whaling was done in larger groups, using the "umiaks". These were skin-covered open boats like the kayaks but up to 9 m long and with a capacity for more than 10 persons. They were also used for transport and were then moved by the women. With great courage the Eskimos attacked the whales, even large bowheads, with lances and harpoons. After capture they climbed on the back of the whales and killed them with lances and knives. The hunters wore watertight skin coats to keep them afloat when sliding into the water. Lances were prepared from driftwood.

The possibilities for fishing in the Arctic, at the border of the distribution of many boreal fish species, are dependent on the varying climatic condi-

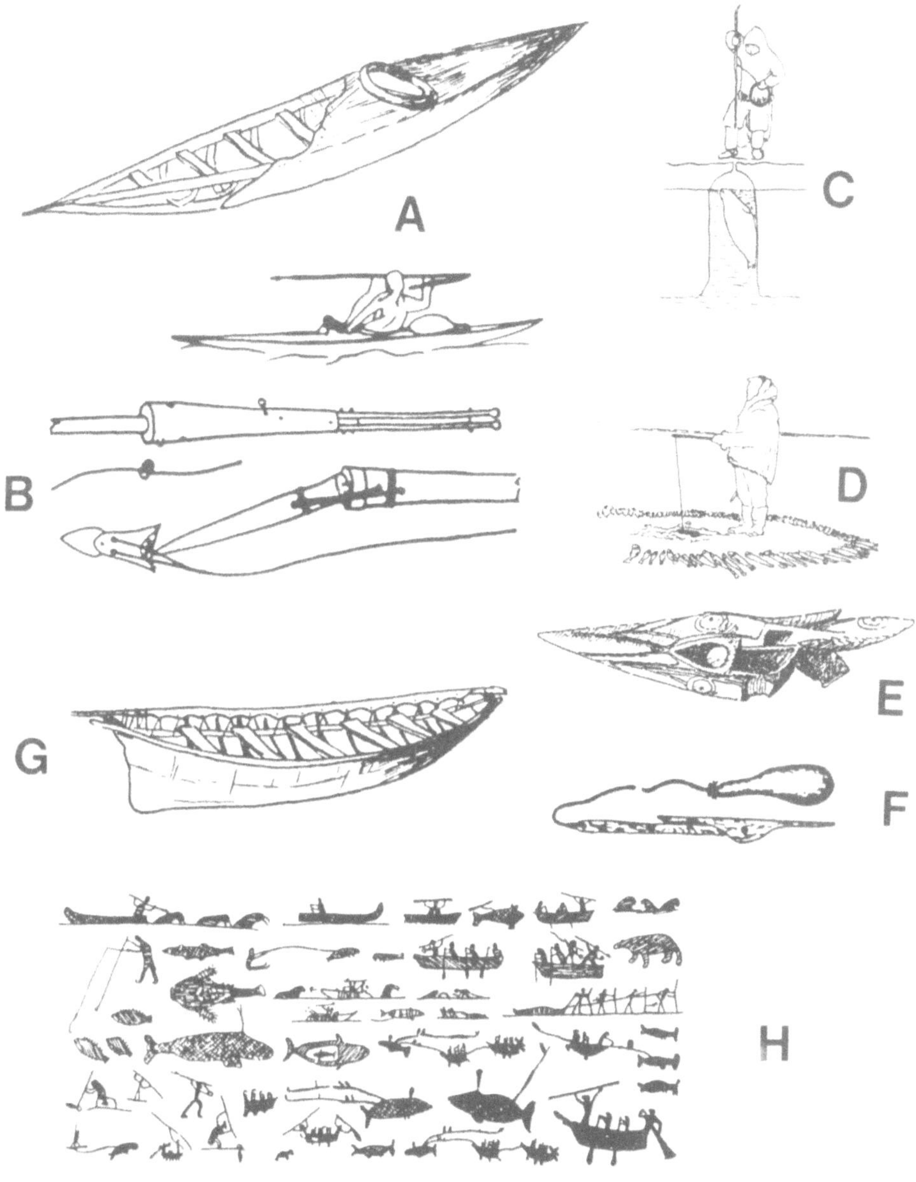

Fig. 8A-H. Eskimo fishing and hunting. **A** Kayak and throwing of harpoon with help of a sling; **B** front and rear ends of bone harpoon, point of stone or iron, with steering mouldings at the end and fastening for bladder; **C** harpooning of seal at breathing hole; **D** fishing for polar cod under the ice; **E** ornamental point of harpoon from ancient Bering Sea culture; **F** gorge with bladder for catching walruses; **G** umiak; **H** portraiture of hunting and fishing from Chukchi area. (**A,E,G** from Mai's Weltführer 35, Grönland, 1st edn, by A. Ludwig Frankfurt/Main 1983; **H** from AE Nordenskiöld 1882)

Fig. 9. Bark canoe of the Naskapi, Labrador. (Turner 1894)

tions. From information on the relative ice-coverage around Iceland large fluctuations can be observed. There are indications that conditions were rather favourable when Eric the Red sailed with the Normans to Greenland in 982. From about 1200 to 1400 it was cooler. After a warmer interim period during the following 200 years living conditions deteriorated greatly from 1600 until the end of the last century. Investigations in modern times have shown that since the 1920s there has been a gradual advance of fish species like cod and salmon in Greenland waters to further north with ameliorating climatic conditions. Such developments have significantly improved the resources available to the Eskimos. In extremely cold waters, where fish resources are poor, the polar cod *(Boreogadus saida)* and ogac *(Gadus ogac)* and, in freshwater, the Arctic char *(Salvelinus alpinus)* are the most important fish species. For fishing, carried out also by Eskimo women, the major types of gear used are hook and line and various types of lance.

Further south, in the thinly populated Subarctic, tribes of the Athapaskans in the west and Algonkins in the east were engaged in hunting and fishing in the numerous rivers and lakes. They had canoes made of a hull of birch or fir bark covering a wooden framework (Fig. 9). The pieces of bark were stitched together and made watertight with pine pitch. The canoes were lightweight and could be transported easily over land.

On the northwest coast, in the region of the mouthes of the rivers Fraser and Columbia and in Puget Sound, a culture of fishermen can be traced back to 4000 B.C.. Almost all Indian tribes along this coast lived nearly exclusively on fish, the Nootka and Makah were also engaged in whaling. The wealth of this culture was based mainly on the resources of several salmon species *(Oncorhynchus* spp.*)* which were easily accessible during the upstream spawning migrations of the fish. Furthermore, concentrations of herring *(Clupea pallasii)*, cod *(Gadus macrocephalus)*, halibut *(Hippoglossus stenolepis)* and other fish close to the coast provided sufficient food for the relatively dense population.

Hooks and lines, spears, traps and dams in the rivers were used for salmon fishing. The Indians built canoes of cedar wood which was hollowed

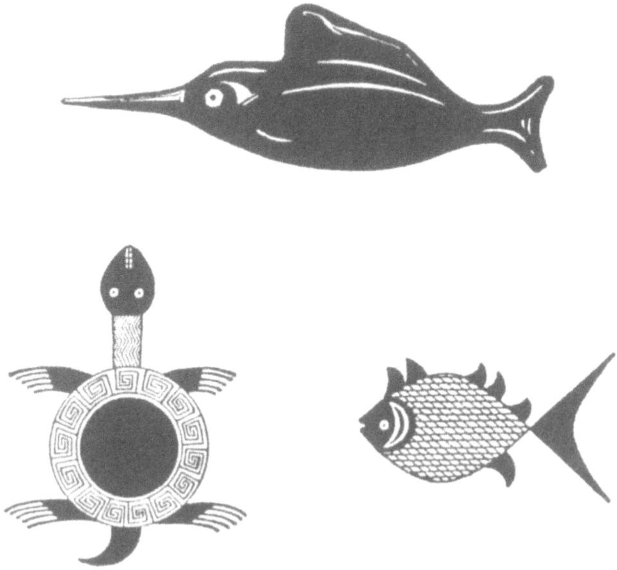

Fig. 10. Indian art from North America. *Above*, swordfish of soapstone with eyes of mussel shell from Gabrielino-Chumash culture, southern California; *below*, naturalistic images, Mimbres black-on-white ceramics from late Mogollon culture, New Mexico. (Haberland 1965; *above*, after Douglas and d'Harnoncourt 1941; *below*, after Cosgrove)

out by fire. The trunks were then forced apart by wooden sticks and cut to measure. The seaworthiness of these boats for fishing at sea was improved by planks to raise both ends of the canoe. Hooks and lines were used in fishing at sea, especially for halibut, and skimming nets for herring. Whales were hunted with lances and harpoons and seals with bow and arrow. The women collected shellfish and large shell mounds can be found in many places along the coast as far north as the Bering Sea.

Fishing localities on the lower Columbia River developed to become major centres of fish processing and fish trade. Fish, especially chum salmon *(Oncorhynchus keta)*, were preserved by drying in the open air and mostly pulverized to make fish pemmican, a very stable food. The products were transported inland across the mountains on trade routes called "grease trails" as fish and whale oil was exported in the same way. Chinoah Indians were the most important traders in this business. The Indian fishermen culture produced mythical sculptures and pictures of animals, especially of salmon ghosts, but also of killer whales and other species.

Interesting remains have also been reported from the southwest of the United States. In southern California the Chumash Indians have lived by the coast in the Santa Barbara area and on the Channel Islands for more than 1000 years. Fishing and whaling was their main activity but it is uncertain how far they went out to sea, though they possessed large seagoing boats of planks, sewn together with tendons and made waterproof with asphalt; these vessels could take up to 30 people. Harpoons were used for hunting seals, dolphins and sea otters. Pearls from mussels served as money. By 300 - 500 A.D., the Chumash had produced beautiful sculptures from soapstone displaying killer whales, swordfish and other marine animals (Fig. 10). Such sculptures were probably used during ritual ceremonies. Highly artistic ceramics of inland fish, tortoises and other animals were found from the Mogollon culture in New Mexico around 1100 A.D..

Also specializing in fishing were the tribes of the Ojibwa Indians living on the shores of the Great Lakes. They mostly fished with nets close to the shore and were engaged in wide-ranging fish trade. Along the North Atlantic coast the Indians were also living on fish to a great extent. In the northern parts of the woodland they fished in the many rivers and lakes with hooks, spears and nets as well as building fish-hedges and stone dams for Atlantic salmon *(Salmo salar)*, trout and other salmonids, eel, perch, pike and other fish.

Indications for the presence of man in South America, in particular in the region of Ecuador, Peru and Chile, date back to rather early times. Artefacts and cave paintings dating from around 7600 B.C. show that these people were hunters and nomads. The rich marine resources along the Pacific coast led to the early formation of cultures of mollusc collectors and fishermen. Excavations of the "Hombre de Paracas" on the Peruvian coast resulted in the discovery of the oldest known fish net in the world with an authentic age of 8800 years, proved by radiocarbon dating. Cultivation of plants had already started but people were still semi-nomads (settlements were promoted by the use of guano as fertilizer around 4500 B.C.). They lived mainly on molluscs and fish. Other coastal groups of collectors and fishermen are represented by the "Hombre de Pucusana" (5300 B.C.) and "Hombre Chilca" (3700 B.C.). The latter were using fish hooks and harpoons of bones but apparently no nets and were hunting seals, pelicans and other birds.

There are many places along the Peruvian coast with large aggregations of mollusc shells ("conchales") which represent the remains of human meals, as do the kökkenmöddinger in Europe. These layers can reach up to 10 m and cover almost 1 km^2. Besides of mussels, particularly scallops *(Pecten purpuratus)* and machas *(Mesodesma donacium)*, they contain

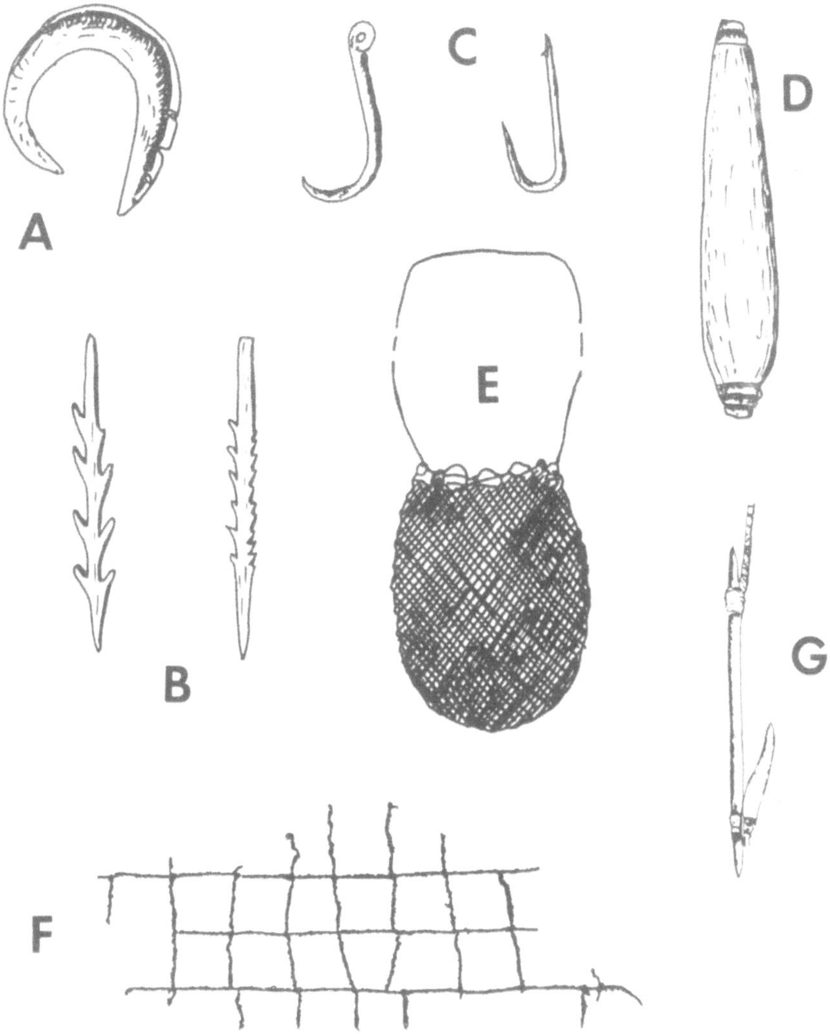

Fig. 11A-G. Fishing equipment from the pre-ceramic era in Peru. **A** Hook of mussel shell; **B** harpoons of bone; **C** hooks of copper; **D** wooden float; **E** bag for collecting molluscs (?); **F** remains of netting; **G** composed hook of wood. (Pesca Peru no date; Museo de la Universidad Nacional Agraria and Museo Amano, Lima)

remains of sea-urchins, starfish, crustaceans, fish, seals and birds. Charles Darwin was the first to investigate these mounds in 1835.

A famous area for excavation is Huaca Prieta which has remains dating from 2500 B.C., i.e. the pre-ceramic era (3800 - 1800 B.C.); by this time

Fig. 12. Reed float, "totora" or "caballito" from Peru; the same type of vessel can be observed on Lake Titicaca. (Pesca Peru)

people had already settled permanently. They were primarily "mariscadores" and fishermen but they also farmed. Cultivation of cotton was started and the fibre used for the manufacture of fishing nets. Stone tools were still in use but were increasingly replaced by bone material, often from seals and whales. Typical of this culture are fish hooks made of mussel shells (Fig. 11). Double-barbed harpoons were made from bones and metal, especially copper (around 5000 B.C.).

In the tidal zone along the coast fish and other animals were caught behind stone walls and fish-hedges. Scaring and tracking of animals to the coast was known and so was the handling of beach seines; this required coordinated work by groups of people.

It is not known at what time people started to go out to sea for offshore fishing but this was certainly also a long time ago. Fishermen used reed floats ("totoras", Spanish "caballitos") of a design which can still be observed in use on the Peruvian coast and on Lake Titicaca today (Fig. 12). In the south of Peru they used "balsas de odres", boats built with two seal skins around a float. These were still in use up to the 19th century. Larger rafts of balsa wood *(Ochroma lagopus)* trunks served mainly for the transport of goods and passengers but were also used for fishing with lines and nets. Both the floats of reed or similar material and the rafts of light trunks belong to the oldest types of vessel and can also

Fig. 13. Bicolour netting with fish design, Chancay style, Peru, 1100 - 1400 A.D.. (Museo Amano, Lima; Pesca Peru)

be found in many other parts of the world, e.g. in India, New Zealand and Micronesia and on Easter Island and African lakes. The floats have to be dried quite frequently to retain their carrying capacity and usually they do not last very long.

The old fishing nets in Peru, consisting of short-fibred cotton in double or triple twines, were several metres long and of splendid workmanship. There were two types of nets: one was without knots but with twisting of the net yarn at the joins; the other was knotted netting with the same type of knots as developed in other parts of the world ("Peru" and "lake-dweller knots" according to von Brandt). Many nets were dyed and around 500 B.C. bicoloured nets appeared. At the Paracas necropolis polychrome nets were discovered from 300 B.C.. It is not known whether the staining was carried out to conserve the nets or to minimize the visibility of the nets to fish. Often the nets contain designs of fish and birds (Fig. 13). 2300 or more partitions with figures have been observed. Obviously such highly artistic nets were used during special ceremonies. The nets used in fishing had large meshes at the borders and were equipped with pumpkin floats and stone sinkers. It seems that fishing nets were already widely in use in Peru 4000 years ago. Fortunately, due to the arid conditions, such nets were well preserved and can be seen today.

Another method that was carried out from balsa-trunk rafts was fishing with harpoons and arrows for fish that were attracted by artificial light from torches. Also, long lines with many hooks were used. In northern Peru cormorants were tamed and trained by the Moche people (100 B.C. - 900 A.D.) to catch fish, a method also developed in China and Japan. Images of such fish-catching birds were painted on beautiful Indian ceramics (Fig. 16). Fishermen of the Nazca culture, at around the same period, were diving for mussels. Unearthed human skulls show damage to the region of hearing; this damage might have arisen from this profession. It has also been proved that a wide range of preservation methods, like salting, drying and smoking of fish, were already in use at this time.

The overwhelming importance of fish and other marine animals, such as crustaceans, cephalopods and mussels, is well demonstrated by the highly artistic pottery and tapestry remaining from the ceramic era, especially from the Moche, Nazca, Chimú and Chancay cultures (Fig. 14).

In the tropical zone of South America remains are scarce because of the humid climate. But shell mounds of the remains of human meals can also be found along rivers and coastlines here, the oldest dating back 7000 years. There is no doubt that fishing, particularly from the many rivers and swamp areas, has always been important for the production of animal protein. A large variety of fish were, and still are, shot with bow and arrow or are harpooned with lances or spears. It seems that fish hooks were unknown until they were introduced by Europeans. As in many other tropical areas of the world, fish are often caught by stupefying or poisoning them with extracts of vegetable poisons. It was estimated that there are about 400 plant species suitable for this purpose, of which 100 grow in South America. Here, the legum *Tephrosia toxicana* and *Lonchocarpus nicou* are the most widely used plants. After toxication the fish can easily be collected and consumed without difficulty. Blow-pipes are mainly used for hunting birds but sometimes are also used for fishing.

Fish hooks were also unknown to the Indians living isolated on Tierra del Fuego in the south. Of the Fuegians the Yahgans were most dependent

Fig. 14. Ancient Peruvian art. *Above*, plate of polished stone with fish design, Chavin culture, around 1000 B.C. (Museo de la Universidad Nacional de San Marcos, Lima); *below*, ceramic fish sculpture from the Chimú culture, 1200 - 1450 A.D. (Museo Nacional de Antropologia y Arqueologia, Lima; Pesca Peru)

on marine resources, especially on molluscs which they could collect along the shores. Numerous campsites with mounds of mussels can still be found. The Yahgans were nomadic Indians who used large pieces of bark from old trees to build their heavy canoes. The boats always carried a fire built on rocks in the centre of the vessel. The women rowed the canoes while the men sat in the bows and watched for animals. They used spears and harpoons to hunt fish and seals. The children were accommodated in the centre of the boat. Another tribe, the Alacaluf, had similar customs but they had sails on their canoes and used bows and arrows for the hunting of birds and guanacos. Most likely the canoes were indigenous inventions by these Indians since their neighbours were only guanaco hunters and did not have such vessels. Charles Darwin observed that the Fuegians fished with simple lines without hooks but with bait. This bait was pulled over the shaft of a bird feather. Such baited lines could hardly have been very effective but they are also found in many other regions.

1.1.4 Asia

Regions of China, particularly the plains along the central Yellow River, were already densely populated by several thousands of years B.C.. The development of their culture was closely linked to the resources of the large streams and numerous other waterways, with the draining of swamp areas and the irrigation of agricultural land. Freshwater fishing was, and still is, the most important sector of fisheries. Fishing along the coasts also has a long tradition but its development was inhibited by frequent typhoons and piracy. The Chinese were always forced to produce a maximum amount of food from extremely small rural areas. This gave rise to the early development of an extensive aquaculture (see Sect. 5.2.3). Fish ponds accounted for a high percentage of the total production of fish. In places, up to one-third of the population lives by fishing and in centres of aquaculture this can extend to two-thirds.

All along the Chinese were not only hard-working people but they were also skillful fishermen, rich in ideas, who developed a wide range of unique and sometimes curious fishing tackle and methods (Fig. 15). Many of these techniques spread to other countries; famous is fishing with tamed and trained cormorants, a rather old invention of the Chinese which produced substantial quantities of fish (Fig. 16). From small boats and rafts the birds dive for fish but rings around their throats prevent them from swallowing the larger pieces of their prey. From China this method spread to Japan and southeast Asia. Another old method is the artificial planting of brushwood in shallow waters to attract fish to the

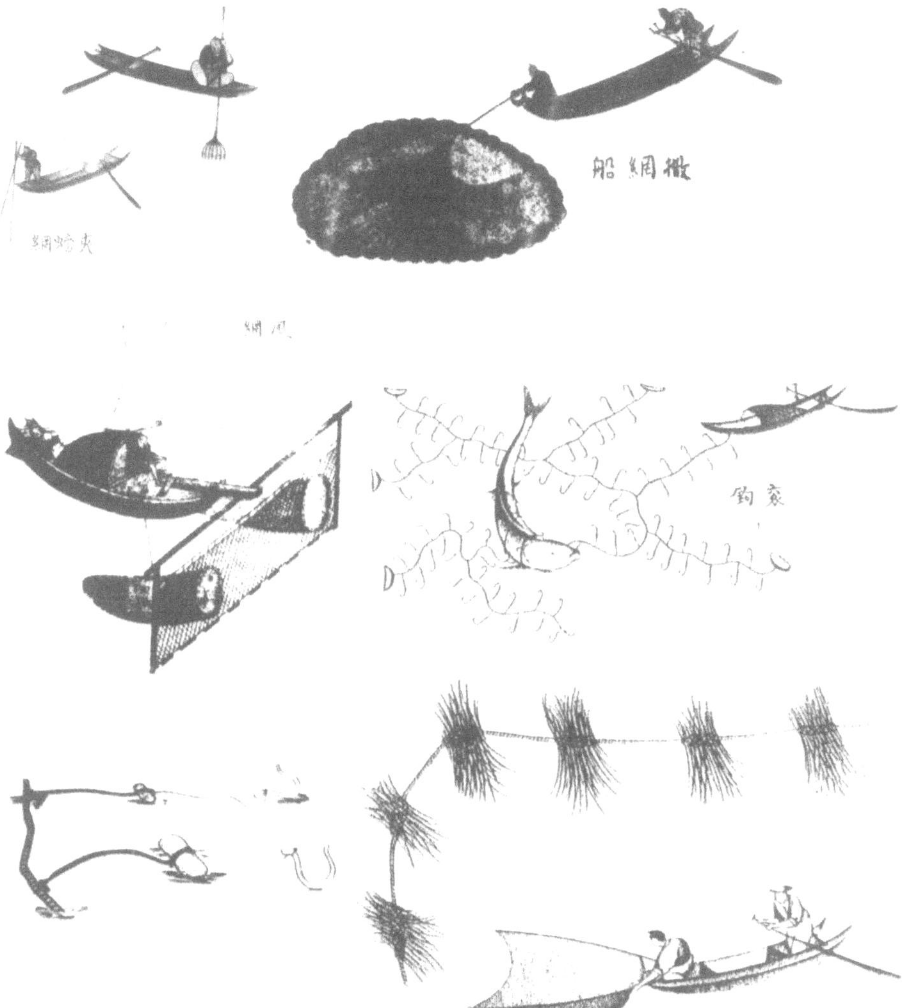

Fig. 15. Old Chinese fishing methods. *Above left*, lances with multibarbed spear and mussel clamp; *above right*, cast net; *centre left*, double bag nets on stakes; *centre right*, multiple line with rip hooks; *below left*, line with rip hooks for catching carp; blunt hooks separated from bait; *below right*, fishing with bundles of twigs on a line. (After Dabry de Thiersant 1872 in Thomazi 1947)

Fig. 16. Fishing with the help of tamed cormorants. *Above*, an old method still in use in China (Dabry de Thiersant 1872); *below*, picture on a ceramic receptacle from the Moche culture, 100 B.C. - 700 A.D. (Museo Amano, Lima; Pesca Peru)

shade and shelter; the fish can be caught with skimming or encircling nets. Again, this method developed independently in a fairly similar form far away in other continents, e.g. Africa and Europe.

Dugouts cannot be found in China. Instead, the junk and sampan were developed as peculiar types of vessels (Fig. 17). The junk is most probably the oldest real plank boat, dating back several thousand years. In early times it was used only on rivers and other inland waters but since about 1000 B.C. it has also been used at sea along the coasts. Its flat bottom and box-like form still show today its suitability for inland waters and probable origin from rafts. Such rafts were widely used to cross shallow

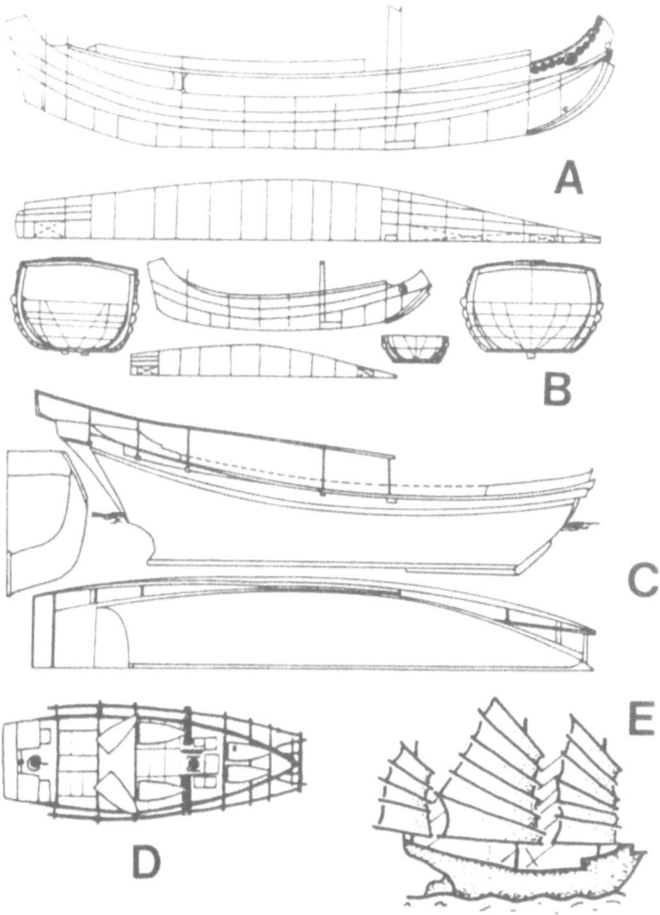

Fig. 17A-E. Chinese fishing junks and sampans. **A** Seine boat from Chekiang; **B** sampan; **C** sea-going junk for trawling; **D** deck plan of long line vessel with six sampans; **E** typical silhouette of a junk

waters and rapids at the upper courses of rivers. Often they were equipped with side-walls and it was not difficult to develop the vessel further by bowing up both ends of the boat. In many inland waters fishing was carried out from rafts that supported huts for accommodation. The location of living quarters under the elevated bridge on the junks reveals the old method of construction (Fig. 17C). Altogether, a wide range of boats were developed.

Since earliest times fish have also played an important role in Chinese culture. Inscriptions on ancient pots from the Shang culture (about 1500 B.C.) hint that fish catches were considered as oracles for rich

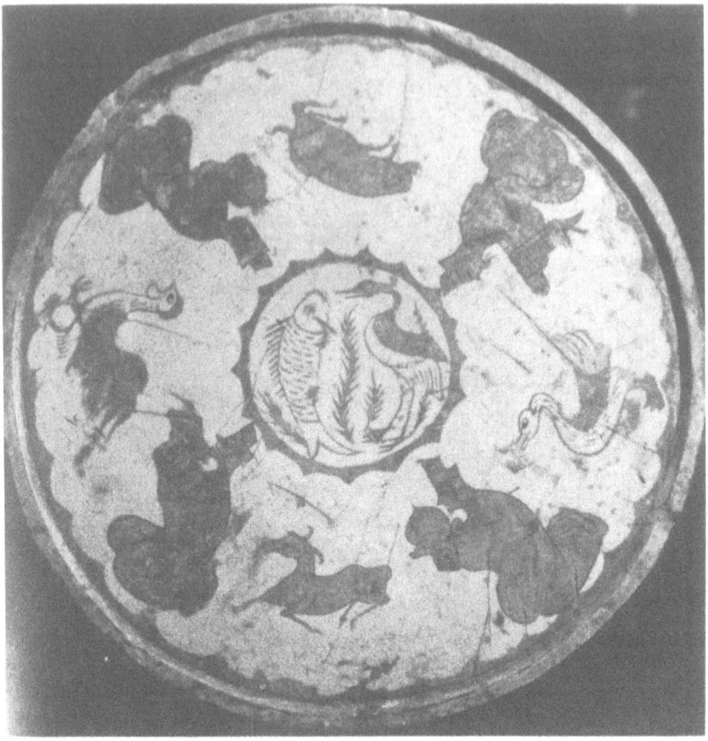

Fig. 18. Chinese plate from the Eastern Han dynasty era (25 - 220 A.D.), fire-clay, painted, diameter 37.4 cm. Four men, wearing the dress of civil servants, present offerings (?) like fish and peaches. (The Nelson - Atkins Museum of Art, Kansas City, Missouri(Nelson Fund))

harvests and prosperity. The same appears in literature and on ceramics of the Han period, 206 B.C. - 220 A.D. (Fig. 18). Fish images also served as money. In the late Zhou era, around 300 B.C., coins of copper or bronze were moulded in various forms, including fish and knives (Fig. 19). Such coins replaced kauri shells which were originally used for this purpose. Later, from the time of the Han dynasty, coins were made in round form.

In contrast to China, people on the mountainous islands of Japan, where there were wide areas unsuitable for agriculture and limited freshwaters, have depended, since the earliest times, on the resources of the sea. Numerous mounds of mussel shells and other remains of marine food were piled up along the coasts by ancient collectors. Generally, the old Japanese fishery can be considered as a branch of the Eurasian-subarctic fishing culture, like the fishery in northern Europe. The wide range of fishing

Fig. 19. Chinese fish coin of bronze from the Zhou period (1027 - 256 B.C.). (After von Erdberg-Consten 1958)

Fig. 20. In Japan Ebisu, one of the gods of luck, is the patron of fishermen; copies of this woodprint are sold at a temple in Mito. (von Brandt 1975)

gear and methods also reflects the influence of Malayan traditions. Furthermore, many inventions from Chinese fishing were imported to Japan. Up to the time when the country opened up to European influence, the Japanese used the Chinese type of junk for fishing and transport. The natural conditions for the development of fisheries were not very favourable as the shelf is mostly rather narrow and the coastline provides shelter and small landing places only in the wider parts. An exception to this is the productive and protected Inland Sea between the islands of Hondo, Shikoku and Kyushu, where open-sea fishing first developed. As in China, fish in Japan have also been considered a symbol of luck and prosperity since early times (Fig. 20).

1.2 Mesopotamia and Egypt

Climatic changes of increasing aridity since the Mesolithic period led people to concentrate in large river valleys, particularly of the rivers Euphrates and Tigris in Mesopotamia and the Nile in Egypt, where cultures of a higher civilization developed in the oases. The dense population required adequate food production and new forms of social structure and organization. The first cities were built and there was increasing specialization into various professions with differences in social standing. High waves from the upper courses of streams often caused dangerous floods, as documented by excavations. The stories of the deluge in the Bible and similar reports in the Gilgamesh epos have their sources in such historical events which might have happened around 3000 B.C.. An extensive system of dams and canals were built to serve for drainage as well as for irrigation and shipping. The extensive waters in these highly productive zones provided excellent possibilities for fishing.

In Mesopotamia archaeological material of remains of mollusc shells, fish and fishing gear can be dated to at least 5000 B.C.. Net sinkers, made of fire-clay, may be even older. Excavations and Sumerian and Babylonian texts of cuneiform characters on clay tablets reveal that there was obviously a distinct difference between the fisheries in freshwater and in the sea, i.e. the Persian Gulf. In the rivers and lakes fish were caught in hedges and weirs or by spearing with harpoons and lances. Since the middle of the third millennium B.C. fish hooks, mostly of copper, were also used (Fig. 21). The most productive waters were the wide marshes in the lower reaches of the rivers Euphrates and Tigris. The catches consisted mainly of cyprinids, especially marsh carp *(Barbus sharpeyi)* and

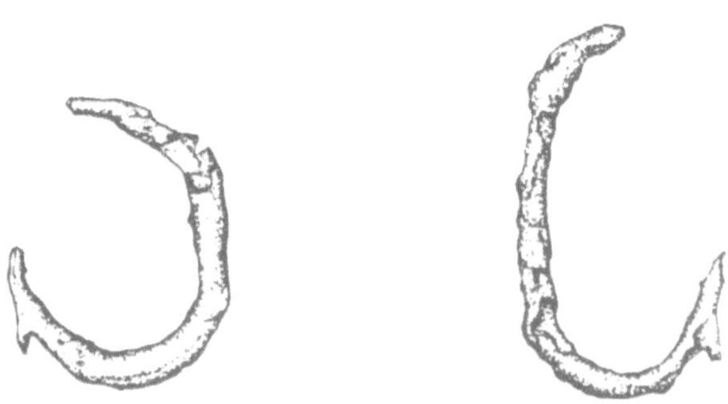

Fig. 21. Fish hooks, made of copper and 4 cm long, from Mesopotamia, about 3000 B.C.. (After a drawing by von Lorentz-König in Strommenger 1980)

Fig. 22. Bowl from Samarra, upper Euphrates, late sixth millennium B.C., showing in the centre four birds catching fish (silurids ?). Fire-clay, diameter 26 cm. (Formerly at Staatliche Museen, Berlin)

giant carp *(B. esocinus)*. In the estuaries and coastal waters of the Persian Gulf fishing was carried out mainly with cast nets and gill nets, and possibly also already with trammel nets. The most common marine fish was the threadfin *(Polydactylus tetradactylus)*. Pearl fishing is also very old in these waters.

Altogether 324 Sumerian and Babylonian fish names are known for about 90 species. Of these, 36 have actually been identified. Four old names are still used in Iraq. The Sumerian and Babylonian texts contain reports on the food and distribution of some of these fish species. They show also that fishing regulations existed with penalties in cases of infringement, and that fishing rights were leased out. Guilds of fishermen were well organized. Some tribes along the lower Euphrates lived exclusively on fish. Fishermen collaborated closely with bird catchers.

Fig. 23. Roller-seal from Mesopotamia with fish being sacrificed by priest. (Musée du Louvre, Paris)

Fig. 24. Assyrian fisherman hooking a fish while riding on a float of goat skin; detail from a relief of Sanherib, around 700 B.C.. (British Museum, London)

Fish was an important food; transport to and marketing in the cities was organized. Fish processing included drying in the sun and larger fish were clipped and hung on lines for this purpose. Also, salting, fermentation and preparation of pickled fish existed and possibly also the smoking of fish by drying them near the fire. Fish roe was processed to a kind of cheese.

Fish also played a major role in myths (Fig. 22). In the temples fish were offered as sacrifices to the gods (Fig. 23). Fishermen were employed by the temple authorities and others had to deliver parts of their catches as duties. Substantial fish consumption, mostly of large carp and marine fish, in the temple districts is documented by remains found in the archaic layers of Uruk. Fish that were not eaten by the priests and their personnel were sold to the public. In certain temples and on specific days fish consumption was prohibited. Fish and turtles were also offered to dead ancestors.

Fishing in Assyria is documented on several reliefs from ancient palaces and other buildings (Fig. 24). Fish were caught with hook and line but it seems that rods were unknown. Besides hand lining, often carried out from floats constructed of animal skins, nets were also used from small boats. As long as 4000 years ago many towns in Assyria had fish ponds where not only freshwater fish but also various marine species were kept as a food supply.

From ancient Egypt many well-preserved remains provide a broad impression of daily life, including fishing activities. Reports by Herodotus, Aelian and other historical writers contain valuable information to complement the overall picture. It appears that fish was a most important and cheap food for most people in the Nile valley since the earliest times. There was no real class of full-time fishermen but all fellaheen and people on the banks of the river Nile used to fish in summer during the flood season, after the end of the agricultural harvest. Sizeable quantities of fish and other animals could also be collected in remaining pools and on the mud of the flood plains after water run-off. The profession of fisherman was regarded socially as only of a low level. Catches usually had to be delivered to the lords of the manor and fish were often used, jointly with other natural produce, as payment in kind for labour. Catches were subject to taxation and fees also had to be paid for fishing rights.

Fish were gutted and eaten raw or grilled. Large quantities were clipped and dried in the sun, often hanging on lines. Also, salted fish had a wide market in Egypt and for export. Mackerel was pickled in brine. Transport

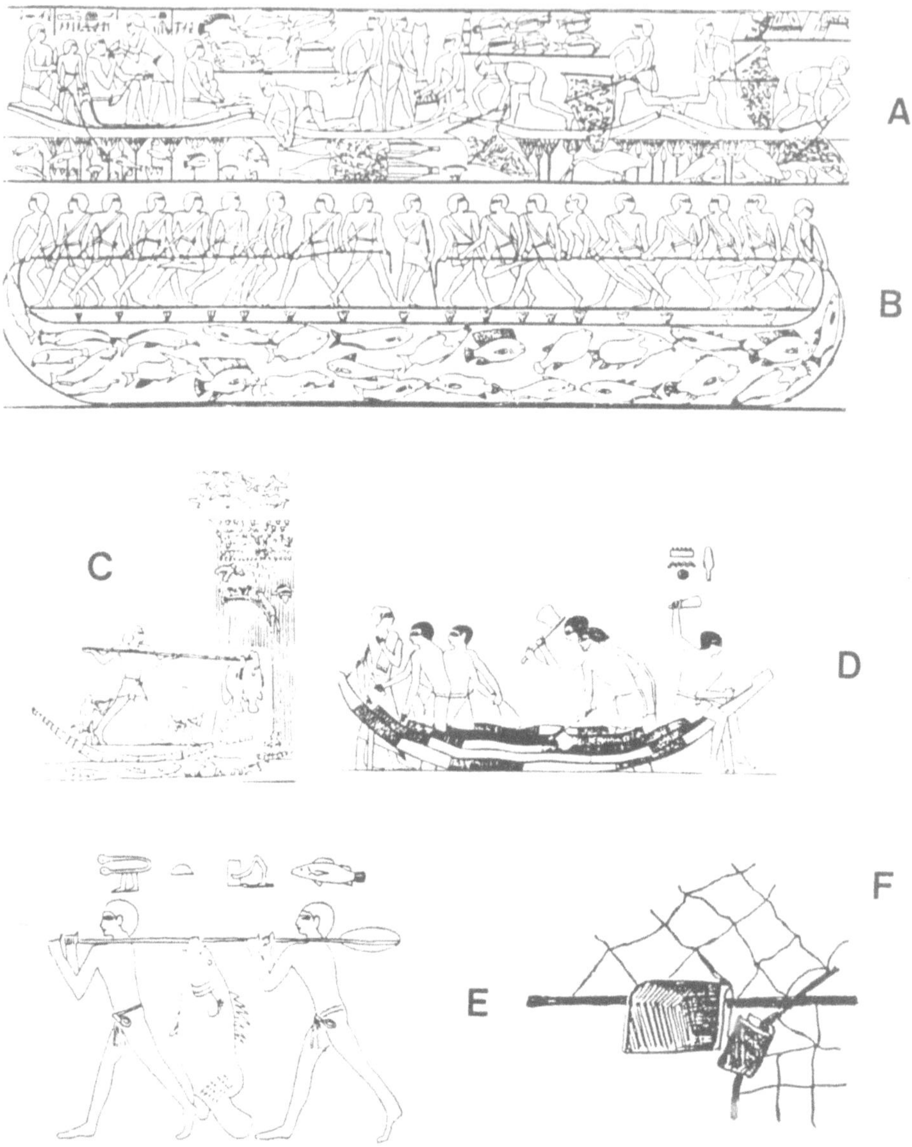

Fig. 25A-F. Fish and fishing in ancient Egypt. **A** Catching fish with hooks and lines, baskets and skimming nets; processing of fish; **B** fishing with seine net carried by nine men on each side under the command of a foreman; also shown are typical fish species from the river Nile. From tomb of Mrrwk, Saqqara, 6th dynasty, around 2300 B.C.; **C** spearing of fish from a papyrus raft; **D** construction of a boat; **E** transport of a Nile perch *(Lates niloticus)*; **F** remains of netting with lead sinkers, 14th century B.C.

of fish in baskets or on stakes shouldered by two people has been portrayed (Fig. 25A,E). This shows great similarity to present conditions. It seems that fish as food consisted almost exclusively of freshwater species, while sea fish was considered unclean. Fish was not only the staple food for the majority of people but was also supplied to palaces and eaten by the nobility. Upper-class people often engaged in sport fishing, mostly using spears and harpoons (Fig. 25C).

Fishing with hand lines was very common. Hooks were first made of mussel shells, bone and horn and later of metal. As documented by excavations, hooks around 2000 B.C. were made of copper and later of bronze, 2 - 6 cm long. In early times they were barbless but from the 12th dynasty (1991 - 1786 B.C.) onwards they were barbed and had a ring or plate for fixing a line to. Lines were fitted with one or more hooks; also, rods were used with one or more lines and several hooks. It is doubtful whether certain pointed bone or ivory pieces can be interpreted as gorges.

Fish were also caught in wicker baskets which existed in a large variety of forms, mostly funnel- or pear-shaped, with an opening at the rear end for removing the catch (Fig. 25A). Smaller baskets were set up side-by-side or opposite each other. Large baskets reached a height of 1.50 m and certainly had to be operated by several people. There are also hints for the existence of fyke nets, i.e. traps with nets. Skimming nets positioned between two crossed sticks, often with transom or linking ropes, are frequently illustrated. Presumably they were used not only for fishing but also for hoisting the catch.

Seine nets were operated by groups of fishermen under the command of foremen from the beach and from boats (Fig. 25B). Most of these nets were of simple construction without bags in the centre. They had floats and sinkers of stone or lead (Fig. 25F). The nets were made of flax or other natural fibres with small mesh-sizes of up to 1.9 cm. Remains of netting found in a tomb from the 12th dynasty proved the existence of weaver knots which also developed quite independently in northern Europe and which nowadays are the most widely used type of knot for making fish nets. Lift nets and cast nets were also used. The latter method, very old and even mentioned in the Bible, was later spread by the Arabs from the Near East to Africa and then by the Greeks to Europe.

The almost 4000-year-old model of two papyrus boats towing a net over the ground is very illustrative (Fig. 26). Fish caught were killed with clubs as shown on the deck of one of these boats.

Fig. 26. Model of two papyrus boats towing a net, found in the tomb of Mektiré at Der-el-Bahari, from the 11th dynasty, around 2000 B.C. Wood, painted, length 91 cm, height 24-26cm. (Egyptian Museum, Cairo)

Old Egyptian illustrations not only demonstrate how fish were caught but also the species that were taken (Fig. 25B) and their distribution. More than 30 of these species could be clearly identified. All of them are still abundant today in the Nile valley waters so that the fish fauna does not seem to have undergone much change since the beginning of the ancient Egyptian culture about 5000 years ago. Fish species most frequently portrayed are the Nile tilapia *(Tilapia nilotica)* and other *Tilapia* species, the Nile perch *(Lates niloticus)*, some mullets *(Mugil cephalus, M. capito)*, the bottlenose *(Mormyrus* sp.*)* and bynni *(Barbus bynni)*.

Evidently the Egyptians already had substantial ichthyological and biological knowledge, based on observation of fish, their distribution, feeding and behaviour. Extracts from various parts of fish were used as drugs for medical purposes. Some ancient pictures indicate that in Egypt, as in Syria, installations existed to allow fish to be kept in ponds.

Also, in ancient Egypt, fish were significant magic or religious symbols. Fish images were found as gifts in numerous tombs (Fig. 27). Nile tilapia, a mouth-breeding species, was particularly taken as a symbol of palingenesis and was carried on amulets. Other species which prevailed in fish worship were the Nile perch and oxyrhynchos *(Mormyrus* sp.*)*. In the late New Kingdom devotion to fish was often combined with a ban on the consumption of certain fish species at certain times and in certain localities, especially by priests in the temples.

Fig. 27. Glass receptacle from Tell el-Amarna, Egypt, about 1350 B.C. Length 14 cm; yellow and white; fins partly blue. (British Museum, London)

Both in Mesopotamia and Egypt, the dearth of wood created special conditions for the construction of boats. The existence of large thickets of papyrus and reed led to the building of rafts. In Mesopotamia shipbuilding remained at a rather simple level. The "gaffa" was a rounded reed bark, made waterproof by animal skins, oakum and pitch. The "kelek" was a reed raft with air-filled skin-bags underneath to increase its buoyancy.

Development went much further in Egypt, where boats played a larger role for both transport and fishing. Simple devices with bundles of reed developed into vessels with thick and firmly laced papyrus which were shaped upwards into bows at both ends (Fig. 26). Such boats were moved by poles or paddles. This original form of the papyrus raft was retained when such vessels were built of wood. Even the old cords along and around the boat were imitated on the wooden vessel by carving or painting. Since it was impossible to produce beams and planks of more than 2 m in length from the endemic acacia and sycamore trees (imports were expensive and palm tree wood was unsuitable), ships were "bound", according to Herodotus, i.e. short but thick blocks of wood were assembled like bricks (Fig. 25D). Soaking in water provided a very durable connection by means of plugs. This type of boat was not only in use for a very long time but it was also developed further into larger seaworthy vessels. Later, the wooden ships were propelled by oars and had a fixed steering rudder. They were rigged with a square sail. Here again, as in China, sea-going vessels were developed from inland-water boats for purposes for which they were neither originally constructed nor especially suited.

1.3 Mediterranean Cultures

In the eastern Mediterranean region Jews were engaged in inland fisheries and Phoenicians in sea fisheries. Both were influenced by developments in Egypt and Babylonia.

The Jewish fishery concentrated on Lake Tiberias (Sea of Galilee) and the Jordan River system. In the lake the main food fish species, the musht *(Tilapia galilea* and some other related species), barbels *(Barbus longiceps, B. canis, Varicorhinus damascinus)* and "sardines" *(Acanthobrama terrae-sanctae)* were caught with seines, cast nets, trammel nets (three-wall gill nets) and hooks and lines. Fish without scales, like catfish *(Clarias lazera)*, lampreys and eels, were considered unclean and were forbidden by law and not eaten by the Jews. Descriptions of these fisheries are contained in the Old and New Testaments of the Bible and the Talmud.

Activities of the seafaring Phoenicians are well documented and as traders they contributed greatly to the diffusion of technical knowledge in the Mediterranean region and beyond but so far little is known about their fishing. Their vessels were similar to the Egyptian ships but showed later influences from the Aegean peoples, who used ships with high bulwarks, a single mast with a haul-up square sail and perpendicular stern. Near the ancient Phoenician town of Sidon large mounds of shells of purple snails *(Murex brandaris, M. trunculus)* indicate that the production of purple dye was an important sector.

In contrast, there are plenty of lively and realistic presentations of fish and fisheries from the culture of Crete (around 1500 - 1400 B.C.). Objects of art mostly show rather stylized pictures of marine animals such as flying fish, tunas and dolphins as well as cephalopods, mussels, snails, starfish and corals (Figs. 28 and 29). Dolphins were highly esteemed as fishermen believed that these friendly and intelligent animals helped them to drive fish into their nets. Killing them was a sin. The presentations seem to indicate a mainly artistic interest, but pictures of food fish are also shown in frescos and on numerous vases (Fig. 30). Cave paintings illustrate various fishing methods such as the use of cast nets, baskets and fish hooks. Doubtless fish and shellfish were an important dish for the Cretans. Salted fish roe (tarichos) was served widely throughout areas of the eastern Mediterranean both as food and an aphrodisiac.

Special production and trade developed from the utilization of purple murex in the east of Crete. This was the basis for the wealth of the har-

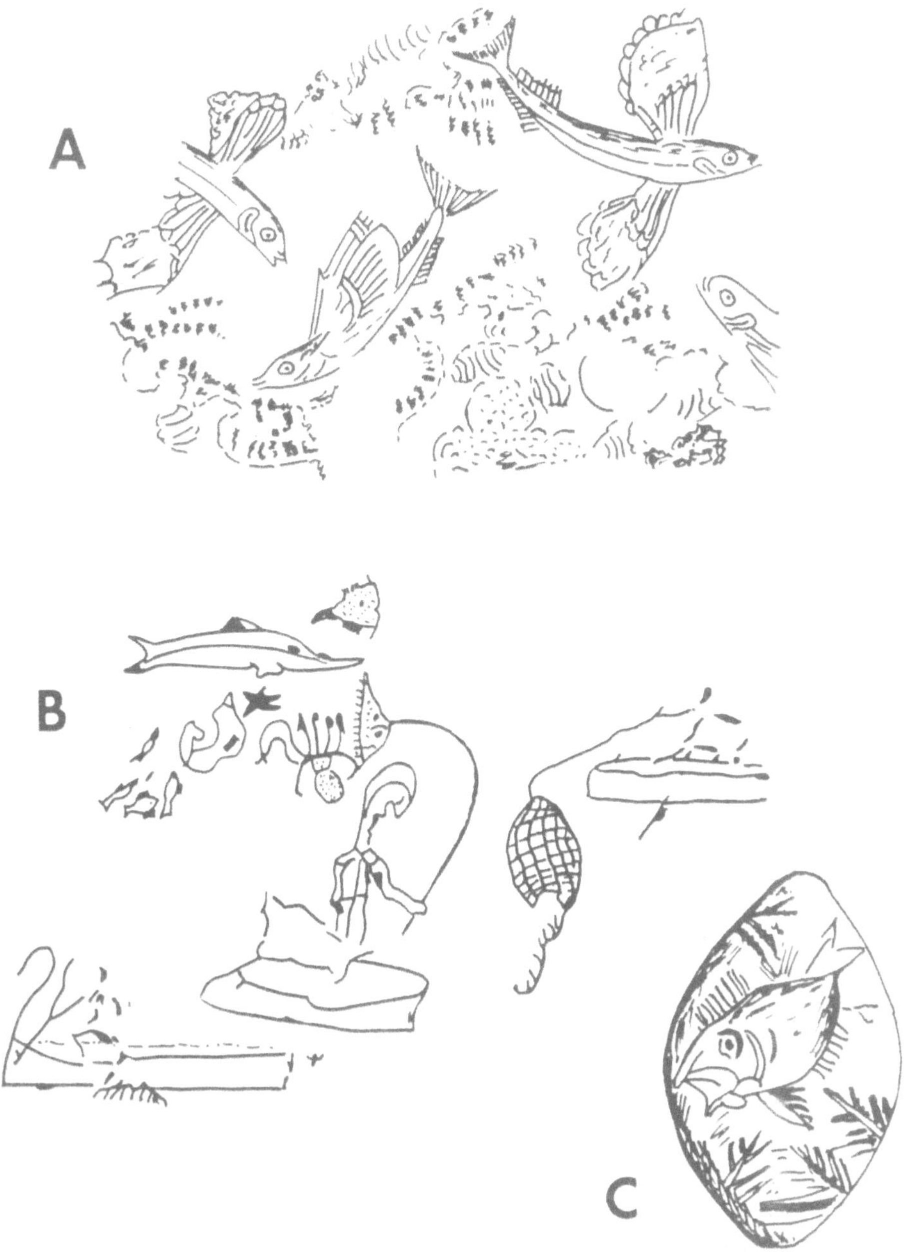

Fig. 28A-C. Fish images from Minoic times in Crete. **A** Parts of a Minoic fresco from Phylakopi (Melos) with flying fish *(Exocoetus volitans)*; **B** painting from the Vornofeto cave at Kato Pervolakia (Sitias), about 1400 B.C., showing fishing with cast net and by other methods; **C** seal of semi-precious stone with fish and water plants. (**A,C** after Matz 1956; **B** after Faure 1973)

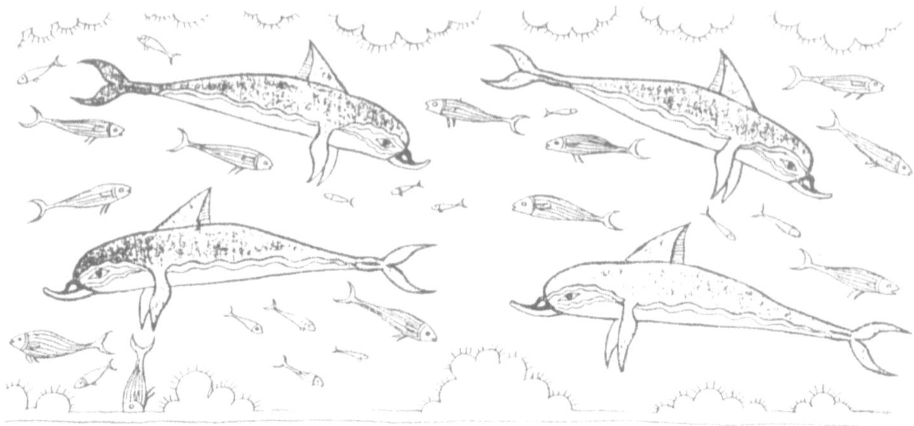

Fig. 29. Fish and dolphins in Minoic art. Fresco from the megaron (living room) of the Queen of Knossos, Crete. (Kreuzer 1974)

Fig. 30. Young man from Thera (Santorin) with bundles of dolphin fish *(Coryphaena hippurus)*. Minoic fresco, about 1500 B.C., Athens. (Photo: Ekdotike Athenon SA)

bour of Itanos (now Erimoupolis) and other places. Aristotle and Pliny the Elder described the process of extracting the stain from the snails. The highly valued purple stain was exported overseas for the dyeing of garmets of dignataries. Purple murex are depicted on several Minoic vases.

The waters of the Aegean Sea have been used for fishing since very early times, as verified by excavations, for instance at Troy from 2000 B.C.. Ceramics of Mycenaean origin depicting fish figures were also found in Syria (Fig. 31). It seems that the Greeks, descending from inland people, the Achaeans and immigrant Dorians, had first to develop their fishery. Homer (ninth or eighth century B.C.) in his poetry on the heroic age before 1200 B.C. expressed disgust at eating fish but his frequent remarks about fishing indicate that he was quite familiar with this subject. In *Iliad* and *Odyssey* he mentioned fishing with hooks, nets and harpoons and also the collection of oysters and other molluscs by divers. It appears that this work was carried out by a distinct class of fishermen. During the early period fish were mostly eaten by the poorer people while the nobility preferred meat.

Later, when fishing had developed further, consumption of fish and other animals from the sea was more widespread, particularly in Athens

Fig. 30

Fig. 31. Drinking vessel (Rhyton) from Minet el-Beida near Ugarit, Syria. Mycenaean, second millennium B.C.; hight 12 cm. (Syria 13)

Fig. 32. Fishmonger cutting a tuna, shown on a vase from the fourth or third centuries B.C. (Museo di Cefalú, Sicily)

(Fig. 32). Fish even became a famous delicacy for luxurious banquets. The life and activities of the fishermen were vivaciously described in Attic comedies, for instance by Antiphanes (around 350 B.C.) and Menander (300 B.C.) and in poems or idylls, for instance by Theocritos (born 305 B.C.). Most of the literature was lost but presentations on vases, frescos and mosaics still provide a lively picture of the habits and customs (Fig. 33). Although fishermen were often mentioned in poetry, they were usually considered poor people of low social standing, e.g. by Plato and Plutarch.

During the second and third centuries A.D., several descriptions of fishing were given in didactic poems. Obviously based on personal experience Oppianos (around 200) in his *Halieutica* gave very detailed and most interesting information on catching fish with hooks, nets, baskets and with trident spears. Knowledge of life habits, reproduction, food, distribution and seasonal conditions of fish was already advanced. Fishing for

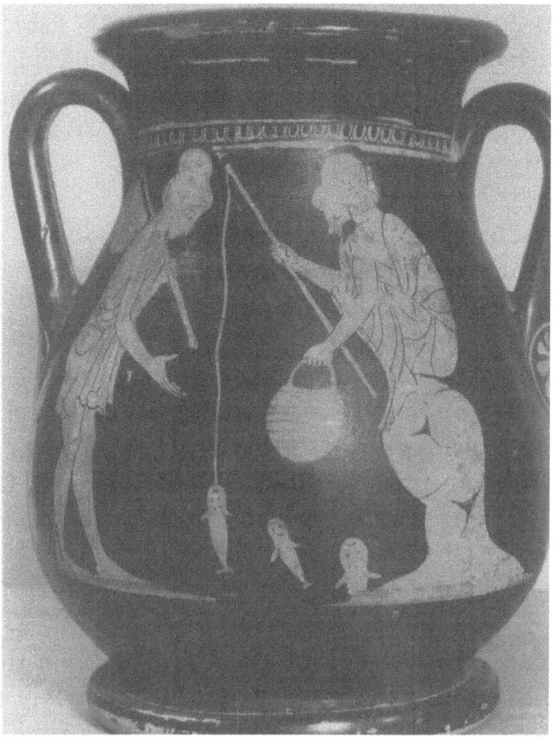

Fig. 33. Fishing with rod, line and hook depicted on a Greek vase, 480 - 470 B.C. (Kunsthistorisches Museum, Wien)

migratory species, primarily tuna, pelamids, swordfish and mackerel, was described. Besides fish corrals, possibly used already by the Phoenicians, there was a fishery with encircling nets operated from four to five collaborating vessels. Fish shoals were detected from high observation points along the coast and were pushed towards the gear by scaring with noise and hitting the water. Furthermore, cast and drift nets were used for the capture of fish near the surface. Also, the attraction of fish by torch light was known, a method still used today in Mediterranean light fishing. During fishing with hooks and lines not only floats and sinkers were used but also tubes of horn to prevent the line being bitten off. Artificial flies were made of cloth or feathers and were also fixed to troll lines that were towed rapidly through the water. Reed bundles were stuck into the ground to attract heliophobous fish. The trident fish spear even became the symbol of the god Poseidon. Dolphins, which were sacred to this god of the water and to other deities, were often depicted.

A most comprehensive report on the biology of fish was written by Aristotle (384 - 322 B.C.) in his zoological publications. He listed 110 fish species, mostly marine, and distinguished between viviparous cartilagenous fish and oviparous bony fish. Based on observations on animal behaviour, he presented useful suggestions for the capture of fish and provided details on the migration of tuna.

From the medical point of view Hippocrates (born 460 B.C.), Galen (129 - 199 A.D.) and others evaluated various fish species as to their contribution to a healthy diet.

A wide-ranging trade of fishery products developed. Even from the mouth of the Dnieper River and the Sea of Azov lumps of dried fish and jars of salted and pickled fish (tarichos) were shipped to Athens and other Greek towns. Byzantium and Sinope on the southern shores of the Black Sea, both originally fishing villages, developed to become centres of salt fish trading.

It seems that in Rome during the early era fish were not a major food source. But, later, fish, oysters and other animals from the sea, and to lesser extent those from freshwater, played an important role. Wealthy people had a special liking for rare fish which were very highly prized. Sea bass *(Morone labrax)* was the most expensive fish during the times of Emperor Augustus. Red mullet *(Mullus barbatus)*, also highly valued, was served at all banquets. Sturgeon was a luxury imported from the East via Rhodes.

In reports from Roman times little mention is made of fishermen besides the remark that they were considered to be poor and deplorable people. In contrast, much is known of the sport fishing undertaken byupper class people. Apart from exorbitancies like Nero's nets of golden filaments and purple ropes, mullet weighed in gold or morays fed with slaves, it seems that further progress was made in fishing methods and tactics. One example is the refinement of fishing with hook and line to an eminent skill by the use of bait soaked in natural or artificial aromatic substances. Fishing of bluefin tuna, bonito and other fish with large iron hooks and in complex installations of traps (almadrabas, madragues, tonnare) in many coastal areas around the Mediterranean was especially well-organized and most profitable.

The importance of fish and fishing is documented by many pictoral and literary objects, for instance in masterpieces of mosaics and frescos (Fig. 34). Paintings in Etruscan tombs prove that also in Etruria fishing

Fig. 34. Part of a Roman mosaic from Tunisia, early third century A.D. (Museum at Sousse)

was important (Fig. 35). Dolphins and fish played a dominant role in mythology. In Roman literature the didactic poems and epigrams of Ovid (43 B.C. - 17 A.D.), Martial (40 - 104) and Ausonius (ca. 310 - 393) were particularly concerned with fishing. In his *Metamorphoses* Ovid described various fish-catching methods and his poem *Halieuticon* contains a catalogue of fish and details of fishing, for instance the operation of large tuna traps, which are most likely based on personal observations. His list of fish comprises 56 species and, like another description by Pliny the Elder (23 - 79) in his *Naturalis historiae*, is much less comprehensive than that given by Aristotle 400 years earlier. Martial portrayed fishing

Fig. 35. Fishing by the Etruscans; part of a fresco in the "Tomba della Caccia e della Pesca" at Tarquinia, Italy, about 520 B.C. (Kreuzer 1974)

by hook and line. Ausonius described the fishery in the river Moselle in Germany with hooks, seines and lift nets. He listed 15 fish species in this river, including river trout, sturgeon, pike, tench, barbel and river lamprey.

Already during the golden age of Greek culture the first attempts at aquaculture were being made. Eels were kept in bins and Aristotle reported that oysters from banks near the island of Lesbos were transplanted to similar hitherto unpopulated grounds off Chios. There is also evidence that Carthaginian prisoners built a fish pond near Agrigentum in Sicily. During Roman times the construction of fish ponds was widespread, especially in the neighbourhood of villas of distinguished people. Much information on fish ponds and aquaculture is contained in publications by Varro (116 - 27 B.C.) and Columella (around 65 A.D.). Varro distinguished between two types of ponds: "dulces" or freshwater ponds kept by commoners for profit from sale of fish and "maritimae", seawater ponds maintained mostly by rich people. Licinius Crassus, later called Muraena, was the first person in 92 B.C. to create in Rome vivaria for feeding fish. Supply of water, its temperature and other conditions were already controlled. Considerable investments were made in this field and transition from agriculture to production of fish in ponds often yielded high profits. Ostarias, installations for the culture of oysters, possibly already existed from the Greek colonization along the Gulf of Taranto and certainly by around 100 B.C. in Lake Lucrinus. Refugees are said to have developed the onset of valliculture in the swamps of Comacchio by the Adriatic Sea; these later became famous as brackish water ponds for culture of eels. According to reports by Pliny the Elder, transplantations of certain preferred fish species as spawners, eggs or young fish were made to new areas.

The overall increase in the population, especially in the region of Rome, resulted in a higher demand for food. It was estimated that during the second century Rome had a population of about 1.2 million. By the existing fairly simple methods of fishing no major increase in production could be achieved. There are even several reports from 200 - 300 A.D. on decreasing yields from coastal waters close to Rome. Under these circumstances a large import of fish and fishery products developed. This was promoted by the general evolution of trade, shipping and basic industries in the Roman Empire. Ostia, Rome's main harbour for the import of grain, also became a trade centre for salt fish and other products. The main fish species on the public markets were tuna (mostly salted), bonito, swordfish, mackerel, eel, conger, mullet and bream. Mullet roe was a favourite dish. Dried and salted Nile perch imported from Egypt fetched good prices. The selling of fish and prices was controlled by the government authorities.

As documented by the large heaps of fragments of earthenware from vessels that were used for the transport of fish products and which have been found in Rome, Pompeii and many other places, imports came mostly from Spain. Fish preserves from Gades (Cadiz) and Malaca (Malaga) were not only sent to Italy but were likewise identified during excavations in the Rhine valley. Sardines in oil were imported from the western riviera (southern Gaul). Also, on the coast of the Syrte and Tunisia in northern Africa installations for the salting of fish developed. The Greeks had already obtained fish from the Azov and Black Seas, partly from their own fishing and processing activities and partly from imports. Now, during Roman times, imports of salted fish and caviar from these waters came via Sinope and Byzantium and gained in importance. Another major source for imports was the northern region of Gaul.

The various fish species and their provenance were valued quite differently. Their consumption varied as a result of trends in fashion. Preservation was, of course, a basic requirement for the transport of this perishable food over long distances. Ice and snow were used in winter for transport from the north, mainly Gaul. Ice cellars may have been of help in the summer. Another method of preservation was the production offermented food. Garon in Greek or garum in Latin was a strong-tasting and strong-smelling spice sauce made from the guts of Spanish mackerel *(Scomber colias)*, tuna and other fish. Black "garum sociorum" of mackerel processed in Spain was considered the best and fetched high prices.

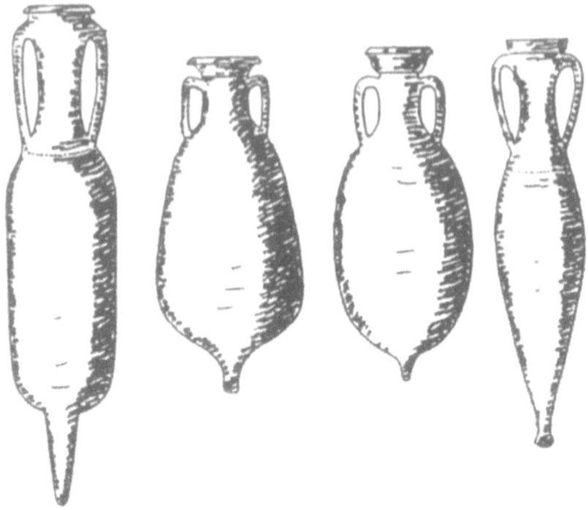

Fig. 36. Various forms of Roman amphorae used for the transport of garum. (After Callender 1965)

Remains of garum were mixed with salt brine and served under the name of alec as food for the man in the street. Muria was also a fish sauce to which other animal food, mostly crustaceans or mussels, or vegetable food had been added.

Factories for the production of garum were traced inter alia in Italy (Rome, Pompeii, Puteoli), Spain (Carteia, Carthago Nova, Barcino, Ibiza), North Africa (Morocco, Leptis Magna, Tunisia), Istria and Bithynia (Turkey). Usually the processing installations with basins for fermentation and salting were located close to tuna traps. In several such places tuna traps were still operated until recent times. As for fish under olive oil and in brine, the fish sauce was transported in amphorae (Fig. 36). Such characteristic earthenware vessels, mostly used in the trade of wine and olive oil, had been produced since the Bronze Age in Mesopotamia, Syria and Egypt, and later in Greece, Italy and other places around the Mediterranean. Thousands of them have been excavated on land or discovered in sunken cargo ships and they prove the wide trade connections that existed in those times. Shipwrecks containing cargoes of fish sauce were found mainly along the Spanish coast of Andalucia, around the Baleares, off southern France, in the Strait of Bonifacio between Corsica and Sardinia, along the Italian coast north of Rome and on the northern coast of Sicily. From these discoveries it can be concluded that the main shipping routes were from Spain via the Baleares and the Strait of Bonifacio to Rome and, as a second route, from Spain all along the continental

Fig. 37. Cameo from the early Christians with fish, pigeons and anchor. (British Museum, London)

coast to Rome. Many of the amphorae carry inscriptions which allow the identification of the manufacturer, his location, the approximate date and the quality of the product.

Many recipes from olden times with garum as an ingredient have been saved over the years. This fish sauce still appeared on menus in the Middle Ages and garum was still eaten in Constantinople in the 16th century. Often garum was used for medical purposes, for instance to ease indigestion, act as an appetizer and also in treatment of many other diseases.

The fish trade was closely linked to the production and distribution of salt required for preservation. Salt pans and mines were operated in many places in the Roman Empire, particularly in southern Spain, Ibiza, southern France (Narbonensis), Sicily, Egypt and in the salt lakes of Phrygia and Cappadocia (Anatolia).

In the early times of Christianity fish and fishing gained a new importance in religious matters. The faithful used the fish symbol as their secret distinctive sign when they were a persecuted brotherhood (Fig. 37). This symbol was painted or carved into the walls of Roman catacombs and early churches where the Christians met for worship. The custom vanished about 300 years later when Christianity was formally accepted under Emperor Constantine. The reason for choosing the fish for this pur-

Fig. 38. Peter the Apostle with his brother Andrew fishing from a boat; mosaic in the church San Apollinare Nuovo at Ravenna, sixth century

pose is open to speculation. Several explanations have been offered; one assumption is that the Greek word for fish "ichthys" stands for the abbreviations of Jesus Christ, God's Son, Saviour (Jesous Christos, Theou hyos, sotér). It seems more likely, though, that the use of the fish symbol was related to the sign of the zodiac pisces. Ancient Jewish astrologists predicted that the manifestation of the Messiah would occur in the time of the conjunction of the planets Saturn and Jupiter in the fish zodiac, an event which actually happened in the year 7 or 6 B.C. and could be an explanation for the appearance of the "Star of Bethlehem".

The Bible contains many references to fish and fisheries, mainly in the New Testament. Most of Jesus' sermons were held by the shores of the Sea of Galilee and fishermen were his earliest disciples (Fig. 38). A comparison of related texts from the Gospels detailing fishing gear and operations used on the lake until the 1950s indicates that fishing with seines, cast nets (Fig. 39) and trammel nets had hardly changed since biblical times. The story of the miraculous draught of fish presents an informative description of fishermen's life. Also, the miracle of the feeding of the 5000 is concerned with fish and there are many presentations of the Lord's Supper with fish and bread.

Fig. 39. Roman oil lamp from Alexandria, decorated with picture of cast net fishing. (Nun 1989)

Pictures of fish were carved on pebbles and into bricks by members of the Cathari or Albigenses in southern France and northern Italy; these people were considered to be, and were persecuted as, a heretical sect in the 11th to 13th centuries. It is not known whether these fish images were meant as a memory of the "ichthys" from the Roman catacombs or whether they served other purposes.

2 Main Fisheries in Europe since the Middle Ages

2.1 Mass Migration and the Vikings

From the time of the fall of the Roman Empire and the unrestful period of mass migration information on fishing is rather scant. Certainly fish, both from freshwater and from the sea, continued to be an important food source. According to Solinus (around 240), the natives of the Hebrides, west of Scotland, lived mainly on fish. When the land along the southern shores of the North Sea was colonized in the first century, fishing was apparently much less important than agriculture. But excavations near the mouth of the river Weser and in Eiderstedt on the German coast proved the utilization of several fish species, like sturgeon, salmon, cod, haddock, plaice and flounder. Records from the sixth century onwards indicate that the English, French and Flemings were actively engaged in fisheries, mainly herring, in the Channel and the North Sea.

In Scandinavia colonization reached northern Norway during the fourth century. People lived mainly as farmers but also hunted, fished and caught whales and walruses. Pictures engraved on rocks show that the light boats covered with skin or bark and used during the Stone Ages had, since the Bronze Age, been replaced by larger seaworthy vessels. These wooden ships were first built with double prolongations on both ends, as exemplified by the Hjortspring boat from the third century B.C., and excavated from the Alsen Peninsula in Denmark (Fig. 40). Later, the typical long boats of the Vikings were developed with clinker construction and high-bowed stems. The oldest ship of this type known so far is the Nydam boat unearthed near Schleswig in northern Germany and dating from about 400 A.D.. The most elaborate boats of this kind are those from Oseberg and Gokstad, originating from the ninth century and exhibited in Bygdö near Oslo. Such vessels were depicted on numerous coins and other objects. Originally the ships were moved by paddles and later by oars. Square sails were only used here since the seventh century, i.e. much later than in the Mediterranean region.

The construction of these fast and flexible vessels, fit for the high seas, enabled the Vikings to undertake long voyages. In the eighth and ninth

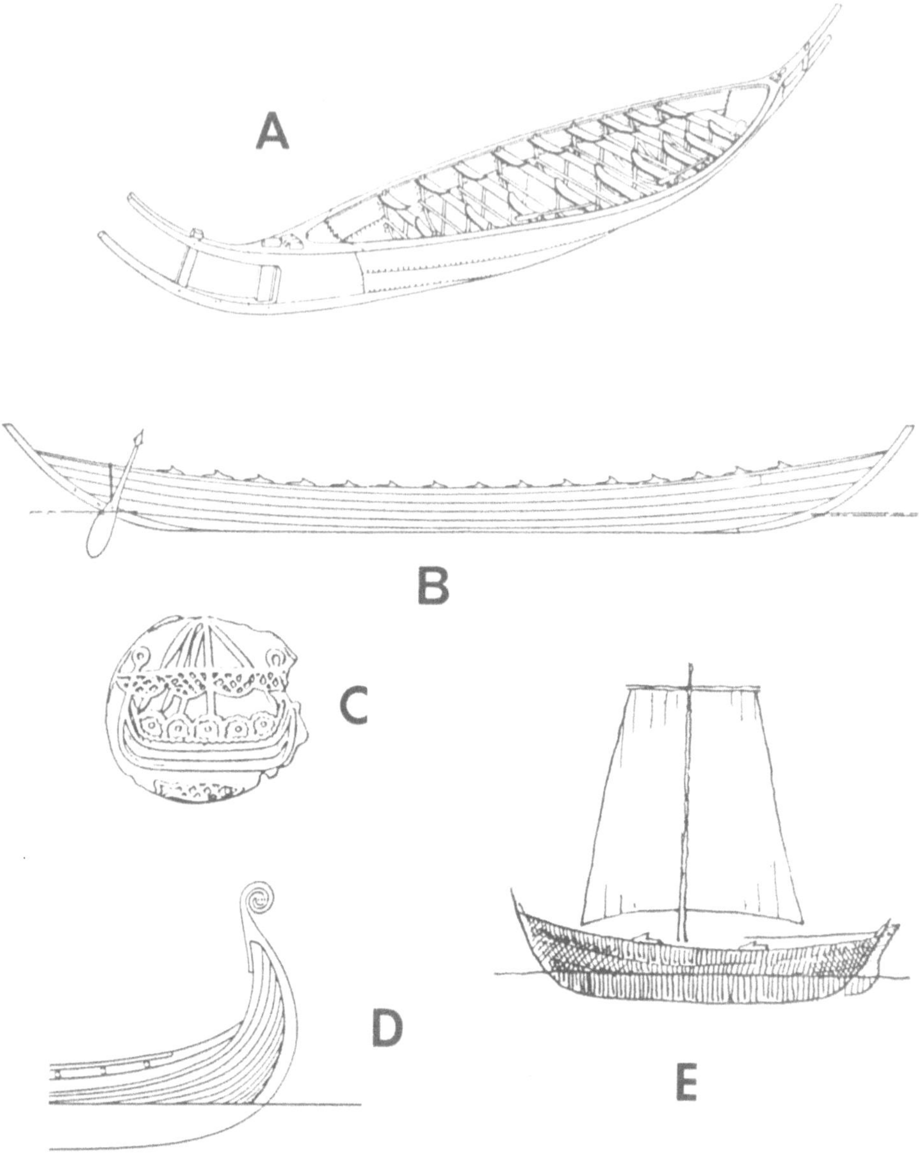

Fig. 40A-E. Viking ships. **A** Boat from Hjortspring, Alsen, Denmark, second century B.C.; length 19 m, width 2 m; **B** Nydam ship from the Nydam fen near Schleswig, Germany, sixth century A.D.; length 23 m, width 3.25 m, clinkered; **C** presentation of Viking ship on silver coin, ninth century; **D** stem of ship from Oseberg, southern Norway, ninth century
E for comparison, sailing boat from western Norway, early 20th century

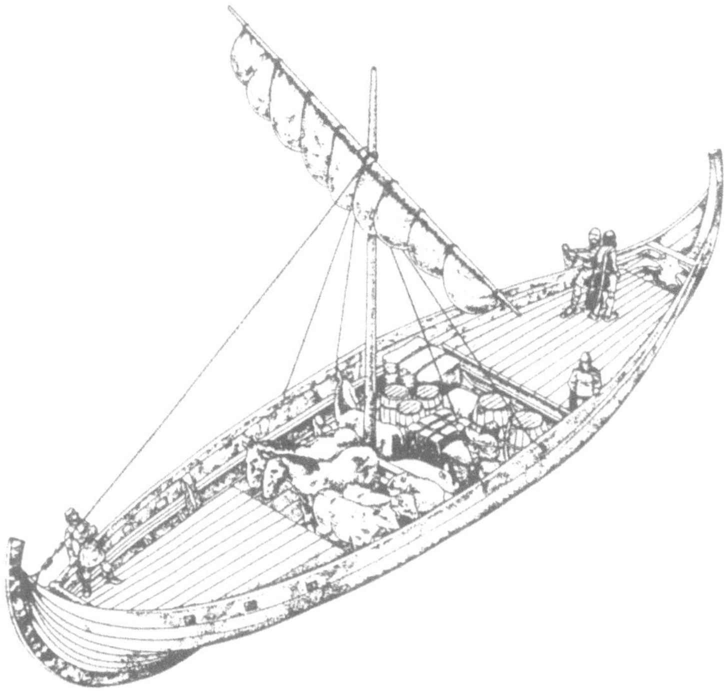

Fig. 41. "Knarr", cargo boat of the Vikings; length 20 - 25 m, width 4.5 - 6 m, loading capacity 35 - 40 t, built mostly of fir timber. (Graham-Campbell 1980)

centuries they sailed to the Shetland, Orkney and Faeroe Islands, to Iceland (874) and to Greenland (985), where they settled as farmers, cattle-breeders, fishermen and hunters of birds, seals and whales. Via Greenland they even reached northern America ("Vinland"). Also, the Germanic Angles and Saxons sailed in such long boats when they migrated to the British Isles in the fifth century.

During the period from 793 (plundering of the monastery at Lindisfarne off the coast of Northumberland) until the occupation of England in 1066, the Norsemen carried war to wide areas along the west European coasts. But they were also engaged in substantial trading activities from northern Europe to the west into the Mediterranean Sea and to the east into the Black and Caspian Seas. These wide trade connections are proved by many archaeological findings. The ships used for the transport of goods, called "knarrs", were half-decked sailing vessels which were shorter and wider than the crank open warships (Fig. 41). In Scandinavia the form of the Viking ships was preserved for a long time and fishing vessels still resembled them until modern times.

Fig. 42. Norse god Thor, fishing with a line and a bull's head as bait; part of a bas-relief on the Gosforth cross, Cumberland, England. (Kreuzer 1974)

There are indications that besides grain, wine, furs and many other goods, as well as slaves, fish formed a part of this trade. An example is the village of Haithabu near Schleswig which, between the sixth and 11th centuries was a major trade centre on the route from the Rhine region to Birka in Sweden and Halogaland in northern Norway. At-Tartúshi, an Arab merchant from Tortosa in Spain, reported around 950 in the golden age of Haithabu that the inhabitants were mainly eating fish which were abundant in nearby waters. Fish remains discovered during excavations were mostly from herring, perch and pike but also from carp and flatfish. It seems that ling, saithe, halibut and large cod were transported to Haithabu dried either as ships' supplies or as merchandise. Well-preserved herring bones may indicate a substantial consumption and trade of salted herring.

It is also known that in the ninth and 10th centuries Danes and Norsemen fished in many places around the British Isles. The famous Domesday Book from 1086 comprises an extensive list of fishing villages in Britain. Fishing with nets for salmon is mentioned in the sagas. Pickled salmon and also charr were early important trade commodities in Scotland and northern England and were a source of income for the monasteries. Already by around 1270 salmon were exported from Aberdeen to England, and in the 15th century this export widened to include ports on the Continent. A fishery for hake existed on the southern coast of Ireland. The fish were dried or salted and exported to southern Europe. Sardines were caught off Ireland and Cornwall and were salted and transported in barrels, as occurred with herring. Since the 10th century La Rochelle was an important place for the fishing and salting of sardines. In Brittany fisheries developed early under Norse influence.

It seems strange that, in spite of the obvious importance of fish as food during these times, effigies of fish and fishing are only rarely found on objects of art and religion (Fig. 42).

2.2 Freshwaters and Fish Ponds

The population increase in Europe resulted in a much higher demand for food. Freshwaters, like rivers and lakes, were used more and more for fishing (Fig. 43). However, possibilities for development in these waters were generally fairly limited. Originally every free man was entitled to fish at least in the area of his rural community. Only some landowners on the banks of rivers and lakes had attained certain customary privileges. It seems that altogether fairly substantial yields were obtained during the times of free fishing. But the strongly increased utilization of the natural resources often led to their overexploitation. Furthermore, due to cultivation of land and its opening up to trade and commerce, fishing areas were reduced.

With the development of the feudal system royal prerogatives and other noble privileges were gradually established so that fishing in public waters was restricted and fishing rights were granted or leased to certain families or persons or to churches and monasteries. Such permissions were usually combined with duties to provide a regular supply of fish to

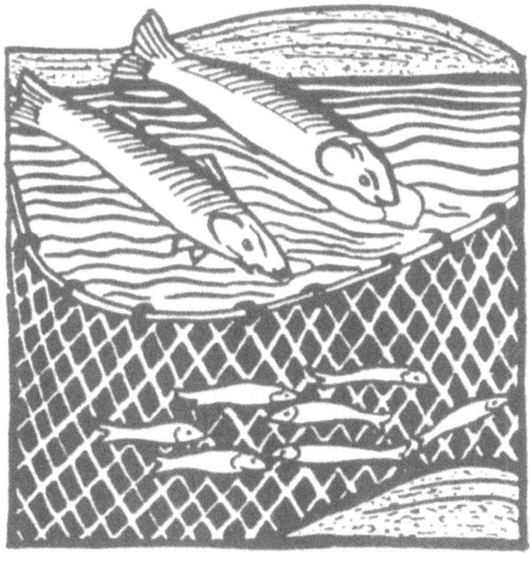

Fig. 43. Woodcut from the *Hortus sanitatis* (1490)

the households of the nobility and to the cloisters or came with financial levies. Often they also contained regulations for the restriction of fishing in time or space, in the gear that could be used and the allowed catch so as to ensure the protection of the resources. The selling of fish was strictly controlled. The earliest regulations date back to the seventh century and from the 12th century a great number of fishing prerogatives and regulations existed, some of which have been maintained until the present time.

The significance of fish in the daily life and economy is reflected by many coats of arms from medieval times. In Scotland it was the salmon which occupied a prominent place, while in Germany trout appeared most often as a fish symbol in heraldry. Fish were also frequently portrayed in the artistic ornamental initials of books written in the cloisters. This was certainly related to the significance of the fish symbol in the Christian religion but is perhaps also an indication of the importance of fish in the diet.

Guilds of fishermen and fishmongers were established, particularly from the 12th century onwards. In Ravenna in 943 a fishing guild already existed and was invested with fishing rights by the archbishop. Traditional festivals of such guilds, often with contests on rivers and lakes, are well known and were depicted in many paintings. It seems that a dilapidation took place in the late Middle Ages and the previously prospering fishermen's guilds changed their character. Active fishing was replaced more and more by mere trading. This development was encouraged by the regulation that fishermen had to sell their own catches. Also, the rule of limitation of the number of members in guilds contributed to the decline of fishing. Most people turned to profitable agricultural, artisan or other urban professions. Fishing was often displaced to remote areas and carried out by poorer people.

An exception was the production of fish in ponds. As described earlier, the keeping of fish in freshwater or seawater ponds was already widely practised during Roman times. With the expansion of Christianity and the foundation of cloisters, the creation of fish ponds spread throughout wide areas of Europe. The cloisters, especially those of the Cistercians, initially played a leading role in the promotion of agriculture and they also had a great interest in the production of fish as Lenten fare. Fish and their offspring were transferred to ponds from rivers and lakes so as to make the cloisters, courts and noble houses more independent from the rather variable catches in natural waters. In this way they could consume fish as food at the right time, e.g. during the period of Lent. The creation of ponds was often invested with the cloisters as royal prerogatives. At a later stage farmers were also permitted to build and utilize fish ponds.

This was not a fish-pond culture in the modern sense. There was no reproductive cycle of fish in the ponds, stocking was only done at discretion and hardly any statistics of inputs and results were kept. In well-managed operations, however, the system of keeping fish of all ages in the same pond was soon replaced by distributing animals of different age-groups to separate reservoirs. Better production was attempted by manuring. Also, through the introduction of new fish species, the cloisters strived to improve the economic returns from their lakes and ponds. Examples include the transfer of perch to Italy around 1400 and of carp from Denmark to Britain in 1514.

Fish culture in cloister ponds was already testified by the time of Charlemagne at around 800. The number of ponds increased markedly during the following epochs and a maximum of production was reached in the 15th century, especially as a result of the demand for fish arising from the Catholic Lent commandments. The production from freshwaters and ponds was often insufficient to satisfy the need so that marine fish, mostly salted herrings, stockfish and salmon, were a most welcome supplement. There are reports that cloisters in southern Germany bought such additional food through the market of Nuremberg. Other monasteries, located closer to the sea, relied more on sea food and were often invested with fishing rights on the coast and were even directly engaged in fisheries (see Sect. 2.3).

The menu of the Benedictine cloister of Maria Laach (Eifel) in Germany around 1550 shows that the friars ate fish almost every day, especially since the partaking of meat was much in dispute. In summer perch and crayfish were caught in the lake close to the monastery. Carp and pike were kept in numerous ponds. During winter fish like barbel, gudgeon, tench, bream, perch, eel, salmon and trout came from the rivers Rhine and Moselle and from other rivers where the monastery had the benefit of fishing privileges. Salted or smoked herring and stockfish were also frequently served dishes.

A region of most intensive fish-pond culture was the district of La Dombes between Lyon and Bourg in France. Here a large number of ponds were built from the early Middle Ages onwards in an area of impervious soil. The system developed almost continuously until the mid-19th century when the ponds reached an overall surface of 18 000 hectares. Thereafter, most of these ponds were relinquished in favour of traffic, settlement and agriculture.

In Germany the fish-pond culture started to decrease much earlier, mainly for economic reasons. Too many ponds had been built, in particular in connection with the creation of corn mills and mill-ponds. This led to overproduction of fish. Another factor was the pauperization of the farmers which gave rise to the Peasant's Wars in 1524. Many pond installations suffered seriously during the Thirty Years War (1618 - 1648). The main reason, however, was the incentive to convert ponds into cornland when cereal prices increased around the end of the 16th century and additional areas suitable for agricultural cultivation became scarce. Furthermore, the demand for fish, usually more expensive than meat, decreased during the 17th and 18th centuries as it became less popular, first for the nobility and later also for the middle class, to eat fish regularly. The old fish-pond culture ceased with the secularization of the cloisters under Napoleon in 1803. It was then not until the mid-19th century that the modern type of fish culture started with the development of artificial insemination of trout eggs by Remy (1850), a method which had already been invented earlier by Jacobi in 1763.

The old fish-pond culture was almost entirely based on the carp *(Cyprinus carpio)*. At later stages tench *(Tinca vulgaris)*, perch *(Perca fluviatilis)* and sometimes orfe *(Leuciscus ide)* were also kept in the ponds. Pike *(Esox lucius)* were added to the carp ponds to reduce the number of undesirable fish. Trials were also occasionally made with keeping trout. The carp was originally widely distributed in Europe, but during the ice ages the northern boundary of this species was shifted southwards so that carp remained only in rivers that flowed to the Black and Caspian Seas. The Romans re-introduced carp to southern Europe from the Near East. Cassiodorus, private secretary to Theodoric, the Ostrogoth king, requested that the provincial governors around 500 provide sufficient quantities of carp from the river Danube for the court banquet. At that time fish caught in natural waters were already kept in reservoirs and this method led gradually to the regular culture of carp. From southern Europe carp was re-introduced to France, Germany and northern Russia, and later also to Scandinavia and to Britain, where eel had been the most valuable species until then.

2.3 Herring Drift Net Fisheries

In western and northern Europe agricultural production and commerce began to flourish around 1000. The population rose quickly and many towns developed as trade centres. It has been estimated that the total population in Europe increased from about 46 million in 1050 to 50 mil-

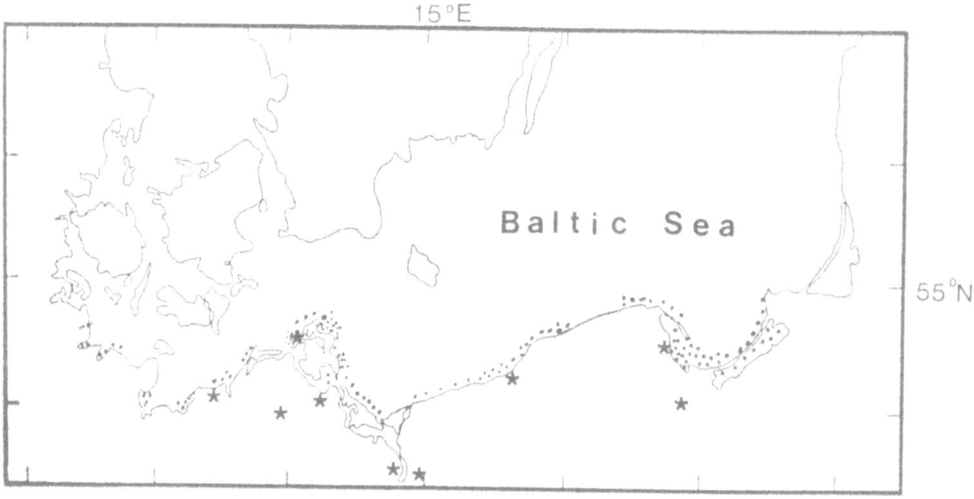

Fig. 44. Documented localities of herring fishing along the southern coasts of the Baltic Sea during the Middle Ages; asterisks mark Cistercian monasteries engaged in herring fishing and/or trade. (After Jagow 1915a; Spiess 1988)

lion in 1150 and that France and Germany alone had around 35 million inhabitants in 1340. Fish production from freshwaters and ponds was insufficient to meet the needs of these people. But rich marine resources in northern Europe, reasonably close to the centres of consumption, provided a good basis for the production of more food. Replacing the Vikings in their trade activities, the Hanseatic League became instrumental in the development of commercial fisheries and fish trading.

The catching and utilization of herring *(Clupea harengus)* was the first fishery to reach the level of a real industry. Herring from the Baltic Sea had already served from ancient times as food for the people along the coasts and in the neighbouring hinterland (Fig. 44). Fishing grounds were close to the land, in bays and river mouthes visited with small open boats, called "schuten" which carried four to six men. Another type of vessel, the "bording", served for the transport of herring, salt and other goods. A large variety of fishing gear was used, particularly seines and fish-hedges.

Fishing was carried out mainly by Wends, Slav fishermen - farmers. The Wends settled along the southern coast of the Baltic after the north Germanic tribes had left these lands before 600. When the Saxons returned around 1200, and German colonists from Westfalia, Frisia and Flanders immigrated a little later, they lived mostly as farmers and left fishing largely to the Wends who became bondmen. A number of names of fish and fishing gear in the German language are of Wendic origin. In several places fishermen's cooperatives were formed, like the "kommünen" in Lower Pomerania, the "maskopeyen" on the lower river Elbe and the "matschappien" at the Hela Peninsula. Already, in the 11th century, Kolberg in Pomerania was an important trading centre for salted herring. The distribution of the fish was in the hands of special carrier guilds. From the second half of the 14th century guilds of German fishermen were also established.

With German colonization in the region from the Elbe to the Baltic provinces, a large number of towns were founded in the 12th and 13th centuries which attracted new settlers. This provided an incentive for the development of fisheries in the Baltic. It was also the basis for the formation of the Hanseatic League under the leadership of Lübeck (founded in 1143).

The sovereigns vested prerogatives for fishing in the Baltic in the cities and also in Cistercian monasteries which were established in considerable numbers in this region. Often fishing rights played a role in the struggle between sovereigns and municipal authorities in their endeavour for more autonomy. In the area of the old Prussians, where fishing was of overwhelming importance, the Knights of the Teutonic Order ruled from 1308. They claimed the fishing rights for their own supply, ceded or renewed existing authorizations and let fishing installations with defined gear to people. This system formed the origin of the later fisheries administration in this region.

While most fishing grounds along the southern coast of the Baltic Sea were only of local importance, the main and most productive area near the island of Rügen attracted people from abroad. Here the trade in herring was testified already by the mid-12th century. On the island there were so-called "vitten", localities where merchants lived temporarily in huts during the fishing season in autumn and where they bought herring from the local Rugian fishermen. Fishing was carried out mostly with set nets overnight. At the "vitten" the herring were salted with salt imported by the merchants from Lüneburg and were put into barrels for transport. Parts of the catches were smoked. Eight trade and curing localities can still be traced on Rügen Island. Some later developed into permanent fishing villages and the name "vitte" has persisted in some places until the present day. Merchants from Lübeck were particularly privileged in this herring trade in return for paying royalties. When the city of Stral-

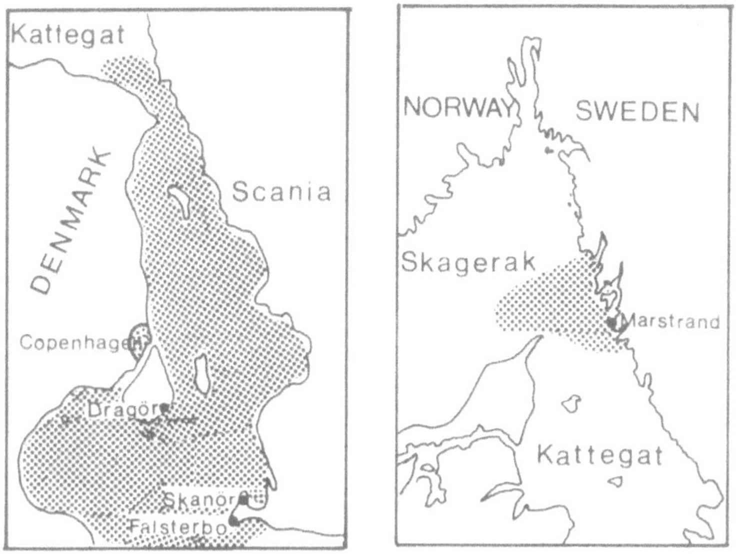

Fig. 45. Fishing grounds of the medieval herring fisheries off Scania (*left*) and Bohuslän (*right*) on the basis of catches in recent times. (After Johansen 1924)

sund, also a member of the Hanseatic League, entered into major competition, this led to war with Lübeck in 1249 which was lost by Stralsund.

Around 1200, merchants from Lübeck and soon also from other Hanseatic cities moved their herring trade activities to the Peninsula of Scania (Fig. 45) which belonged to Denmark. Here, in the coastal waters of the Oere Sound, herring had already been fished since the second half of the 10th century, or even earlier. Old reports speak of very dense herring concentrations at certain times. It seems that Dutch traders had already visited the place in the 12th century. Now a substantial herring fishery developed with trading centres on the spit between Skanör and Falsterbo and at Dragör on the island of Zealand.

The annual fishing season ran from the end of July until the end of October. During this time Danish fishermen caught herring in set nets and drift nets overnight, and in bays beach seines were also used (Fig. 46). They used open boats with oars or sails manned with three to six people. Crews were admitted to fishing only as enduring groups under the command of headmen. The catches were sold daily to the foreign merchants who lived on separate land in "vitten" with wooden huts and installations for the salting and smoking of herring (Fig. 47). As on Rügen, the mer-

Fig. 46. Herring fishery with beach seine in Scania

Fig. 47. Salting and smoking of herring on land. (After a painting in the Alfred Wegener Institute for Polar and Marine Research, Bremerhaven)

chants imported the salt. Coopers built on the spot the barrels in which the salted herring were packed for transport. Originally the herring was salted and left ungutted but later the intestines were removed. The quality was subject to strict regulations.

The "Scania Fair" was soon a major commercial event which was not only concerned with fish trading. Various duties provided the Danish Crown with substantial financial income. Through strict regulations it attempted to facilitate the social life of people from many nationalities who gathered there temporarily but life was certainly very hard, especially when seasons were bad and herring catches low. It seems that such failures occurred frequently. In Hamburgian shipping regulations it was specified that higher wages had to be paid to the sailors when there was no herring to be loaded in Scania so that the owner would continue the voyage to Norway.

By 1520 about 7500 fishing boats participated in this fishery. It was estimated that in Skanör and Falsterbo between 33 000 and 76 000 barrels of herring were packed in productive years during the period from 1368 to 1494. In 1537 the production in Falsterbo alone was in the order of 96 000 barrels. Taking a barrel as 100 kg, and considering that the landings in Skanör and Falsterbo made up more than half of the overall production of the fishery in the Oere Sound, it can be estimated that the annual catches were in the order of 10 000 metric tons (t). In comparison it is interesting to note that in more recent times, during the productive period between 1916 and 1920, the annual herring catches in these waters amounted to 5000 - 10 000 t. Today such a quantity does not appear large but for medieval times this was certainly a very important production.

Favourable conditions for the development of this fishery were created not only by the rapidly rising demand for food but also through the existence of an effective trading organization. The Hansa grew in the 12th century out of cooperative groups of merchants travelling together to important markets. After a federation between the cities of Lübeck and Hamburg, which was soon joined by Wismar, Rostock, Stralsund, Greifswald, Danzig, Lüneburg and many other cities, the Hanseatic League became commercially and politically very powerful. More than 160 cities belonged to this League at different times. The Hansa maintained a wide system of trading posts of which the ones in Brügge, London (Steelyard), Bergen (Tyske Brygge) and Novgorod were the most important. Generally, natural products were shipped from the east to the west in exchange for manufactured goods transported in the opposite direction. Trade of

Fig. 48. Hanseatic cog, cargo ship for herring, stockfish and other goods; length about 25 m. (Tesch and de Veen 1933)

salted herring and stockfish (see Sect. 2.4) became one of the props of Hanseatic prosperity. Over the centuries the Hansa managed to maintain a monopoly in herring trade for the German merchants, called the Easterlings. Hanseatic fish barrels were distributed throughout Europe in the 13th and 14th centuries. During the 15th century pressure from merchants of other countries, particularly from the Netherlands and England, grew to take their share in this trade. The Hansa frequently used trade embargos to prevent such developments but finally their power declined. The Thirty Years War then put an end to this era.

One of the reasons for the success of the Hansa was the superiority of their shipbuilding. They mainly used the "cog" which was developed during the 12th century to become the main cargo vessel in northern Europe (Fig. 48). The cog, having its origin in the shallow waters of the Wadden Sea, was a clinker-built wooden ship with flat bottom, straight keel, steep stems and high poop, and equipped with square sails. With typical loading capacities of 100 - 200 tons these vessels were highly superior to the earlier long boats of Viking and Frisian traders which had only a capacity of 20 - 25 tons. After the mid-15th century the cogs were gradually replaced by "holks" and "carvels" which were larger.

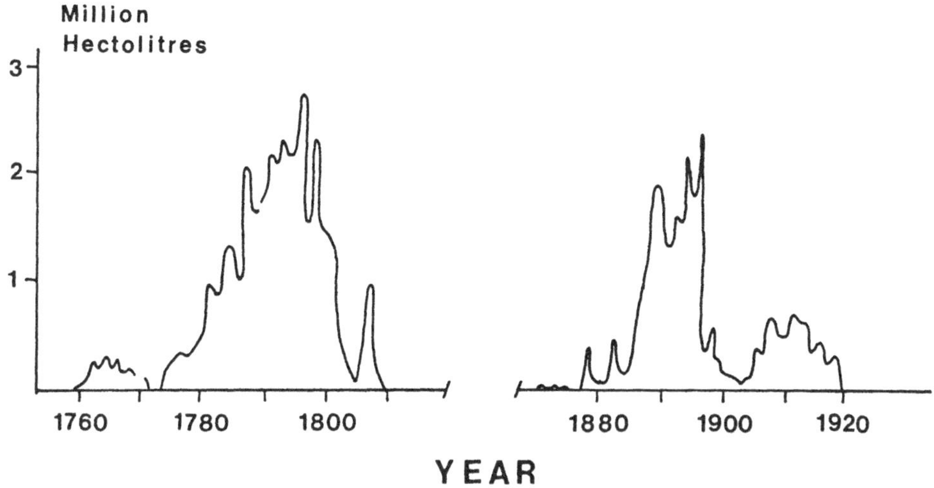

Fig. 49. Catches during two periods of herring fishing on the Bohuslän coast. (Devold 1963)

The herring fishery off Scania declined in the 15th century and ceased in the middle of the 16th century. Development in Scandinavian countries and substantial taxes, particularly the Sound toll raised by the Danish kings, contributed to the decay, as did frequent hostilities.

Another important herring fishery which, in the long run, produced higher yields than the Scania fishery was the so-called "vigen" fishery further north in the waters extending from Marstrand on the Swedish Bohuslän coast to the Skagerrak (Fig. 45). Here, the abundance of herring was always very variable. From the end of the 10th century a total of nine periods of productive herring fishing were reported: (1) from the end of the 10th century until the early 11th century; (2) from the end of the 11th century until the early 12th century; (3) from the end of the 12th century until the mid-13th century; (4) from the end of the 13th century until the mid-14th century; (5) in the mid-15th century; (6) from 1556 to 1590; (7) from ca. 1660 to 1680; (8) from 1747 to 1809; and (9) from 1877 to 1918. The lack of herring between these periods caused great hardship for the local population (Fig. 49).

Several attempts were made to explain the phenomenon of these fluctuations in herring numbers. One hypothesis was that the abundance of herring was periodically greatly enhanced by Atlanto-Scandian spring spawners entering the bays and sounds of the Bohuslän area at times when the herring fishery off western Norway showed low productivity. Based

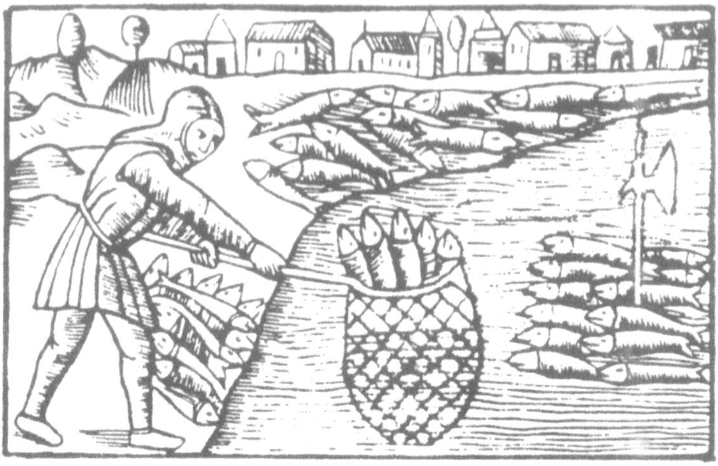

Fig. 50. Presentation on herring fishing in *Historia om de nordiska folken* from 1555 by the Swedish archbishop Olaus Magnus (1490 - 1557)

on the investigation of remains of herring skeletons excavated from the ponds of old oil factories, this hypothesis was rejected and it was assumed that the fluctuations were caused by increased abundance of autumn-spawning herrings from the North Sea, the Skagerrak and Kattegat. A further explanation was offered by relating the great periods of Bohuslän fishing to changes in environmental conditions, specifically to phases of less frequent westerly winds, relatively low winter temperatures and higher salinities in the Bohuslän waters. In spite of these attempts, it may be fair to state that the cause of the extreme variation in abundance of herring in these waters is still unknown.

Similar to Scania, the fishing season in Bohuslän lasted for 2 - 3 months in autumn. Between the skerries herring were caught with set nets or seines and further offshore with drift nets. From the end of the 13th century until 1660, when the province became Swedish, German, English and Dutch merchants participated in this fishery (Fig. 50). Production reached 600 000 barrels of salted herring during the best years.

Around 1760 a method for the production of herring train oil was discovered and all herring, except the leanest, were then utilized. In the mid-1790s there were almost 500 oil factories along the Bohuslän coast. The herring were boiled in water in large copper pots and the fat that floated on the surface was skimmed off and used as oil. The rest, the so-called trangrums, was regarded as waste and dumped in the sea. As pollution problems arose such dumping was prohibited and the remains were buried or tipped into specially constructed fishwaste ponds, when it was impossible to use them as manure or fodder for livestock.

Also, in the eastern Limfjord in Denmark an important herring fishery developed in the 12th century which lasted until 1825. This fishery was carried out as a seasonal event in a rather similar manner to that off Scania. Here, the herring were spring spawners of lower quality than the herring in the Oere Sound. Aalborg served as the centre of the herring trade.

Further west, in the North Sea, herring fishing also has a very long tradition. There are records that fishermen from England, France, Flanders and other countries were already fishing in the sixth century in the waters off the East Anglian coast. In the ninth century Dutch merchants bought salted herring from Scottish east coast fishermen. Even before the Norman invasion numerous lords of the manor and institutions of the church in England and Scotland were invested with fishing rights and many were directly involved in fisheries. Norwich, formerly situated on the coast, was the most prominent fishing harbour until the 10th century. Then Yarmouth was established on the spit of a sandbank in the Yare River where foreign fishermen had, for a long time, regularly gathered each year from September until November to participate in the herring fishery in inshore waters. Later, a new generation of local fishermen settled here; they shifted their activities from the mudflats, bays and estuaries to the open waters along the coast. By around 1350 about 300 foreign ships and 40 English vessels participated in this fishery. Until the 14th century Yarmouth remained the centre of this flourishing herring fishery. The importance of herring for the economy is reflected by the fish shown on the coat of arms of this city. Later, Yarmouth suffered from the silting up of its port and its importance declined.

Initially, fish production from coastal waters around Britain was not large enough to meet the demand. Considerable quantities of stockfish and salted herring were, therefore, imported from Scandinavia and through the Hanseatic cities. Later, the trade changed to exports from England and Scotland to the Continent. In the beginning these exports consisted of catches processed on land by foreign fishermen and carried with them to their home countries at the end of the fishing season.

Also, on the coasts of Cornwall and Devon, and on the French and Flemish coasts, the existence of herring fisheries can be traced back to the times of Charlemagne or even to the Merovingians. Fecamp, Dieppe and Calais were already established as fishing ports but the fisheries suffered after 1340 from the long war between England and France. The catches were mainly consumed in the densely populated provinces of Flanders and in France, especially in Paris. When the English and French fisheries declined, this was an incentive for fisheries in Flanders and the Netherlands to fill the gap. In Flanders Grevelingen and Nieuport developed into major fishing ports and Brügge was an important centre for the herring trade.

In the long run, the herring fisheries in the North Sea proved to be superior to those in the Baltic Sea, Oere Sound and Skagerrak, because they had much larger resources. When the Hanseatic League concentrated their activities initially on the development of the Baltic area as a supply region for central and western Europe with food and other natural products, the main fish trade connections ran from the Baltic in a westerly direction to Flanders, England and France. With the decline of the Hansa, the main herring fisheries shifted to western Europe. The leading Hanseatic cities near the Baltic ceased to participate in the herring trade. Hamburg retained its central role only by entrusting its "Schonenfahrergesellschaft" (Scania Farers Association) with the trade of herring from the Netherlands and later from Scotland.

Also, Norwegian herring, so-called Bergen herring, was already playing an important role. As early as 1100, Hamburg held fishing privileges for this herring which was shipped via Hamburg to France and other countries in the 12th and 13th centuries.

Dutch people became the leading group of foreign fishermen in England. Their regular fishing from Yarmouth and the maintenance of land installations can be traced back to the 13th century. In 1285/1295 and again in 1494 equality with local fishermen for the rights to fish in British coastal waters had been conceded to the Dutch by English kings. At times fees were also claimed and paid. But from the end of the 14th century the English attempted to impede the activities of the Dutch in favour of their own fisheries. Furthermore, the port of Yarmouth silted up. As a result, the Dutch began to fish with their larger boats which were previously used for transport purposes, and to salt the herring on board these ships at sea. This gave rise, in the 15th century, to the development of the first real high-seas fishery in history.

The port of Brielle was initially the centre of this "Grote Visscherij". Later, it was concentrated at Vlaardingen and a great number of other harbours in the northern Netherlands, in the region of the river Meuse, as well as in Zeeland and Flanders. Means and methods of fishing were

Fig. 51. Dutch buss, oldest type of vessel in the "Grote Visscherij". (Tesch and de Veen 1933)

adapted for work on the open sea. The vessel was the "buize" or buss, a ship of heavy construction with a flat bottom, especially suited to the difficult shipping conditions in the Waddensea area of the southern North Sea (Fig. 51). This type of vessel was first mentioned in 1410. Used originally only in coastal waters, the ships were built larger and fitted with a full deck and a narrow high poop. Around 1630 they were mostly 22 m long but could reach sizes of up to 30 m and they usually carried 14 - 15 men, although sometimes up to 30 men, including picklers and coopers. Their loading capacity was around 90 - 100 tons. The vessels had three masts rigged with square sails. During fishing operations the two foremasts were laid down to reduce the wind pressure and the movement of the ship. Herrings were caught with drift nets, long walls of netting often combined from 40 - 55 separate pieces (Fig. 52). In early times each net was about 18 m long and 5 m deep. The same type of gear, which is said to have been constructed for the first time at Hoorn in the Netherlands in 1416, has been used in the herring fisheries up to modern times. The net wall was kept straight by the vessel, connected with the gear at one end and drifting with the wind and currents. Mesh-sizes of the netting were selected to ensure the capture of herrings of the most desirable size, trapped with their gills in the net.

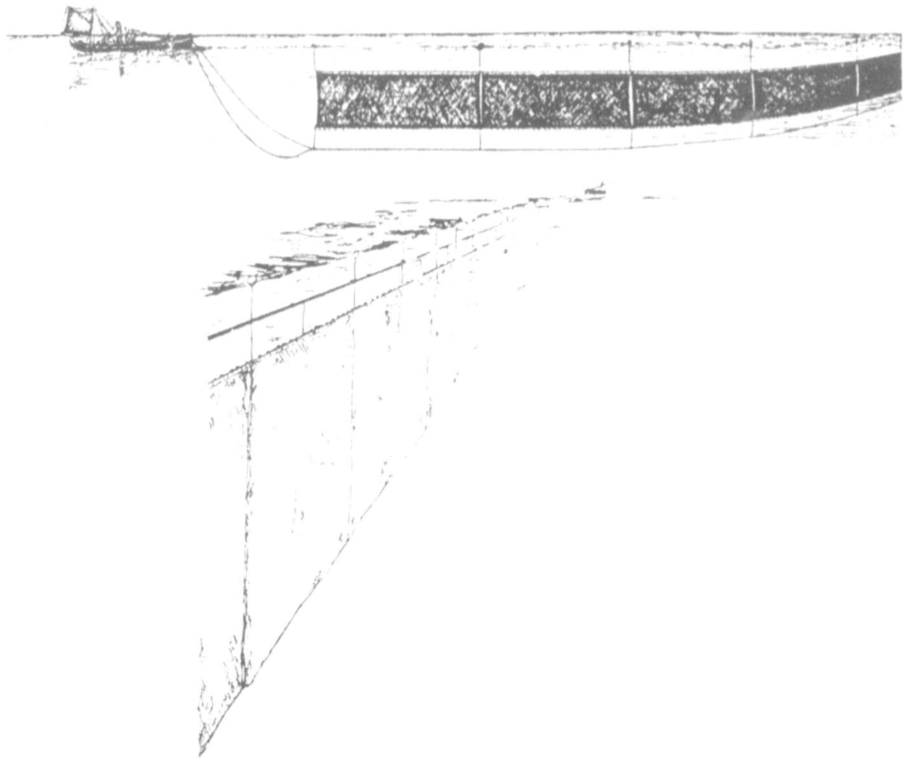

Fig. 52. "Fleet" of herring drift nets, operated from a Scottish steam drifter (*above*) and Dutch type of drift nets (*below*). (Kyle 1929)

While the East Anglian herring fishery was limited to the late autumn, the Dutch fishery using busses was mobile enough to cover a much wider area and to stay at sea for longer in order to follow the herring migrations. The beginning of the fishing season shifted to June when the Dutch vessels fished around the Shetland Islands. During the following months the fleet moved further south and the season ended in December in the English Channel (Fig. 53). Special quick-sailing vessels, the "ventjagers", were invented as tenders to receive the herrings from the busses at sea and to transport the catches ashore. Very high prices were paid for the first fresh herrings caught in the season and appearing in July at the markets. During a time when fresh meat was extremely scarce (almost all cattle were slaughtered in autumn due to the lack of fodder and most people ate pickled meat all winter), freshly caught herrings were a highly welcomed delicacy.

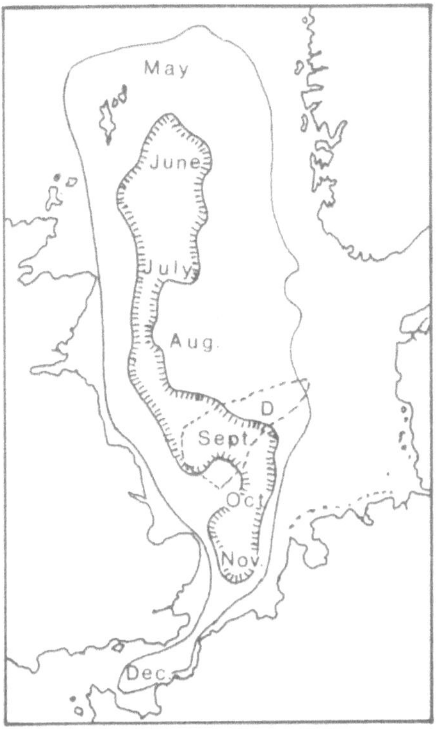

Fig. 53. Herring fishing grounds in the North Sea and the Channel, indicated by the distribution of the German drift net fishery during the years 1926 - 1937. *Thin line*, overall area, *thick line*, most regularly visited grounds; *D* Doggerbank

The superiority of the Dutch herring fishery was based mainly on three factors. Firstly, they salted the fresh herring at sea immediately after catching it. Secondly, they applied an improved method of curing and packing. Possibly it was Willem Beukelsz or Beukelzoon who, around 1400, invented the "kaaken" (hollowing) of the herring with a special cut in the throat to remove the intestines and blood vessels. It is, however, not certain whether Beukelsz was a historical figure. Thirdly, the Dutch established an efficient system of sales management with strict regulations for the catch, processing and shipment of herring; this helped to maintain a high standard of quality. Rules for the type of salt and fabrication of barrels were even laid down and each barrel was branded for identification. With their superior product the Dutch suppressed the Hanseatic herring trade and in 1614 they exported around 150 000 t of herring.

Fig. 54. Dutch herring drift net fishery off Scotland in the 17th century under protection by warships

During the heyday of the Dutch herring fishery in the early 17th century, 1000 - 2000, possibly even 3000, busses were engaged in fishing mainly around the British Isles. Each buss made an average of three voyages per season and landed about 70 t of salted herring. Operational bases were established on the Shetland Islands, the Hebrides and on the northern coast of Ireland. Also, the waters off Norway were fished at times. The ships were manned by 30 000 - 50 000 fishermen and, together with all the auxiliary personnel engaged in curing, trading and producing barrels, nets, etc., about 450 000 people worked directly or indirectly in the sector of fisheries. This was one-fifth of the population of the Netherlands. The saying that "Amsterdam was built on the bones of herring" is a reflection of this situation.

In Britain stronger attempts were made from the time of Queen Elizabeth I (1533 - 1603) to develop their own fishery. Dutch vessels were increasingly inhibited from fishing close to the British coasts. In 1609 the English Crown claimed part of the North Sea as British territory. Subsequently, the Dutch fishing fleet suffered from English naval warfare and was, therefore, escorted by warships for protection (Fig. 54). Nevertheless, Robert Blake captured 100 Dutch busses off Buchan Ness in 1652

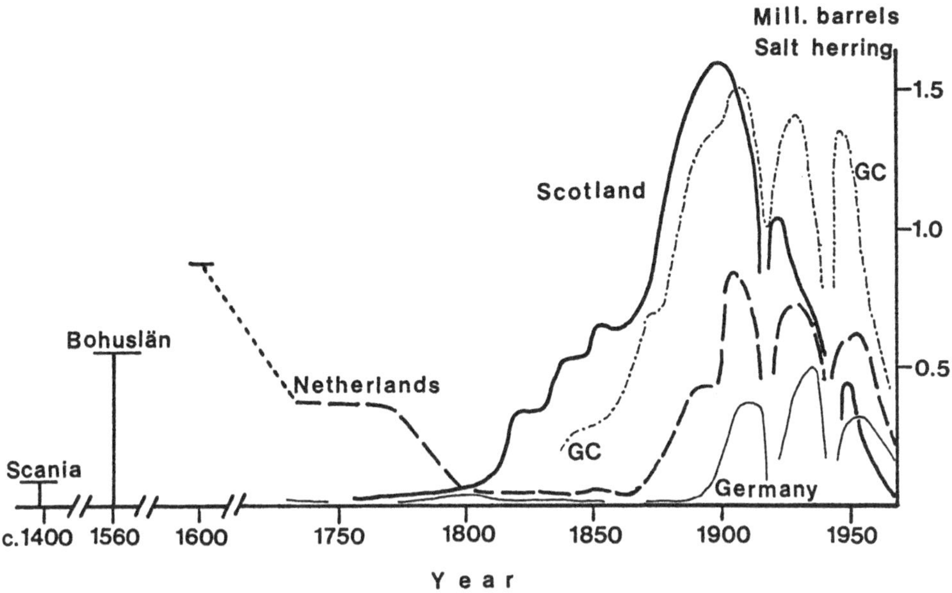

Fig. 55. Estimated average production of salted herring in some of the major drift net fisheries and average consumption of salt herrings in Germany (*GC*)

and many other vessels were also lost to privateers during these times. A large part of the Dutch fishing fleet was destroyed by French warships during the War of the Spanish Succession in 1703. Productive herring seasons in the Bohuslän waters during the periods 1556 - 1590 and 1747 - 1809 had detrimental effects on the Dutch herring trade by providing serious competition. Thus, the number of Dutch fishing vessels declined sharply in the 17th and 18th centuries and, after further difficulties during the Napoleonic wars, there were, in 1815, only 137 busses left, landing about 7000 t of salt herring. It was then not until 1850 - 1860 that the Dutch fishery received a fresh impetus.

From the end of the 18th century the herring fisheries in Britain expanded substantially (Fig. 55). The development was promoted by government subsidies. The Scottish herring drift net fishery in the Firth of Forth, the Clyde and neighbouring coastal waters, which was originally stimulated by Dutch activities, had the advantage of short distances from the ports to the fishing grounds. Small open boats could participate and fishing trips usually lasted only one night. In contrast to the longer distance fisheries, the catches here were processed on land. The port of Wick in the northeast of Scotland was developed in 1810 to become a centre of herring fishing, salting and smoking, and Lerwick on the Shetland Islands, Peterhead and Fraserburgh became main fishing ports. Scots also participated extensively in the East Anglia herring fishery, which revived with Yarmouth and Lowestoft as the main harbours. In the beginning there were considerable quality problems but exports overseas, particularly to the plantations in the West Indies, provided an outlet for lower grade salt herring products. Later, the land-salted herring was widely accepted and exported in large quantities throughout Europe and to America. The Scottish herring largely replaced the more expensive Dutch cure and was mainly consumed in Russia and other eastern European countries but also in Germany (Fig. 55).

Herring fisheries in other countries remained at much lower levels. In France, the harbour of Boulogne was a centre of this fishery and in Belgium it was Ostende. The German herring fishery developed in close connection with the Dutch fishery but on a much more modest scale with varying success and several interruptions. It started in 1552 when some Dutch herring fishing companies emigrated temporarily to Emden to escape the difficulties of war. German citizens became acquainted with the industry and built their own fleet which consisted of 25 busses in 1598. However, 100 years after its inception this first German high-seas fishery ceased for economic reasons. Also, several other ventures in Hamburg (1603 - 1614), Altona (1767 - 1807), Emden (1770 - 1811) and, later, in some other places were only of relatively short duration. Finally, after the foundation of the German Empire, a drift net herring fishery developed from 1872 onwards at the ports of Emden, Leer, Vegesack near Bremen and Glückstadt near Hamburg; this lasted until the end of the 1960s when this type of herring fishing declined internationally.

Like other branches of the economy, the traditional herring fishery benefited from the general increase in the demand for food, the increased purchasing power of many people and better transport facilities. A most prosperous step was the replacement of the busses by a new type of vessel, the lugger. During more than 200 years of herring fishing the old type of buss had gradually been modified to facilitate the operation of long "fleets" of drift nets. The number of masts was reduced from three to two, and the main mast made reversible during fishing. Now the lugger proved to be more seaworthy, faster and more efficient than the buss. It has its origins in France, where the "lougre" served in the 18th century as a quick-sailing tender in the Breton fishery. In England the lugger was first mentioned in 1781 and was used as a fishing vessel off Cornwall and further east in the English Channel (Fig. 56). Around 1865 the lugger was introduced to the Netherlands and was considerably modified here.

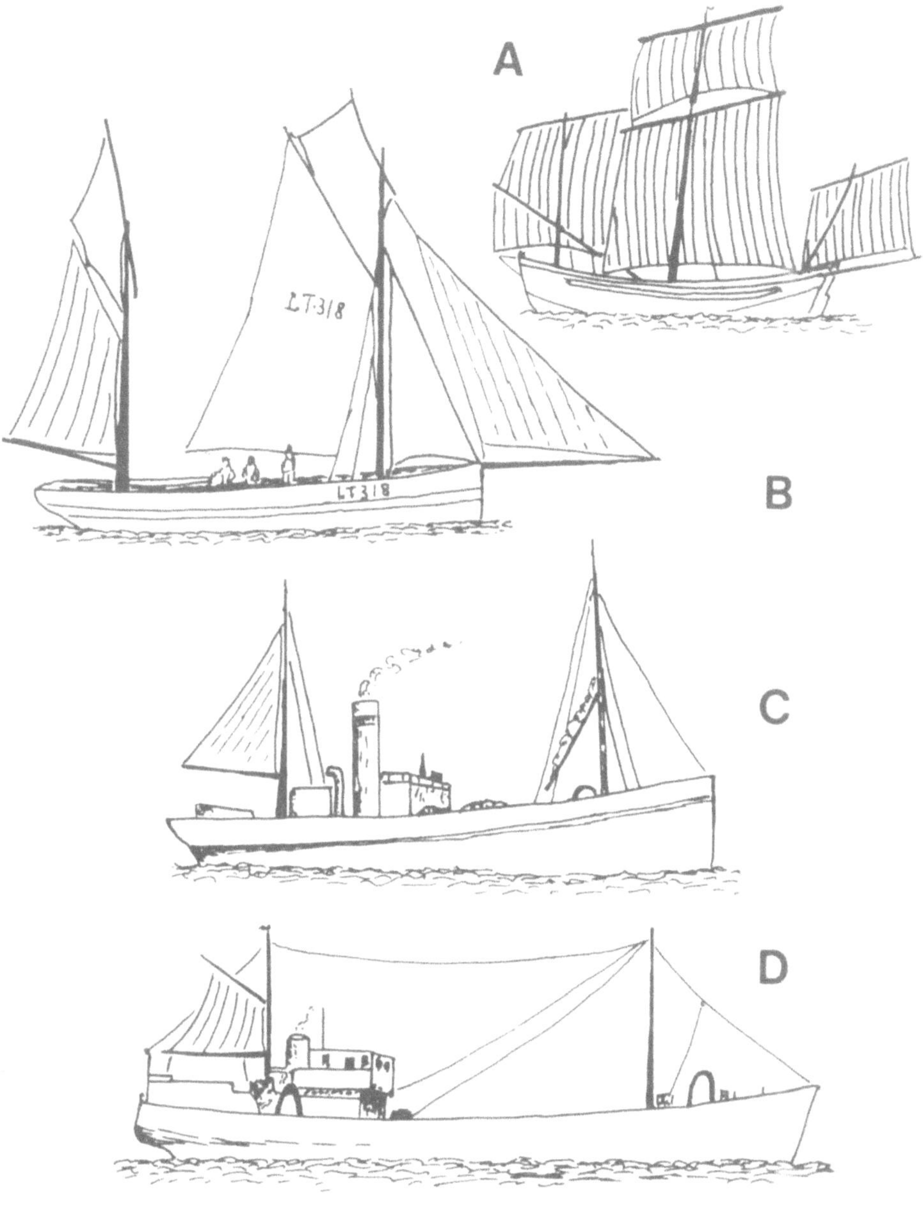

Fig. 56A-D. Developments in the construction of luggers or drifters. **A** Cornish lugger of 1814, keel length 11 m (March 1952); **B** sailing drifter from Lowestoft in 1883, length 18 m (March 1952); **C** English steam drifter, length 26 m (Kyle 1929;) **D** German "combination" motor lugger for drift netting and trawling in 1957, length 38 m, engine 750 hp. (Sahrhage 1957)

The reversible main mast and mizzen were fitted with sails fore-and-aft in ketch rigging. Initially about 22 - 24 m long, with a displacement of around 100 gross tons and not larger than the late busses, the size of the sailing luggers was gradually increased. Around the same time the first decked drifters were constructed in Great Britain (Fig. 56). Most of these ships were built in Scotland and belonged to the types of "scaffies", "fifies" and "zulus". The last busses sailed from the Netherlands in 1886.

Mechanization of fishing began during the late 19th century, making the vessels much less dependent on the wind and other weather conditions than before (see Chapt. 3). In 1882 the first steam drifter was introduced to the Scottish herring fishery and many luggers were equipped with steam capstans to facilitate the hauling of the long "fleet" of drift nets. After 1897, when the first steam drifter had also been put into operation in Lowestoft, sailing luggers were increasingly replaced in all participating countries by vessels with steam engines as their main form of propulsion. Especially around 1914 the old sailing drifters disappeared very rapidly from the fishing grounds. At that time steam drifters were usually about 24 - 27 m long with a capacity of 90 - 100 gross tons (Fig. 56). From early in the 20th century diesel motors were also installed in drifters.

Yields increased not only with more efficient vessels but also with improvements in the fishing gear. The old drift nets made of hemp were replaced by lighter cotton nettings which were, since the 1820s, produced with net-making machines. The gear consisted mostly of at least 100 nets, each measuring about 30 x 15 m (Fig. 52).

The economic success of this fishery led to a considerable increase in the number of drifters. In Scotland the fleet of steam drifters grew from 41 in 1898 to 943 in 1908. About 600 luggers operated from the Netherlands and nearly 300 from Germany. As a result, the production of salted herring rose steeply after 1860, especially in Scotland and the Netherlands. A peak of more than 500 000 t was reached in the herring landings during the period 1900 - 1914. Thereafter, the statistics show a more or less steady decline (Fig. 55).

The decrease in the salt herring production of Scotland was, to great extent, caused by the loss of the Russian market after World War I. The Soviet Union stopped the large imports and started to develop its own fisheries. During recent times a sharp decline is also evident for the herring drift net fisheries of luggers from the Soviet Union and other socialist countries, whose fisheries developed considerably during the 1950s but had already dwindled 20 years later. During the peak of this fishery in the mid-1960s luggers from the Soviet Union caught almost 500 000 t of herring per year in the Norwegian Sea, up to 115 000 t in the North Sea and smaller quantities off Iceland and the Faeroes. A substantial salt herring production also existed in Norway but drift net fishing was never important there. Instead, herring were mainly caught by purse seines.

The general decline in the herring drift net fisheries was due to economic reasons. The demand for salted herring decreased when the use of ice on board the ships and faster transport connections, with the construction of rail links, promoted the distribution of fresh fish, preferred by the consumers. Furthermore, the traditional drift net fishery had a number of disadvantages. Firstly, a high input in manpower and material was required to operate the very large gear. Secondly, the herring could be caught in the drift nets only during the night when the fish migrated upwards to the surface waters. Thirdly, the fisheries concentrated on the catch of sexually mature herring in the western North Sea with a season limited to the period from June, later May, until December. Thus, the crews of the ships and auxiliary personnel on land could only be employed for part of the year and their families had to carry out supplementary agricultural work in order to secure a minimum of earnings.

It was uneconomic to leave the vessels and installations inactive during the off-season. For this reason many attempts were made to extend the fishing season. Such endeavours were finally successful when "combination luggers" were built at the end of the 1930s. These ships were equipped to fish alternatively with drift nets or trawls (Fig. 56). This required more powerful motor vessels and had higher costs but it ensured a much greater mobility in fishing operations throughout the entire year. Then the herring fishery also expanded to the northeastern North Sea to catch the spring-spawning herring in the early months of the year. However, salted herring in barrels were less and less in demand and the luggers could not compete with the trawlers and seiners which were landing large quantities of fresh herring and keeping them on ice. Therefore, after 1970, only a few luggers continued with the old traditional fishing.

2.4 Old-Time Cod Fisheries

Besides herring, cod *(Gadus morhua)* was the most commonly available fish species in Europe during medieval times. Also, the ordinance of the Holy Roman Church to eat fish every Friday and on holidays created increasing demands for this species and incentives for fishing and trade. In the early Middle Ages productive fishing for cod and also for hake *(Merluccius merluccius)*, ling *(Molva molva)*, haddock *(Melanogrammus aeglefinus)* and whiting *(Merlangius merlangus)* was carried out close to the shores of the English Channel, especially off Devon and Cornwall. A large part of the catches was dried on sticks or poles to yield stockfish, a durable product suitable for transport over large distances. Part of this production was exported to the Iberian Peninsula. At times, Spanish fishermen obtained rights to participate in the fishery in English waters.

With an increasing demand for food and better technological abilities the fishery expanded along the coasts of the North Sea. In England, Cistercian monasteries became engaged in fishing in the 12th century and obtained fishing rights in the Humber area in 1160. Regulations for the fish market at Grimsby date back to 1258.

Already by the 12th century English ships reached Iceland and started trading in stockfish. Old sagas prove that at that time the drying of fish was known in Iceland. Cod, ling and other lean fish were caught by the local people from large Norse-type rowing boats equipped with a mast and square sail. The English fish trade with Iceland intensified considerably after 1412, when the economic links between Norway and Iceland were interrupted by the Hanseatic League in the interest of their monopoly in Norway. The main ports of this English trade were Dunwich, an ancient port now under the sea, King's Lynn and Bristol. The English competed with merchants of the Hansa who came mostly from Hamburg, where an "Iceland Farers Association" already existed before 1500. English and German traders sailed in summer to a number of places along the Icelandic coasts, particularly in the west but also in the north and east of the island. In some places they built huts on land where the Icelanders delivered their stockfish, train oil, wool and downs in exchange for salt and manufactured goods. Gautavik in southeast Iceland was such a trading place which has been excavated. Another site was Hafnafjördur near Reykjavik, mentioned since 1391 and in the 15th and 16th centuries as serving as a main trading centre for Hamburgian merchants. The latter sold much of their cargo in London through their Hanseatic trading post, the "Steelyard".

Fig. 57. Dutch "hoeker" buss, fishing with hand lines. (Tesch and de Veen 1933)

As early as 1436, English fishing vessels visited Icelandic waters to catch cod and ling. In 1490 a treaty was concluded with Denmark, giving English ships the right to fish off Iceland under licence and against payment of customs. During the early 16th century about 150 English sailing smacks participated in the fishery annually between April and August using hand lines mainly off the east coast of Iceland. The fish were cleaned, split and salted for transport back to England. The English ships had the advantage that they were able to operate further offshore while the Icelandic long boats were restricted to coastal waters. The English fishery declined during the second half of the 16th century and temporarily ceased in 1665. The reasons were that, on the one hand, there was a decrease in the demand for salted fish when the Anglican Church was formed in 1534 and the Lent ordinance was lifted but, on the other hand, probably more detrimental was the increasing competition by the Dutch fisheries.

From Flanders and the Netherlands fisheries had spread from coastal waters into the open North Sea, even as far as the Shetland Islands. Many of the Dutch busses had taken up fishing with lines during the periods when no herring drift net fishing was possible. Vessels equipped for this type of alternative fishing were called "hoeker buss" (Fig. 57). Usually these ships made one trip to the productive Dogger Bank area in spring prior to the herring fishing season and fished with hand lines; these were single lines with one or two hooks. After the herring season they mostly returned to the Dogger Bank from October to the end of January to catch cod and haddock with long lines, trains of lines with many hooks attached to the main line by snoods (cf. Fig. 62). During winter they continued their fishing on the Great Fisher Bank and Jutland Bank. The cod were salted and later dried to a product called "labberdan". People and ships working in this way were called "doggevaarders".

From the end of the 16th century Dutch hoeker busses, coming mostly from Maasluis and Vlaardingen, ports on the river Meuse, competed with the English in the line fishery off Iceland. The level of participation in this "Groote Abberdaan Visscherij" was somewhat inversely related to the economic situation of the Dutch herring drift net fishery in the North Sea. After 1786 the Dutch cod fishery off Iceland declined quickly due to political difficulties with Denmark and to the introduction of prohibitive custom fees in the Austrian part of the Netherlands. Several attempts to revive this fishery after the end of the Napoleonic wars were unsuccessful.

During the late 16th century and the first half of the 17th century the Dutch fishery in the North Sea and off Iceland suffered greatly under naval warfare from English and French privateers. Therefore, the fishing fleets sailed in convoys escorted by warships. Despite such protective measures many ships were taken and destroyed.

At the same time a considerable export of salt-cured cod developed from the Dutch republic to countries in southern Europe. This trade connection is reflected by similarities between the Portuguese word "bacalhao" for salted cod, in Spanish "bacalao", and the name "bakeljauw" in ancient Dutch for cod (now "kabeljauw" in Dutch and Flemish). It is, however, not known in which direction the word was transferred.

Several other types of vessels were also used in the Dutch line fishery, for instance the "schuyt" and "bom", heavy clinker-built ships with flat bottoms and drop-keels on both flanks, which made them suitable for towing on to the beach, the "smak" with two drop-keels and a mizzen, the "kwee", a modified buss, and later the "beugsloep", a sloop for long lining (Fig 58). An increasing number of ships were fitted with a "well" to allow for the transport of live fresh fish which found a ready market not only in the Netherlands but also in Britain, with London as the major centre of consumption. The wells were large watertight compartments amidships, containing fresh seawater which circulated through holes bored in the hull.

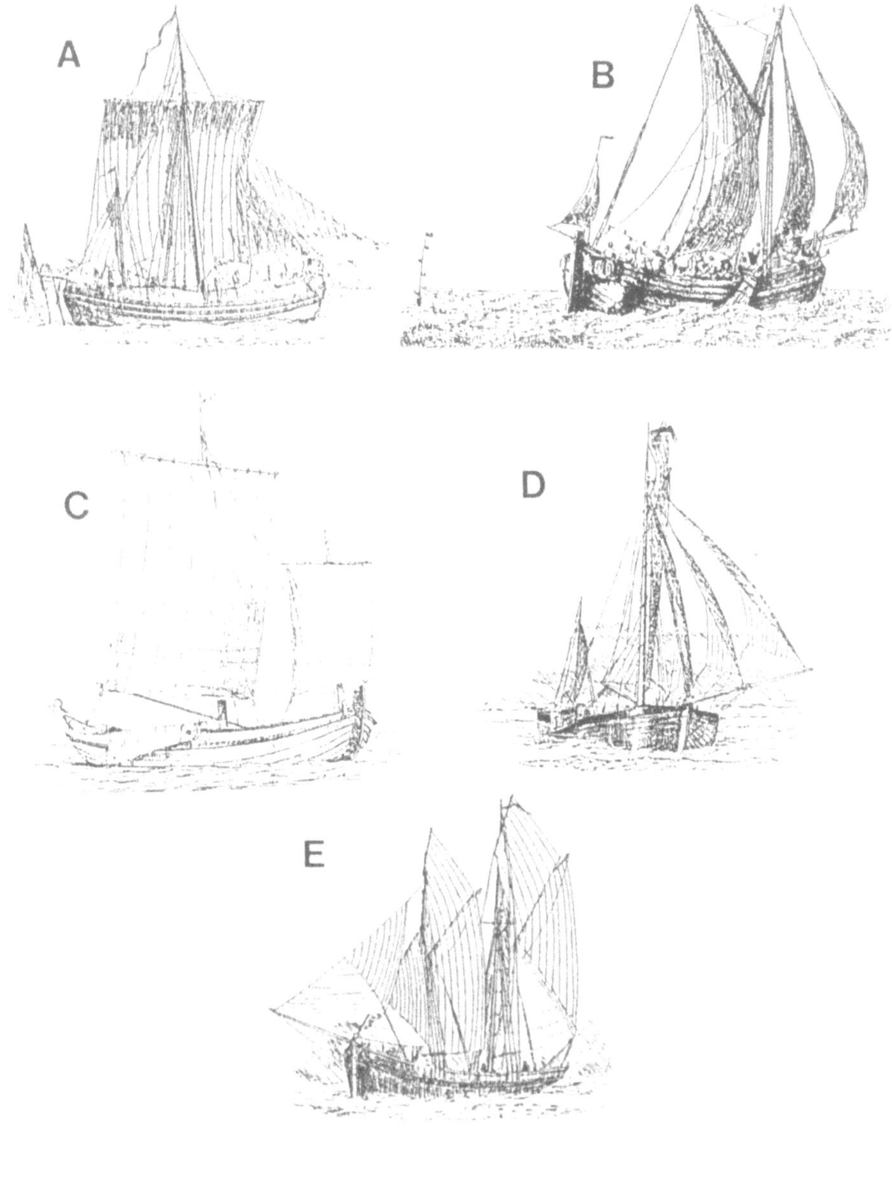

Fig. 58A-E. Various types of vessels in Dutch line fishing.
A "Kwee" from 18th century, equipped with well; **B** "smak", shooting a long line ("beug"); **C** "schuyt" from Scheveningen, 17th century; **D** "bom" from Katwijk, 18th century; **E** "beugsloep", sloop for long lining, end of 19th century. (Tesch and de Veen 1933)

During periods when ships with wells participated in the herring drift net fishery, the holes could be closed with wooden plugs and the wells be used as holds. Similar wells for the transport of live fish to the markets have been discovered in boats from Roman times excavated at Fiumicino near Rome close to the ancient port of Ostia. Such boats were called "vivariae naves".

The landing of fresh fish instead of salted cod was of great economic advantage for the Dutch. However, the English succeeded in breaking this monopoly. From 1712 their fishing smacks were also built with wells. With government promotion of the fishing industry in the mid-18th century the number of smacks increased considerably. Many of these smacks were engaged in trawling in coastal waters (see Sect. 3.1) but, in 1766, fishing with long lines in the Dogger Bank area and on other grounds was also taken up, following the Dutch example. Such liner smacks were called "doggers". Barking near London and Harwich were the leading ports for this fishery. After 1815 cod smacks from Barking fished not only in the North Sea but also near the Faeroe Islands and off Iceland. From around 1850 the water of the river Thames was so polluted that the well smacks were forced to land their live fish in Harwich from where they were transported to London in wooden chests. With more than 1 million inhabitants London was a steadily growing large market. After the construction of the railway from London to Grimsby in 1848 this harbour developed rapidly to become the main centre of fleet operations.

The development of line fisheries created a great demand and market for bait. Long lines often reached lengths of more than 10 miles and were fitted with up to 12 000 hooks which had to be baited by hand. Mainly herring, lamprey and whelk *(Buccinum undatum)* were used for this purpose. Great efforts were necessary to obtain the required quantities. The Dutch procured part of their bait in Britain. Around 1750 restrictions were imposed on the export of lamprey from England in an attempt to impede the competing Dutch fishery.

During the second half of the 19th century yields from line fishing started to decline and more and more ships engaged in trawling and in the herring drift net fishery. In Scotland, however, lining remained the main sector for a long time. Requiring less capital than gill netting and trawling, it was particularly suitable for small- and medium-sized ships which could operate with their lines on coastal grounds where the sea bottom was rough and not suitable for trawling. In the mid-1920s the Scottish liner fleet consisted of about 950 motor vessels, and almost the same number of herring drifters were engaged part-time in line fishing. The total landings in Scotland and England from this fishery at that time amounted to 50 000 t annually. Besides cod, the catches consisted of ling, halibut, tusk *(Brosme brosme)*, conger eel and skate. Small long lines were used for haddock. With the use of ice for the preservation of fish after 1850 the use of wells in vessels was less common.

Another major source of fish for the European market was the Norwegian cod fishery around the Lofoten Islands and Vesteraalen. The cod was processed into air-dried stockfish. Norwegian middlemen transported the cod from the north to Bergen. This city developed to become a most important market and trading centre. German merchants from Bremen and later from Lübeck were already visiting Bergen in the 12th and 13th centuries during the summer and from 1259 they also stayed over winter. After the formation of the Hanseatic League the German merchants forced the Norwegians to cut their existing trade links with Scotland, England and the Netherlands in order to accept a monopoly by the Hansa. In 1343 privileges were confirmed for the Hansa to maintain a major trading post in Bergen, the "Tyske Brygge". The most prosperous period for this post began around 1350.

The power of the Hansa in Norway was based mainly on imports of grain on which the country was absolutely dependent. Furthermore, the German merchants imported many other goods, like manufactured products from the Rhine area and cloth from Flanders. In exchange, they bought not only stockfish but also train oil, herring, furs and sheepskins. From Greenland came hides of seal and walrus ivory. A special sector was the production of cod liver oil. The best grade was clear oil extracted from the livers and brown oil of lower grade was produced by boiling the livers. The liver oil was transported in barrels and used all over Europe for illumination and for the impregnation of fabrics and leather. Since most of the products at that time were shipped in barrels, coopery was a very important profession.

The hegemony of the Hansa in Bergen was broken in the mid-16th century as a result of the decay of the League and of increasing competition by British and Dutch traders. The last house of the Tyske Brygge was sold in 1764.

The traditional Norwegian cod ("skrei") fishery in the Lofoten and Vesteraalen area with a centre in the Vestfjord has continued until the present time. Each year in late January, when the cod aggregated for spawning, a large number of vessels of all sizes, from small open boats (Fig. 40E) to steamers and motor ships, appeared on the fishing grounds with thousands of fishermen, mostly coming from other places further south. In the 1920s and 1930s around 15 000 to 25 000 people participated on more than 13 000 ships. The cod were caught with hand lines, various types of long lines and gill nets. Long lines had already been introduced in 1580 but inhibitions and quarrels between fishermen using different gear delayed the final official permission for their use until 1786. The fishing season usually lasted until mid April and in these 3 months between 50 000 and 100 000 t of cod were taken annually in the 1930s. Fishing in such rough seas was always difficult and often dangerous, especially in earlier times when small sailing boats prevailed. In the period 1892 - 1901 alone, 170 boats sank and 168 lives were lost.

The fishery expanded to the Finnmarken coast further north between Hammerfest and Vardö, where adolescent cod were fished from April to June. Trawling was also started in the late 1920s. Production here increased to 50 000 t in 1936.

Along the Finnmarken coast most catches were dried on racks in the open air to yield stockfish while, further south in the Lofoten area with less favourable climatic conditions for drying, the production of klipfish dominated. The application of salt for curing cod was introduced to Norway in the 18th century from England. The Lofoten cod was split and salted near the fishing grounds, usually under the control of buying agents who then transported the fish further south to the province of Möre for finalizing the drying process before the product was marketed via Bergen or other places.

Estimates for the stockfish and klipfish production in Norway during the last 350 years are given in Fig. 64. The whole production was exported and specific trade connections developed which are still maintained today. Stockfish is exported mainly to Italy and Nigeria and klipfish mostly to Portugal and to Latin America, especially to Brazil, and to Angola.

After this early period the most prominent cod fishery developed in the Northwest Atlantic off Newfoundland. Its origins remain obscure since the reports of early explorers were kept secret in order to maintain an advantage. Probably Basque sailors had already reached these waters in the early 14th century while searching for whales which had become scarce in the Gulf of Biscay. Portuguese explorers crossed the North Atlantic in 1450 and 1472 and again in 1495-1498 as far as Labrador. During their first voyage they fished on the Grand Banks and during their second voyage they discovered the island which was later called

Newfoundland. John Cabot, a Genoese who sailed on behalf of the English king Henry VII to this area in 1497, and who is often considered to be the discoverer of Newfoundland, called this island "Isla dos baccalaos", a term that referred to the rich cod resources. Cabot confirmed what the Basque whalers and fishermen knew and may have taken advantage of for a long time: "The sea there is swarming with fish which can be taken not only with the net but in baskets let down with a stone so that it sinks in the water."

The presence of European whalers and fishermen can be reliably detected from 1504. At that time the fishing fleets were already sizeable so that the fisheries must have developed rather rapidly. It was estimated that in 1518 about 15 000 Europeans participated in these fisheries. During his expedition in 1534 Cartier observed around 100 French vessels. Other reports speak of 149 English ships from Bristol and other west English ports. A report from 1578 refers to about 330 fishing vessels, of which 150 were French, 100 Spanish, 50 Portuguese and 30 English. The wealth of the cod resources in Newfoundland waters and the possibility to sell the fish for good prices on the European market led to harsh competition and even to wars between the nations.

The participation of Spanish vessels was of relatively short duration. The "pesca de gran altura" off Newfoundland, carried out mostly from Basque ports in the Gulf of Biscay with up to 6000 men and more than 100 ships, peaked around 1575 but had already diminished by the end of the 16th century because of economic problems and as a result of the decay in the political power of Spain. With the decline in the Spanish fishery, French and English merchants started to compete strongly for the delivery of dried salted cod to the Spanish market.

The struggle between French and English fisheries was part of the overall fight for superiority and colonial territories in North America. During the early period of the fishery the English were less interested in the Newfoundland waters than the French and Portuguese since they concentrated their activities more on fishing and the fish trade in Iceland. But, like the French and Portuguese, they fished in summer from the coasts of Newfoundland and erected temporary land installations for the processing of their catches. French fishermen, from St. Malo and other ports on the Channel, worked mainly on the northern coast from Bonavista Bay to Belle Isle ("Petit Nord") while Basque fishermen, from the Gulf of Biscay, worked on the west and south coasts, particularly in Placentia Bay, where a permanent settlement was established. Soon the French fishery expanded also to southern Labrador, Cape Breton Island, the Acadie

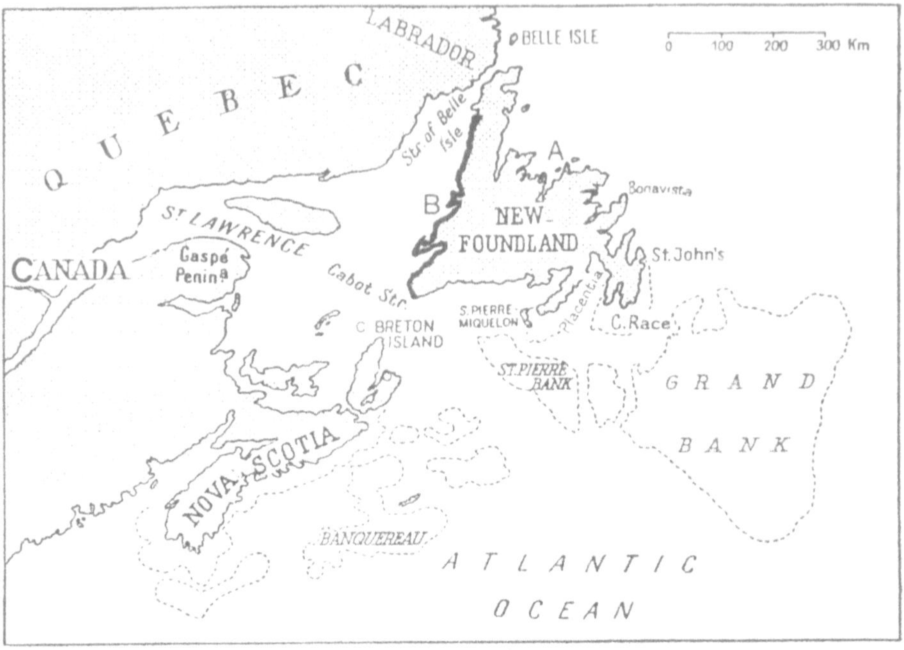

Fig. 59A,B. Cod fishing areas around Newfoundland and the Grand Banks. **A** "Petit Nord", French fishing rights 1713 - 1783; **B** "French shore", French fishing rights 1713 - 1904

Peninsula (later called Nova Scotia) and to the Gaspé Peninsula on the mouth of the St. Lawrence River (Fig. 59). Thus, the French controlled most of the coasts and took possession of Newfoundland in 1524. The Portuguese fishery concentrated on the coastal waters of the Avalon Peninsula in the southeast of Newfoundland between Cape Race and Bonavista Bay. But soon the Portuguese were dislodged as the same coast was the centre of the English cod fishery and the nucleus of the British colonization of Newfoundland. As a result of political difficulties the Portuguese fishery ended around 1600 and was not resumed until 1835. All cod were now imported, mostly by English ships.

After 1580, when English vessels were forced to pay licence fees for fishing in Icelandic waters, the English fishery intensified in Newfoundland. About 300 ships participated in this fishery by around 1620. The ships sailed mostly from Bristol, already a centre for cod trade at that time, to Portugal for loading salt before crossing the Atlantic. In Newfoundland cod were caught from small boats close to the coast and processed ashore. At the end of the season the vessels carried most of the lightly salted dried cod to Portugal and returned to England with a cargo of wine, olive oil and salt. Companies in London competed with fishing enterprises in western English ports for the trading of dry cod from Newfoundland to Mediterranean countries. In view of the growing economic interests the English Crown claimed Newfoundland in 1583.

In contrast to the French, the English established a greater number of permanent settlements on Newfoundland and more and more colonists travelled to settle on the island. St. John's grew to become a central trading place. This development gave rise to conflicts between the old rights of the French fishermen, the demands of the new inhabitants who also depended much on the fish for their livelihood, and the English fishing interests. Over the years many French ships with fish cargoes were captured by the English during war operations. But, also, the English trade suffered during wars with France and Spain. Furthermore, many ships were taken by pirates. The struggle lasted for several centuries and ended with the almost complete dispossession of the French.

With the Treaty of Utrecht in 1713, to end the War of Spanish Succession, the French lost the southern coast of Newfoundland, including Placentia, and the Acadie Peninsula (renamed Nova Scotia by the British colonists) to Britain. They could still maintain Cape Breton Island as an important fishing centre. On Newfoundland the French fishermen were restricted to the so-called French shore and the Petit Nord (Fig. 59). At times they were even here obliged to pay fees for their fishing rights. During the Seven Years War (1756-1763) the English fisheries advanced into the Petit Nord. With the Treaty of Paris in 1763, the French territory in Canada was relinquished and reduced to the small islands of St. Pierre and Miquelon. In spite of these unfavourable developments the French fishery was still carried out by about 300 - 400 vessels, but most of them were engaged in offshore fishing. At the same time the English fishing fleet comprised about 250 - 350 ships. On Newfoundland the French fishermen lost the northern coast ("Petit Nord") with the Treaty of Versailles in 1783. At the same time fishing rights were conceded to the fishermen from the now independent United States of America. French fishing rights along the "French shore" on the western side of Newfoundland had been confirmed in the Treaty of Utrecht in 1713 but were nevertheless under dispute between France and Great Britain for almost 200 years. Finally, in a treaty of 1904/06, France renounced all rights along the coast in exchange for other concessions.

Although Great Britain succeeded with territorial claims against France, the English fishery was already suffering economically from the 18th century with competition from the Newfoundlanders and fishermen from New England. Difficulties between the local inhabitants of Newfoundland and the interests of the fishing and trading companies in England developed into a major issue which contributed to a movement for independence. In 1765 about 15 000 colonists were living on the island and 9000 fishermen stayed temporarily during the summer fishing season. One hundred years before this the situation had been the reverse. The number of boats belonging to Newfoundlanders and the related catches increased substantially. On the other hand, less and less fishing vessels were sent to Newfoundland from west English ports after the mid-18th century. They were replaced by larger cargo ships which carried numerous "bye-boat keepers" on board. These people fished at their own risk during summer and sold their catch to the trading companies. Impediment by the Newfoundlanders resulted, in 1806, in the cessation of bye-boat fishing. Finally, by around 1830, only a few English vessels remained in the Newfoundland fishery, fishing offshore on the Banks.

With the colonization of New England since the early 17th century a commercial fishery for cod also developed further south (see Sect. 5.4). In early years this fishery served the needs of local consumption but it soon started to expand to northern waters as far as Labrador and the Newfoundland Banks and to produce salted cod for export. While cod of lower grade were sold to the West Indies as cheap food for the slaves on plantations, good quality dried cod were shipped to the south European countries. In both fields the American products competed strongly with the English fishery.

French vessels for the Newfoundland fishery sailed first from ports on the Channel, mainly Dieppe, Fecamp, Honfleur, Le Havre, Granville and St. Malo. They were mostly heavy-built ships, fairly small with 50 - 70 tons capacity (Fig. 60). The crew of these "voiliers" for salt cod production consisted of about 24 men. Later followed larger vessels with up to 250 gross tons from ports on the Gulf of Biscay, such as La Rochelle, Nantes, Bordeaux and Bayonne. These ships carried around 150 men for the dry cod production. In 1720, schooners, vessels of American origin with a crank hull and three or four masts with sails in fore-and-aft rigging, were introduced to the offshore cod fisheries. Originally these ships, called "goelettes" by the French, had a size of about 50 tons and a crew of 10. In later years the schooners were gradually built larger and at the end of the 19th century most of them displaced 300 - 400 tons. They carried a crew of about 30 and their loading capacity reached about 350 - 400 t.

The cod fishery started with the crossing of the Atlantic between January and March. Upon arrival, the ships anchored in bays where land instal-

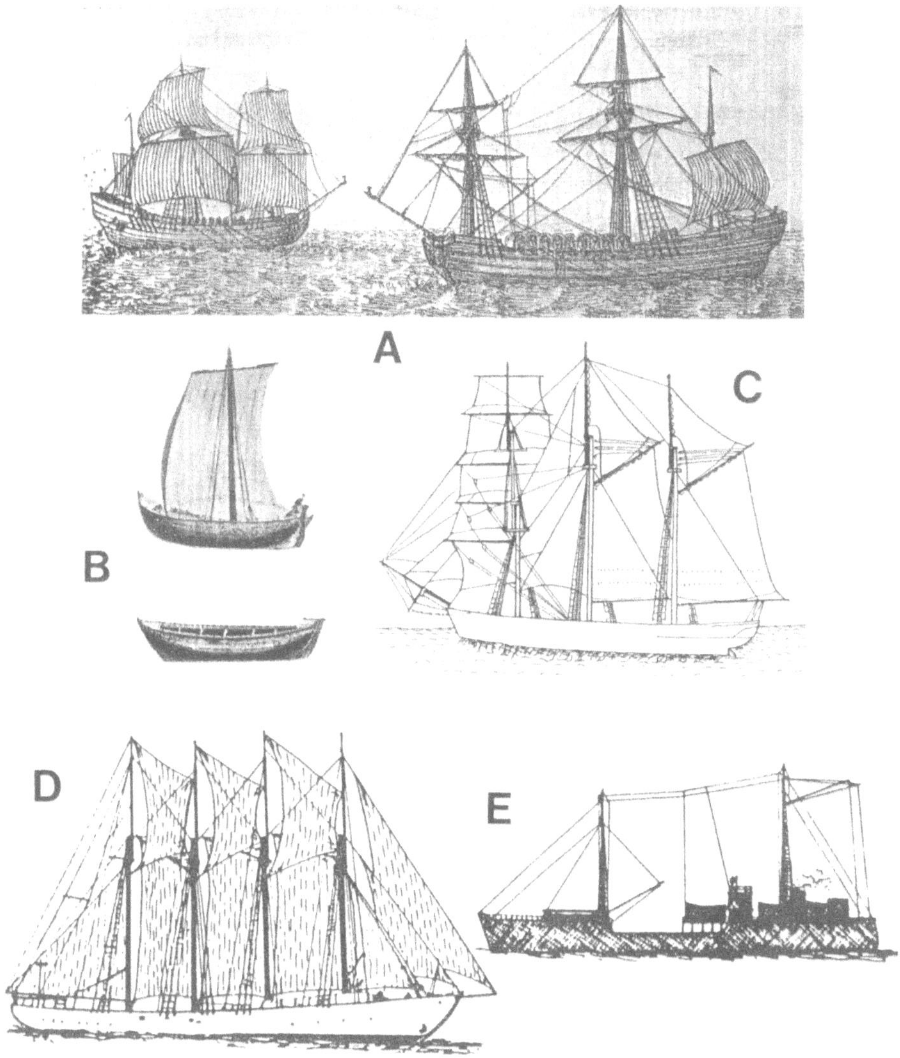

Fig. 60A-E. Ships in the cod fisheries around Newfoundland.
A French "morutiers" around 1680; **B** chaloupes for cod hand-lining, length about 6 m; **C** French "goelette", three-mast schooner in the Grand Banks fishery; **D** Portuguese four-mast schooner, built 1938, 696 GRT, length 52 m, 53 dorymen; **E** Portuguese motor ship for cod fishing with dories, 1950

lations were usually already available from previous seasons (Fig. 61). In these "sedentary" fisheries fishing was carried out close to the shore from open boats (chaloupes). Each ship coming from Europe transported up to 17 chaloupes (Fig. 60B). These little sailing boats were 6 - 7 m long and carried three men who fished with hand lines during the daytime, using mackerel, herring or capelin as bait. Some of this bait was caught with seines. The cod were landed and processed in stages,; their bodies were split, the backbones taken out, and the flesh salted. After washing the cod were laid out on the beach for several weeks to dry. The main fishing season ran from June until August. Return voyages to Europe commenced mostly in September. Occasionally ships managed to make two trips per year and the final return could be as late as November. Smaller vessels often participated first in the herring fisheries in the English Channel during winter before sailing to Newfoundland, from where they returned in September.

As a result of competition for the best fishing grounds and shore facilities, some people were left behind to take care of the land installations during winter and to prepare things for the next season. Certain rules developed which were later laid down by the authorities. In the French fishery the captain of the vessel which arrived first was called "maitre de pêche" and other captains arriving later had to report their plans to him and, if necessary, to respect his decisions. Later the oldest captain acted as "prud'homme de la baie". Similar regulations developed in the English fishery where "fishing admirals" were appointed.

Increasing difficulties and hostilities with the local English inhabitants led to the development of the so-called pêche errante on the Grand Banks. This fishery was carried out without shore bases in America. It had certain disadvantages, particularly the loss of drying facilities on land, but, on the other hand, it gave the opportunity to choose freely the most promising fishing grounds. There is evidence that in 1719 out of 500 participating French vessels 100 were engaged in this new form of high-seas fishing on the outer Grand Banks and in the Gulf of St. Lawrence, particularly around Magdalen Island. Often fishing had already started further offshore on the Flemish Cape, where large cod were abundant. Fishing was then continued along the southern edge of the Grand Banks, on the neighbouring smaller banks, also beyond Cabot Strait off Nova Scotia, and later on the Grand Bank proper at depths of 50 - 100 m.

During fishing in open waters the crew stood in wooden barrels, fastened to the deck on the weather side of the drifting vessel, and they were protected against the cold by leather aprons and against the hooks by straw

Fig. 61. French cod fishing in Newfoundland waters. *Above*, shore installations of the sedentary fishery with stage for fish processing and drying of cod on the beach; *centre and below*, hand lining on the Grand Banks (see text).(Duhamel du Monceau 1769)

shields (Fig. 61). Fishing was done with hand lines, each catching between 25 and 200 fish per day, sometimes as many as 400. Three men were engaged in processing the catch and stapling the cod below deck in layers with salt. Less frequently the cod were stored in barrels as "laberdan", a product esteemed in Dunkerque. The drying of fish was accomplished at the home ports. In autumn the surroundings of these harbours, for instance at St. Malo, were widely covered with fish exposed to the air for drying.

The heavier salted "green" fish ("morue verte") from the Bank fishery were predominantly landed in the Channel ports and, after drying, were consumed in France, mainly in Paris. Lighter salted cod from the sedentary fishery, dried ashore in Newfoundland, on the Gaspé Peninsula and other places, were transported to the ports along the Gulf of Biscay for export to Spain and other Mediterranean countries and to the West Indies. Marseilles was a trading centre for dried cod for export to Italy.

The English fishery competed with the French in these exports of lightly salted dry cod as the home market was rather limited. The English had the advantage of having at their disposal in Newfoundland and neighbouring territories the wide areas of land necessary for the drying of fish. When space along the Newfoundland coast became insufficient, English fishermen emigrated to Nova Scotia which developed into a second fishing centre. After a poor fishing season in the inshore waters of Newfoundland in 1713-1714, the English fishery also started to participate in fishing on the Grand Banks.

The Portuguese cod fishery was resumed in 1835 after an interruption of more than 200 years and, from 1885, a fleet of schooners sailed from the ports of Oporto, Aveiro, Figueira da Foz, Viana do Castelo and Lisbon regularly every year to the Grand Bank and adjacent banks. The cod were salted on board, as on the French vessels, and were dried on racks in Portugal after the return of the ships.

Major improvements in gear technology were introduced rather late. In 1778 French fishermen from Dieppe started to replace the hand lines by long lines, a method already used in Europe (Fig. 62). Successful long lining led to a rapid introduction of this gear although it had been forbidden for some decades because of loss of boats during its operation. By around 1865 each schooner was equipped with long lines of a total length of about 30 - 40 km with 15 000 - 20 000 hooks. The long lines were operated from long boats, each carrying 4000 - 5000 m ropes which were exposed in 24 radial lines of 120 - 130 m length. At each fathom there

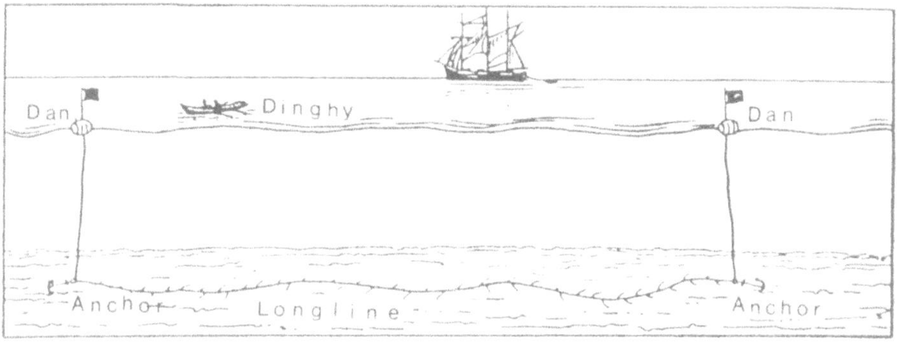

Fig. 62. Three-mast schooner with long line on the Banks; by means of floats and sinkers the long line can be set at various depths. (After Le Danois 1937)

was a snood with a hook. Instead of hauling the lines in the morning and setting the newly baited gear in the evening, American fishermen kept the gear in the water for longer times, controlling the lines only three times daily to collect the catch and replenish the bait.

The introduction of long lines resulted in a much larger demand for bait. Since preserved bait, like salted fish, was not well taken by the cod, the long line fishermen bought fresh herring from the local fisheries. Later, in summer, they also obtained capelin *(Mallotus villosus)* and squid *(Illex illecebrosus)*. In 1886 the selling of bait fish was legally forbidden in Newfoundland in an attempt to inhibit the foreign fisheries. In spite of this handicap the foreigners managed to cover their bait needs on the high seas. Besides the species already mentioned, whelks *(Buccinum undatum)* were also used as bait since they were abundant on the sea floor.

The second innovation, around 1855-1858, was the introduction of dories, simple flat-bottomed wooden plank boats which were more handy than the heavy long boats used so far. The origin of the dory is not quite certain. Some sources refer to it as an American invention. The Portuguese claim that the name dory stems from "pescadores", the Portuguese word for fishermen, and the design of these boats resembles that of boats used in southwest European coastal fisheries. On the decks of the schooners the dories needed little space as they could be piled up on top of one another (Fig. 63). With the replacement of two to three long boats by a dozen dories, fishing could be made much more efficient. Later, around 1950, Portuguese schooners carried 40 - 70 dories.

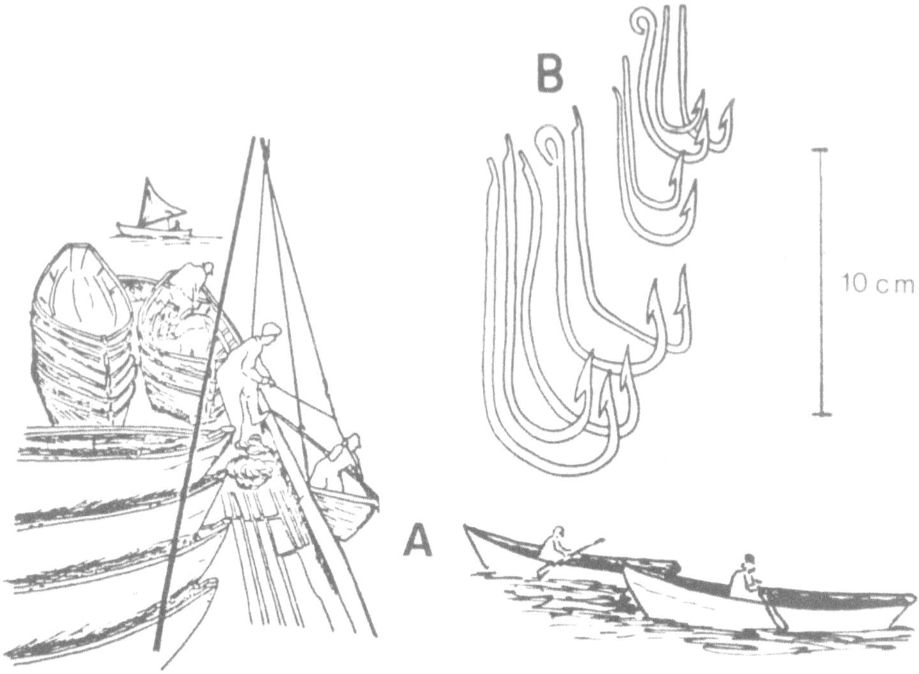

Fig. 63A,B. Cod fishing from dories in the Portuguese fishery. **A** Dories, stacked on deck, lowered over the side and rowed or propelled by triangular sails; **B** hooks for hand lining (*left*) and long line fishing (*right*)

Cod fishing from dories was certainly one of the most adventurous and dangerous activities. On French ships the dories were about 5 m long and had a crew of two men, while in the Portuguese fishery each boat was manned by one doryman. The latter were often highly skilled fishermen from the Azores. Whenever the weather permitted, the dories were watered and spread out from the anchored schooner to catch cod with long lines, hand lines and also jigging equipment when bait was lacking. The dories were rowed but also had a small triangular sail. Fishing lasted for at least 12 hours per day before the dories returned to their mothership, often overloaded with fish. Frequently stormy weather, ice and fog made this fishing a perilous profession and many dorymen never returned. When the catch and the dories had been boarded, the cleaning of the cod took further hours of work. The fish were headed and gutted, the liver removed for producing train oil, the body split along its length, ripping out the backbone, and stored in salt below deck. A full load for a Portuguese schooner could consist of up to 750 t of salted cod.

The French fleet also fished off Iceland on their way to and from the Grand Banks, often to complete their cargo. Their fishery was without contact with the Icelandic coast and occurred as "pêche derivante" with hand lines from drifting vessels. The cod here was smaller but of higher quality than on the Grand Banks. Since these fish arrived earlier in the year on the fishing grounds, voyages could commence in February. On the way to Iceland larger cod were occasionally caught near the Faeroe Islands. Finally, a part of the fleet concentrated their fishing entirely in Icelandic waters. During the peak of this fishery in the mid-19th century, the French fleet engaged in this fishery comprised about 250 vessels, i.e. slightly more than half of the total fleet. The landings of 12 000 t in 1878 almost reached the level of production by the fishery off Newfoundland (16 000 t).

From the 1920s steam trawlers started to replace sailing vessels, first off Iceland, then also in Newfoundland waters. Of 11 sailing ships which still operated in the French fishery in 1939, none was re-activated after the war. The old tradition of cod fishing, however, was continued by Portuguese ships. In Portugal the bacalhao was further consumed in large quantities. At the end of the 1950s the Portuguese liner fleet still consisted of more than 30 sailing ships, three- and four-mast schooners of up to 700 GRT, now equipped with auxiliary motors which were mainly used in the fishing grounds (Fig. 60). They carried altogether about 3000 dories; 4000 men participated in this fishery.

When concentrations of cod were detected off West Greenland in the 1920s and first fished by French vessels, the Portuguese fleet also turned their fishing more and more from the Grand Banks to the Greenland waters. Usually the voyages began from Lisbon in April, with fishing on the Grand Banks lasting until June, and then the ships sailed north to the Store and Lille Hellefiske Banks and Fyllas Bank off Greenland where fishing was very productive. If ships could not complete their cargo there, they returned to the Grand Banks. In September or October the schooners sailed back to their home ports. From the 1950s motor vessels were put into operation as motherships for dories and they increasingly replaced the old schooners. But soon large and efficient trawlers caught most of the fish in these waters.

The expansion of the trawl fishery to Newfoundland waters led, for some years, to a considerable increase in the production of salted cod, particularly in Spain and Portugal (Fig. 64). However, with the development of distribution of fresh fish on ice and later in a deep-frozen condition, the demand for salted and dried cod declined and, subsequently, production in all countries decreased to low levels in the 1960s and 1970s.

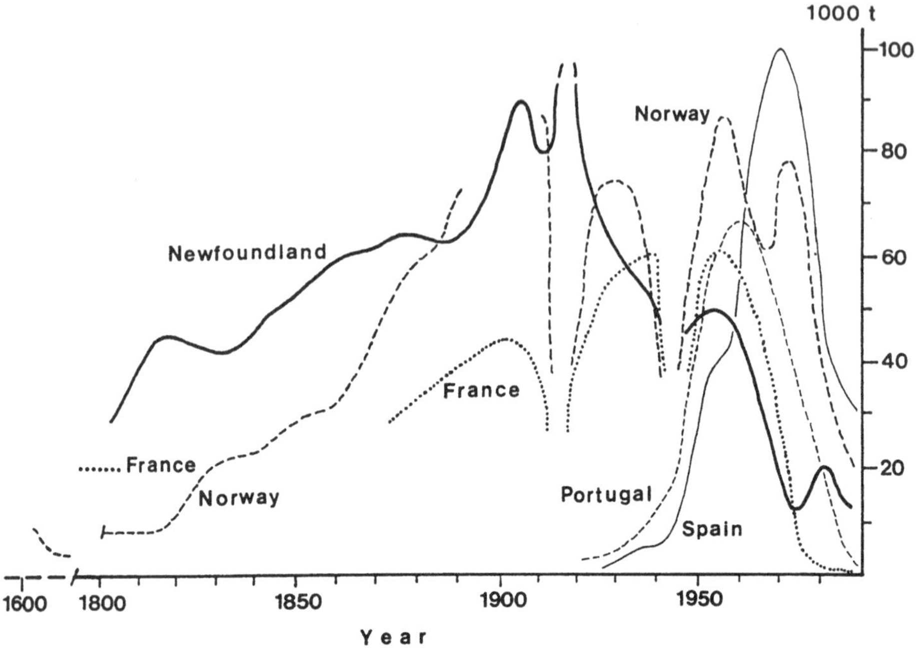

Fig. 64. Estimated production (metric tons) of dried and salted cod in various countries; since about 1900 an increasing part of this production came from cod caught with trawls

For over 500 years the waters around Newfoundland and adjacent banks have been fished with remarkable success. From very incomplete statistics it has been estimated that in the old line fisheries up to 250 000 t of cod may have been caught annually from about 1580 to 1750. For the 19th century the annual catch was estimated at 400 000 - 500 000 t.

3 Mechanization of Fishing

The invention of machinery at the beginning of the industrial era led, in fishing, as in many other sectors, to revolutionary developments and caused a complete change to the structure of fisheries.

In the field of fishing technology a basic change in tactics took place, from the outwitting of fish, used since earliest times, to the application of mechanical power for collecting the catch. With steam engines and, a short time later, motors, instead of sails, vessels became largely independent of wind conditions. This not only increased tremendously their mobility and catching efficiency but it also allowed the transport of fish in a fresh condition over larger distances and facilitated regular landings for a steady supply to the markets.

After the construction of the first serviceable steam engine by James Watt in 1769 it still took 100 years until such machines were applied to the propulsion of fishing vessels. The reasons for this rather slow introduction lay in the special requirements of the fishing operation by relatively small ships in remote areas where there were often rough weather conditions and, in particular, the need not only to propel the vessel but also to drive the capstan and trawl winch. In fact, the development began with the application of steam to hauling nets and long lines around 1860. At the same time steam ships were often used as tenders for the transport of fish from the fleets of sailing trawlers and liners to the ports. Steamers were also used as tugs for towing the sailing ships to and from the fishing grounds or harbours when wind was lacking or blowing from unfavourable directions. When the initial problems with steam engines in fishing vessels were eventually overcome, a most dramatic development began.

The construction of railway lines since 1830 facilitated the distribution of fish to the areas of consumption. In Great Britain, which played a leading role, all important ports were already connected to railroads by around 1850. The development of the distribution system, firstly in Britain and then also on the Continent, opened up a wide field of new markets and further promoted increases in fisheries production. Fishing enterprises started to concentrate more than before on certain ports which developed rapidly to become major centres of fish landing and distribution. At the same time coastal fisheries from numerous smaller localities retained their importance. To a great extent they concentrated on the production of the most valuable coastal species which were landed in prime quality and often packed in specially treated boxes or even kept in wells on board for landing them alive. On the other hand, the main fishing ports were equipped to handle quickly large quantities of landings, mainly from the high-seas fisheries.

Mechanization also led to rapid improvements in the preservation of fish and the development of fish products. The application of artificial ice and, later, deep-freezing, and the production of tinned food and fish meal were milestones on this way.

3.1 Trawling: From Sailing Ships to Steamers

It was the trawl fishery which received the largest impetus from mechanization. The origin of the trawl is not known but it is assumed that it developed in western Europe out of a very old gear, the seine ("sagena" from the ancient Greeks), which was used from the beach and later offshore from boats. Probably the seine was modified to a towed net to increase the catch. The "dreige", a trammel (three-walled) net of up to 500 x 2 m, moved by two watersails in tidal zones and widely used in the 16th and 17th centuries along the French coast of the Channel indicates such development. The wings of the trawl may hint at such an origin. Another possible source is the dredge, a heavy armed net bag, used since olden times in the oyster fisheries. This source seems to be reflected in the beam trawl.

After the development of trawling to an industrial scale this method was transferred from western Europe to all over the world and was further adapted for local conditions (see Sect. 5.2). It is certain that at least in Japan, possibly also in some other countries, various types of trawls were invented quite independently.

The earliest references to the trawl relate to gear used in the Netherlands and in England. The "wonderkuil", a small-meshed net for the capture of little fish and towed between two boats in Dutch nearshore waters, was prohibited in 1341. Likewise, efforts are documented in England in 1376-1377 to ban a gear called "wondyrchoun" which was extens-

Fig. 65. Beam trawl. *Above*, sketch of a beam trawl from the back of a state paper from 1635 (*The Fish Trades Gazette*, 1921); *centre*, "chausse", towed by a French smack; *below*, beam trawl with iron trawl heads, towed on one warp. (Le Danois 1937)

ively used in the estuary of the river Thames. It was described as "made in the form of a drag for oysters which is of unusual length ... with so small meshes that not even the smallest fish could escape". Reports on these fisheries indicate that there was already concern for the preservation of fish stocks. Numerous attempts were also made during the following centuries to suppress trawl fisheries as being destructive to young fish and to life on the sea bottom. However, such opposition did not prevent the further expansion of this type of fishing.

Trawl fishing can also be traced back to as early as 1302 in the waters of the Baltic lagoons of East Prussia. Here the trawl was known as "kütel" or "keitel", names which persisted until modern times. It may be assumed that German colonists, immigrating from the west (see Sect. 2.3), introduced the gear to this region where local conditions with a flat bottom in shallow waters were ideal for such fishing.

The large variety of gear which existed along the French coasts and in the Netherlands, particularly the Zuider Zee, seems to indicate that these waters were a nucleus for the development of trawling. It is likely that from here the method of keeping dragged net bags open with a beam was transferred to England, where the trawling technique was then further developed.

From the beginning of the 17th century the term "trawle" was used for drag nets in the fishery along the southern coasts of England and in the Thames estuary. A sketch from 1635 shows clearly the design of a beam trawl (Fig. 65). Originally the beams to which the nets were attached were oak trunks of about 4 m length. They were supported at both ends by wooden trawl heads bound with iron to keep the beam about 60 cm above the bottom. It seems that for neither the trawls on the Continent nor those in England is it known when and why the round knitted net bags were replaced by upper and lower pieces of netting mounted together. Originally the nets were made of hemp dipped in tar and around 1850 manila was introduced.

In England beam trawls were towed by fishing smacks. These were initially small boats, pointed at the stem and deep and wide at the stern, with one or two masts and a square sail. Little is known of the origin of this vessel type but it certainly dates back beyond the Middle Ages. Similar ships fished along the French coast (Fig. 65). In the Netherlands trawling was first carried out by poor coastal fishermen with small flat-bottomed boats like the "schuyt" and the "pink" (Fig. 58). Here trawling concentrated on plaice while in England sole, turbot and brill were special targets. Cod, ling, hake, whiting and other species were mostly salted until, in around 1850, ice was introduced to preserve the fish. During the 17th century on most ships the old square sails were replaced by cutter rigging with sails in fore-and-aft direction; this greatly improved the freedom of manoeuvre. In 1822 the English type of beam trawl was exported to Oostende and from there it was soon introduced to many other places along the coasts of the Continent.

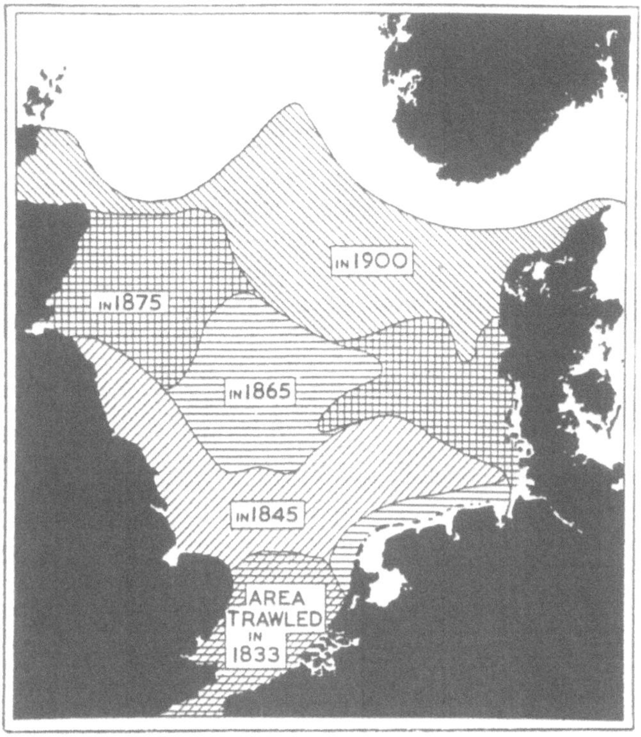

Fig. 66. Spread of English trawling in the North Sea during the 19th century. (Wimpenny 1953)

A centre for the early trawl fishery was Brixham in Devon. Here, where the smacks were called sloops, 112 vessels had already been registered by the 1830s. By this time the trawl fishery had expanded from southwest England and the Thames estuary to the waters off the Dutch coast and along the shores of Norfolk (Fig. 66).

The rapidly increasing demand for fish stimulated a further expansion of trawling in the North Sea along the English coast and even far to the east. Vessels from southern ports like Brixham and Ramsgate began to work temporarily from North Sea harbours, especially Scarborough and Hull, and ships from Barking on the Thames moved to Yarmouth and Lowestoft, thus reducing the distances between the fishing grounds and the landing places. Moreover, the expanding ports were suitably located to enable quick delivery of the catches to London, where Billingsgate had developed to become the largest fish market in the world. Finally, in

Fig. 67. English trawling smacks and reloading of fish at sea by small boats to a steamer serving as carrier in the North Sea fleet fishery. (March 1953)

around 1855, fishermen from ports in the south settled permanently in the North Sea harbours. The fishing industry began to boom.

Good catches and profits led to the construction of many new fishing smacks. Development accelerated strongly when the ports were connected with the railroad system. The number of smacks increased from about 200 in the early 1850s to 1000 in the 1870s. A peak in the size of the fleet was reached in the 1880s with around 2500 smacks. More than 800 of these ships operated alone from Grimsby, a port which was developed dramatically after 1860 with new docks and a railway connection to London.

With very intensive fishing yields declined in nearshore waters. But new productive fishing grounds were discovered in the open North Sea farther and farther from the ports. This led, in around 1840, to the introduction of a fleet fishing system. The smacks fished in smaller or larger groups of up to 200 boats together under the command of an "admiral". The trawl catches of each day were transferred at sea by small boats to carriers, mostly fast-sailing cutters and later also steam vessels, which transported the fish to the markets (Fig. 67). The sailing smacks stayed at sea for 8 or even 12 weeks before returning to the port for a week of rest and refurbishment. In this way fishing was made more economic by avoiding voyages of all ships to and from the ports. But the life of the fishermen became even harder than before. The transfer of fish at sea was especially risky and many sailors lost their lives during such operations. After several years of fleet fishing, some mission boats were put into service to provide a minimum of medical care and re-supply for the hardworking fishermen. When steam ships came into operation, they were used, from 1864, as fish carriers and also as tugs to tow the sailing vessels.

In the early days of the beam trawl fishery the smacks were mostly ships of 18 - 22 m length and 30 - 60 GRT. With the extension of fishing to off-shore waters and with increasing competition more and more vessels were, in the 1860s to 1880s, enlarged or new ships built to larger sizes of 60 - 80 GRT. At the same time the fishing gear was made larger for bigger catches, and the beam finally reached a length of 15 m. On the larger ships the main sail became so huge and the boom on the mast so long and unhandy that the cutter rigging was changed to ketch rigging by addition of a mizzen aft. This arrangement was called dandy-rigging and was also introduced to ships in the Netherlands and other countries. The smacks were very seaworthy ships with hulls of heavy oak timber (Fig. 67). With a crew of five to six they could sail at speeds of up to 12 knots. Hauls with the trawl were usually of 5 - 6 hours duration. The fish were initially packed in baskets but, later, wooden boxes were used and the fish were preserved in natural ice which was imported to England in large quantities from Norway and also produced locally near British fishing ports. A number of smacks were fitted to alternate between trawling and long lining, particularly for cod and for turbot *(Rhombus maximus)*.

The end of this era came with the introduction of steam engines to the fishery. There is some uncertainty as to when this development actually started; many experiments were made in various countries but often failed. The earliest reports on vessels specifically constructed for or converted to steam trawling are from Arcachon in 1836 and 1838 and from Bordeaux in 1864 and 1866. In France the first steam trawler making regular voyages was inaugurated in 1876. The first trials in the USA date back to 1866 and in Britain attempts to develop steam trawlers were made from 1856. This included unsuccessful experiments with paddle tugs towing a beam trawl. The final breakthrough was not achieved until 1881 with the *Zodiac*, a ship 34 m long and built in Grimsby. From 1882 steam trawlers were also put into service in Scotland, particularly at Granton but, generally, the implementation of this fishery developed at a slower pace than in England. In Scotland the existing fleet of sailing ships turned their activities entirely to the line and drift net fisheries whereas in England the sailing vessels were rather rapidly replaced by steam trawlers. This can best be demonstrated by the number of vessels in Grimsby, the leading English fishing port (Table 1).

Table 1: Number of fishing vessels registered in Grimsby during the transition period from sailing ships to steam trawlers

Year	Sailing ships	Steam trawlers
1882	623	2
1890	769	35
1900	69	471
1909	29	608

Table 2: Number of trawlers in various European countries

(after ICES *Bulletin Statistique*)

Country	1906 - 1910		End of 1930s		
	Steam trawlers	Sailing trawlers	Steam trawlers	Motor trawlers	Sailing trawlers
England	1326(+16)	917(+68)	1078(+10)	77(+47)	26
Scotland	302	55	320(+510)		
Germany	210(+20)	143	355	173	-
Netherlands	75(+13)	14(+29)	95(+2)	52(+441)	-
Belgium	26	123	(16)	(471)	-
France			(274)	(81)	-
Denmark	3				
Sweden	14		19	(148)	-
Norway	-	-	9	-	-
Faeroe Islands			12	(275)	-
Iceland			50		

Data are incomplete. Figures in parentheses indicate vessels temporarily participating in trawl fisheries

Fig. 68. First German steam trawler Sagitta (1885).
(Archiv "Nordsee", Deutsche Hochseefischerei GmbH, Bremerhaven)

Altogether, the English fleet in 1895 consisted of 465 steam trawlers and about 2200 sailing smacks. At the same time in Scotland there were 112 steam trawlers, of which 76 were stationed at Aberdeen. In Scotland the development of trawling was hampered by various prohibitive regulations so that, at certain times, to ensure a supply of fish, landings by foreign vessels, mostly English and German, were necessary.

With competition from the more efficient new steam trawlers the English fishing smacks were soon abandoned or sold to the Netherlands, Belgium, Norway or elsewhere. Several were converted to coastal trading vessels. Most of the crews found new employment in the steam trawl fishery.

Britain also took the lead in the new fishery. After several years the steam trawler fleet reached more than 1600 ships (Table 2). Rapid development was also observed in several other countries with a coinciding decrease in the number of sailing vessels. In Belgium steam trawling started in 1884 with five ships bought in England and the fleet reached its maximum size in around 1925 with 57 trawlers. In Germany the first newly constructed steam trawler, with a size of 146 GRT and an engine of 260 hp, was commissioned in 1885 (Fig. 68). Favourable results led to a rapid increase in the fleet's size, mostly with newly built ships. At the beginning of the 1920s the number of German trawlers exceeded 350. The Dutch trawler fleet grew from two ships in 1892 to 210 in 1929, mainly through purchases from England. In the Netherlands the first steam lugger was put into operation in 1897 for herring drift net fishing. In France, after the early attempts already mentioned, the construction of a trawler fleet began in 1894. Besides small steam trawlers (200 - 450 GRT) for the fishery in the North Sea and Celtic Sea, which numbered 220 vessels by 1930, larger trawlers (500 - 1200 GRT) were soon built for the salt cod fishery off Newfoundland and Iceland. Participation from Scandinavian countries in this type of fishing was not very large. In Sweden the first steam trawler was bought in 1901 and a maximum fleet size was reached in 1914 with 58 ships. Later the number decreased gradually until, in 1952, the last steam trawler was laid up. For Norway trawling was too expensive to produce cheap stockfish and klipfish for export. Not until 1935-1936 were nine steam trawlers bought which could operate only under special licence. The Danish fishery concentrated almost entirely on fishing with Danish seine. In Iceland steam trawling began in 1899 and the fleet grew gradually to about 50 vessels. Altogether, the European fleet of steam trawlers was of remarkable uniformity, based on the original English model which was used by other countries. Steam trawling was also introduced in many other waters of the world (see Sect. 5.2).

With the introduction of mechanical power trawlers were now able to fish further offshore on deeper grounds to depths of 400 m while the sailing smacks were usually limited in their trawling to 100 m. The range of operations was also widened by the preservation of catches with ice, which was now produced artificially in increasing quantities by ice factories erected in the ports since the 1890s. On average the catches made by steam trawlers were six to eight times larger than those by the sailing ships.

The German steam trawlers measured about 40 x 7 x 4 m, were of 200 GRT and had steam engines of about 300 - 350 hp. They were manned by a crew of 10 - 13. The English trawlers, built slightly earlier, were mostly a little smaller but their distance from the fishing grounds was usually shorter. In contrast to the British vessels, the German trawlers initially concentrated their fishing entirely to the eastern parts of the North Sea. Gradually their main fishing areas shifted from the southeastern and southern North Sea to more northern waters. The trawlers were equipped with insulated fish holds where the catches were stored in ice in layers with wooden partitions between to avoid pressure. Including the ice, added to the fish in a ratio of 1:3, German trawlers had a loading capacity of 500 - 800 baskets or 25 - 40 t. During many voyages this capacity was not fully used since there was a fixed time limit by which the fish had to be landed to keep its quality. Thus, the fishing time and, con-

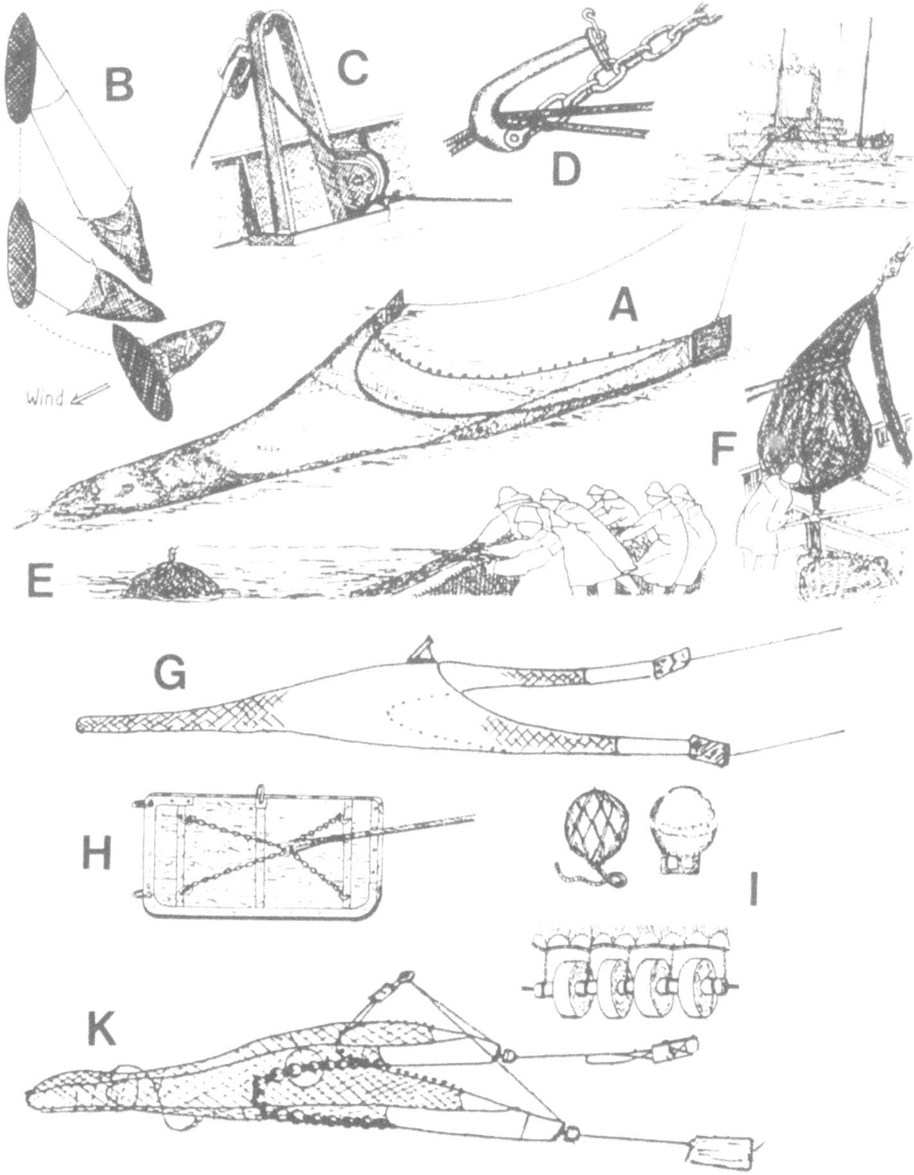

Fig. 69A-K. Otter trawling and further technical improvements. **A** steam trawler towing a Granton trawl, which is kept open by two otter boards at the wings of the net; **B** shooting of the trawl; **C** one of two gallows with blocks for the running of the warp; **D** slip hook for joining the warps outside the ship's side; **E** hauling the catch; **F** opening the net bag at the codend; **G** Vigneron-Dahl trawl with legs between the net and otter boards; **H** otter board with chains and warp; **I** floats for the headline, bobbins at the ground rope; **K** high-opening trawl with kites over the net and sweep lines to the otter boards

sequently, the quantity of the catches were dependent on the distance between the fishing grounds and home port; the whole trip did not usually exceed 10 days.

The efficiency of trawling was greatly enhanced when, in 1892 in Granton (Scotland), otter boards were attached to the net to spread it open by water pressure (Fig. 69). Such wooden boards with a metal keel had already been used for some time in Irish and Scottish trolling fisheries for salmon. Experiments at this time demonstrated that such devices could replace the beams of trawls. In 1894-1895 the whole British trawler fleet implemented the new Granton trawl and soon it was also introduced to the fisheries of Germany and other countries. From experiments it was estimated that the Granton trawl increased the catch efficiency by about 35% as compared to the beam trawl. A further advantage was the easier operation of the Granton trawl during shooting and hauling. The trawlers were fitted with the typical gallows on the side fore and aft to facilitate the handling of the otter boards and the two warps. A great step forward was the introduction of the double-drum steam winch installed on deck. The size of the trawl was now no longer limited by the length of the beam so that larger nets could be constructed.

Soon steam trawling became the main method for the capture of demersal fish in the North Sea, and also in the Skagerrak and Kattegat and in the waters west of the British Isles. Even with the foundation of the International Council for the Exploration of the Sea (ICES) in Copenhagen in 1902, the statistics which became available to demonstrate the extent of this fishery were, at first, rather scant. They show that during the period 1904 - 1910 about 3400 steam-driven fishing vessels operated in the north European sea fisheries. About 2000 of them were engaged in trawling (88% in the North Sea), about 1000 smaller ships were engaged in the herring drift net fisheries and approximately 400 carried out line fishing. The number of high-seas sailing ships at that time can be estimated at 1300 each for trawling and for the herring drift net fishery. Of the sailing trawlers 75% were English, while about 800 of the sailing drifters were luggers from the Netherlands. The above figures do not include 876 Scottish steam vessels which operated partly in the drift net fishery and partly in the line fishery.

From these early statistics it can also be concluded that during the period 1906 - 1910 the average total fish landings from the North Sea were in the order of 1 million t per year. Of this amount about 250 000 t of demersal fish were caught by the trawler fleets, both steamers and sailing ships. Another 620 000 t were herring caught by the drift net fisheries

(430 000 t by British vessels) whilst line fisheries produced about 30 000 t of demersal fish per year. The remainder is unspecified and may have been landed by coastal fisheries using a variety of methods. From the Skagerrak - Kattegat area trawlers landed about 16 000 t per year and from the waters west of the British Isles more than 87 000 t were caught.

The statistics also provide some information on the species composition in the trawl catches during that period. In the landings from steam trawlers haddock made up the largest part of the catch at 46%, followed by cod (19%), plaice (11%), whiting (6%) and rays (2%). The landings from sailing trawlers were different with plaice constituting 53% of the catch, dab (8%), whiting (7%), rays (6%), flounder (6%), sole (5%) and cod (4%). In comparison the landings from the line fisheries consisted mainly of cod (38%), haddock (34%) and ling (8%). These differences result from fishing on different grounds. Steam trawling was carried out mostly in the open North Sea, whereas the sailing trawlers fished in shallower waters nearer to the coasts, where there was an abundance of flatfish. In contrast, the line fisheries caught mainly species that took larger bait.

With the dramatic expansion of trawling in the North Sea, the average catch per vessel started to decline and the catch composition changed. Fish of the "prime" group, highly prized flatfish like sole, turbot and brill and John Dory *(Zeus faber)* became less abundant, plaice and haddock were now the main targets. Then, in spite of further increased efforts, the overall yields began to decline. This was largely due to heavy exploitation of fisheries as indicated by the strong recovery of fish stocks in 1914 - 1918, when only a little fishing could be carried out during World War I. However, biological changes also contributed to the decline, particularly to the gradual disappearance of haddock from the southern parts of the North Sea, observed since the 1890s. After the end of World War I, by which time many fishing vessels had been lost, the fleets were soon rebuilt and intensive fishing during the 1920s again reduced the accumulated commercial fish stocks so that the yields declined once more. At the end of the 1920s the fisheries and fishing industries suffered greatly from the world economic crisis.

The construction, maintenance and operation of steam trawlers of increasing size required substantial capital and adequate technical and administrative services. More and more companies were, therefore, established for such enterprises. In Germany, for example, 14 trawler companies in 1929 owned 68% of the 400 steam trawlers. The largest company alone had 85 of these vessels.

Competition and decreasing yields resulted in greater efforts to further improve the efficiency of trawling. Fishing on rough grounds, which occupy about 17% of the North Sea, was made possible by the invention of the roller ground rope with bobbins, a series of wooden, later iron, rollers in front of the net. Then, in the mid of the 1920s, a major achievement was the Vigneron-Dahl trawl invented in France (Fig. 69). In this gear single or double lengths of steel cable ("legs") were fitted between the net and the otter boards, thus widening the spread of the otter boards and enlarging the zone over which fish could be scared into the net. It was estimated that this modification produced a one-third increase in the catch as compared to the simple otter trawl. Many further technical improvements followed in later years. New trawlers were made bigger and with more powerful engines, enabling them to fish with larger gear (Fig. 70). The usual towing times per haul were reduced to 3 hours or less to improve the quality of catches. During the 1930s the last sailing trawlers were abolished.

Herrings had been caught, to some extent, by trawl since Fleetwood fishermen began this fishery near the Hebrides in 1904. With the experience that floats on the headline and a small otter ("kite") above the net could cause an increased opening of the net, a special herring trawl was developed in the 1920s. This gear was of lighter construction with finer twine and smaller mesh and could be towed at a higher speed. The high opening of the net was later increased by fitting a second kite above the net. A specialized herring fishery with steam trawlers in the North Sea, following the herring shoals from the Fladenground to the eastern Channel, developed in Germany. The landings from this fishery increased to around 180 000 t per year by the end of the 1930s. Soon after World War II it was resumed and, in 1955, reached a peak with 200 000 t. With the construction of factory trawlers during the 1960s and fishing in more distant waters, this fishery became insignificant.

For the German trawl fishery the herring catches counterbalanced the losses caused by the withdrawal of haddock from the German Bight and southern North Sea. The German demersal fish trawling concentrated more and more on saithe *(Pollachius virens)*, a species which had meanwhile been introduced to the market. In contrast, the British trawl fishery remained largely unchanged.

Knowledge of the rich fish resources in northern waters, acquired during the Great Fisheries since the Middle Ages, led to encouraged endeavours from the early days of steam trawling to expand this fishery to waters beyond the North Sea and adjacent areas. Already, since the early 1890s,

British, French and German trawlers had fished in waters off Iceland and from 1898 on the Faeroe Banks (Fig. 71). Off Iceland trawling concentrated initially on haddock and flatfish caught during the summer in shallow waters along the south coast but expanded gradually to the southwest and west for a spring fishery of cod. After 1930 vessels fished all year long. Cod and haddock were also the main targets on the Faeroe Banks. A Russian expedition to explore the Murman coasts (1898 - 1908) stimulated the development, from 1905, of a trawl fishery by English and German ships in the southern Barents Sea and White Sea; this was directed initially on plaice and later on cod and haddock. Russian trawlers did not fish here until the 1920s. German trawler explorations in the 1880s to near the Lofoten in Norway and south of Bear Island were unsuccessful at that time. Generally, steam trawling in northern waters initially developed at a slow pace since most vessels were really not large enough and powerful enough for such long-distance fishing. The catches were sufficiently large, sometimes too large for the small ships, but the sale of old big fish, some of little known species, created difficulties. By 1910 the annual landings from Iceland (90%) and the Barents Sea (10%) reached only slightly more than 120 000 t, i.e. half the quantity landed by the North Sea trawlers.

When, at the end of the 1920s, the bottom trawl fisheries in the North Sea stagnated with discouraging yields, new interest grew to develop fishing in the north. Within a few years, by around 1930, the size and engine power of newly built trawlers surpassed the usual dimensions. The increase in vessel size in the German trawler fleet may demonstrate this development (Fig. 70). In 1910 there was already a differentiation between the North Sea trawler, of about 32 m length, 140 GRT and a fish capacity of around 800 baskets, and larger Iceland trawlers, of 40 m length, 250 GRT and a fish hold capacity of 1500 baskets. Then, new trawlers were built to between 280 and 500 GRT with more than 500 hp. At the end of the 1930s trawlers had reached a size of 70 m with 650 hp (besides steam engines, motors were also in use) and could hold 6000 and more baskets of fish. Internationally, several types of trawlers were distinguished: small steam trawlers (including motor ships) of 40 - 150 GRT with nine men for voyages of 3 - 4 days, larger North Sea trawlers of 150 - 225 GRT with 11 - 15 men for voyages of 1 - 2 weeks, Iceland-Barents Sea trawlers of 250 - 650 GRT with crews of 15 - 20, and the Iceland-Newfoundland trawlers of 500 - 1500 GRT, mostly fitted with diesel engines and with a crew of about 50.

With the operation of larger trawlers new fishing areas were developed in the North Atlantic; for example on the shelves off Norway from 1927, in

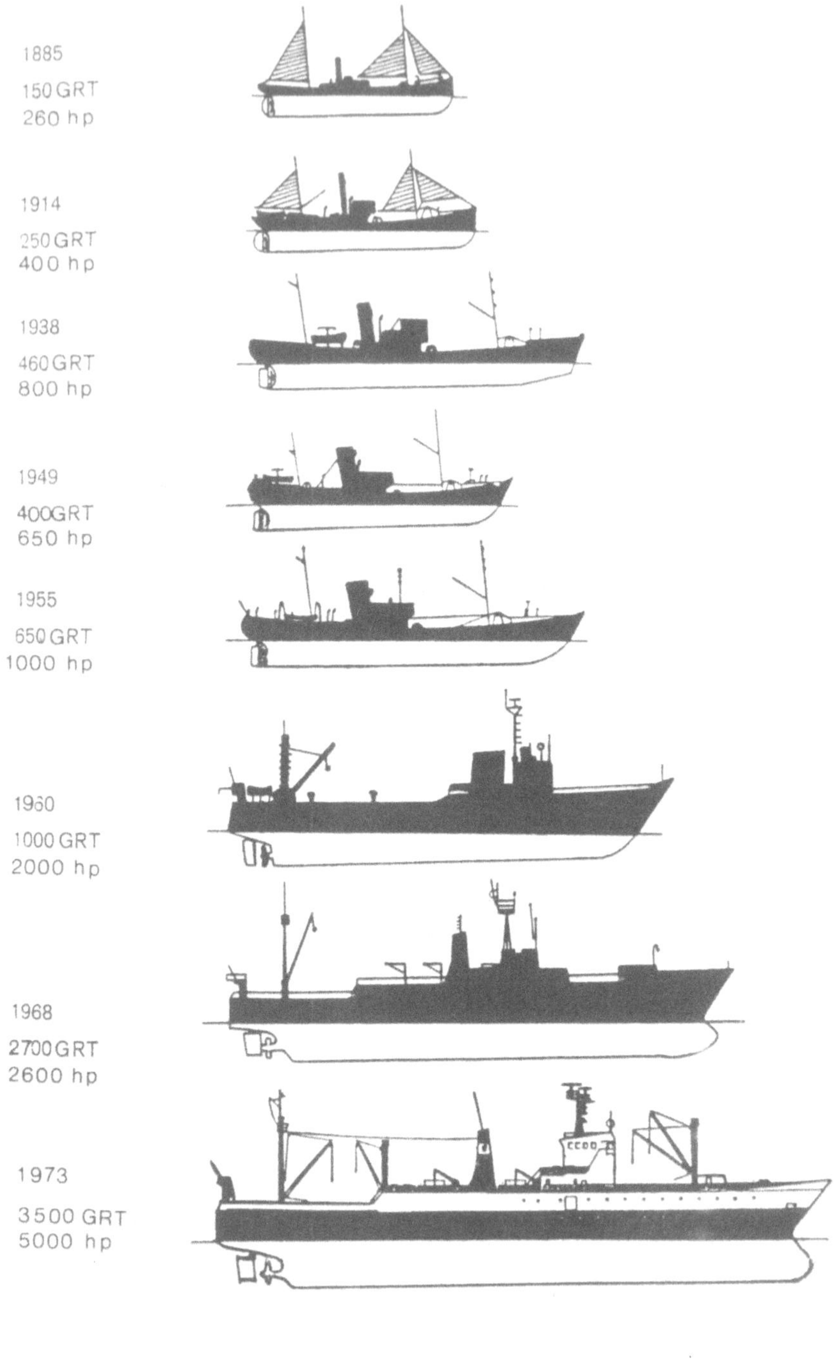

Fig. 70. Development of German trawlers from 1885 until 1973. (Anon 1969, supplemented)

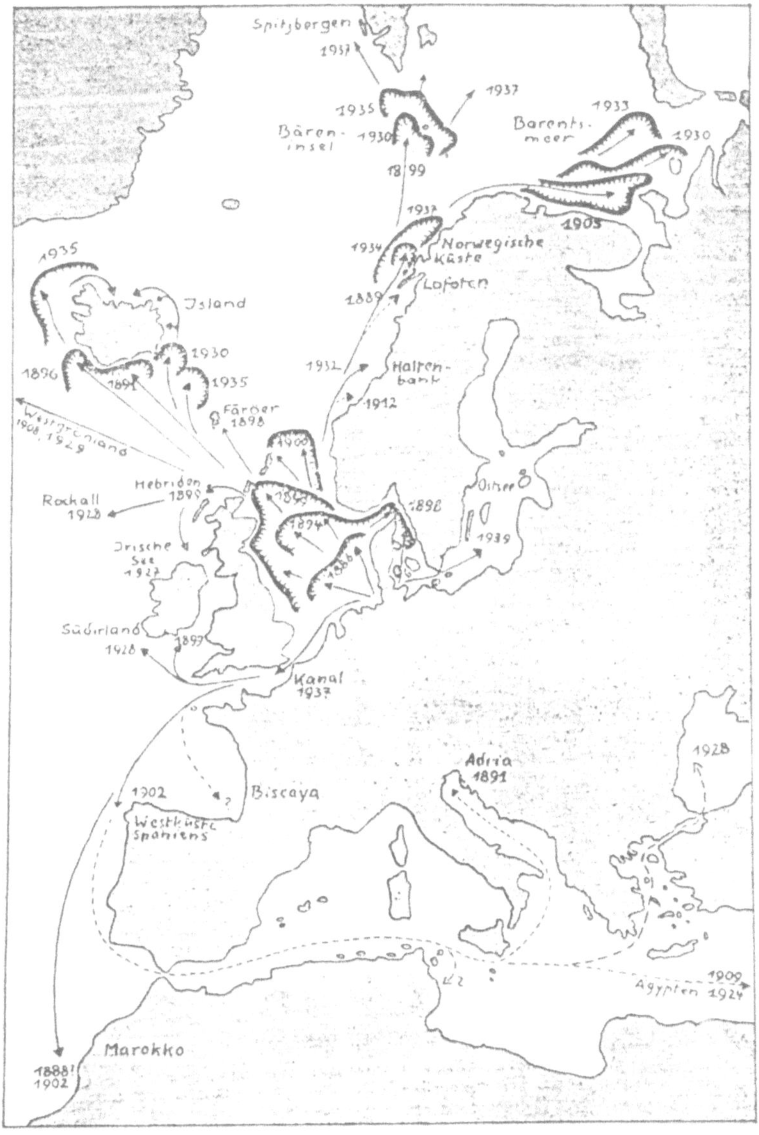

Fig. 71. Expansion of German steam trawling, 1885 - 1939

the Bear Island region from 1928, and around West Greenland and the Grand Banks from the mid-1930s (Fig. 71). The more powerful vessels could also fish in deeper waters further offshore and this influenced greatly the composition of catches. German trawlers especially now landed increasing quantities of saithe and redfish *(Sebastes marinus* and *S. mentella).*

At the end of the 1930s, when the European steam trawl fisheries were at their zenith, landings from northern waters had already surpassed those from the North Sea. The total landings, without including herring catches, can be estimated at around 1 million t per year, of which 300 000 t were caught near Iceland, 400 000 t in the Barents Sea-Bear Island-Norway region, and 200 000 t from the North Sea.

Since the early days of steam trawling efforts were also made to expand fishing to southern waters. The first trials of a German trawler off Morocco date back to 1888. Around the turn of the century a trawl fishery developed in the Gulf of Biscay and along the Portuguese coast. The fishery extended to the waters off Morocco where English trawlers fished mainly for hake. Also, German ships participated in this fishery and landed most of their catches in Portugal, until this was ended in 1910. Mention should be made also of sporadic activities of German steam trawlers in the northern Adriatic (1891), off Egypt (1909) and in the Black Sea (1928). Altogether, fish landings from the south were rather small compared to those from northern waters.

Although an increasing number of trawlers were equipped with motors, steam engines remained the main type of propulsion in offshore trawl fisheries for a remarkably long time. They outlasted World War II (1939 - 1945) when many fishing vessels were lost. New ships were often fitted with oil burners instead of coal furnaces or with steam turbines but in the 1950s and 1960s steam engines finally gave way to motors. The north European steam trawler fleet, which comprised nearly 2000 vessels in 1938 (Table 2) consisted, in 1962, of only 340 ships (England 201, Scotland 14, Germany 86, Iceland 35) and during the following years these vessels were replaced by motor ships.

3.2 Motorization

The installation of motors in fishing vessels was initially concentrated on coastal fisheries since early attempts to install steam engines in small ships had failed. Such engines proved to be too heavy and bulky for smaller boats. In contrast, motors needed less space not only for the machine but also for the bunker of oil required instead of coal. The first trials with motor propulsion were made in Germany in 1891 with the cutter *Matador* (44 GRT). Since 1894 experiments were undertaken in Denmark with spark-plug motors, developed in England. After 1900 such motors were used more generally and became of great importance in the fisheries until they were replaced by diesel engines in the 1920s.

Motorization progressed most speedily in north European countries where steam trawling was not an important sector. In Denmark 200 motor fishing vessels already existed in 1900. In Norway the first motor was installed in a fishing vessel in 1901 and only 2 years later there were already 30 motor skoiten besides more than 2000 decked sailing ships. In 1904 the number of motorized vessels exceeded 100. Also, in Sweden, where motorization began in 1903, more than 400 small motor trawlers and 250 motor purse seiners were operating only a few years later. Vessel motorization initiated in Scandinavia a major intensification of fishing and an expansion to waters outside the fiords and skerries. Long lines, Danish seines and purse seines were the most important gear while trawling was rather limited until, after World War II, industrial fishing for fish meal developed in the North Sea.

In the region of the southern North Sea, where the motorization of small coastal vessels had been promoted since the first years of this century, a new type of small motor trawler was generated in the 1920s and this soon went into competition with the larger steam trawlers and replaced them to a great extent when they left the North Sea to fish in more distant waters. The first of these wooden or iron trawlers, constructed in Belgium, had a size of 50 - 80 GRT and was equipped with a motor of 100 - 200 hp. By around 1930 new ships were 31 m long, 150 GRT and had 250 - 300 hp motors. Such vessels operated quite successfully and were soon built in considerable numbers in the Netherlands, Germany and other countries. In Germany in the 1940s the small trawlers ("cutters") usually had a length of 24 m and 150 hp motors but were, after the war, also built to lengths of up to 35 m. From this type of small motor trawler the Dutch have developed, since the 1960s, a fleet of about 600 larger ships which now reach lengths of 60 m and carry very powerful motors of up to 4000 hp; these are suitable for fishing with large beam trawls for valuable flatfish.

In the fleets of large steam trawlers there was little incentive to change the mode of propulsion as long as fishing was still profitable. However, when catches in the North Sea declined and economic conditions deteriorated in the 1920s, the transition also began in these fisheries.

On small trawlers and ships for drift netting or line fishing a slow-running direct motor propulsion was used and a reversing gear or pitch propeller was sufficient to vary the efficiency of the engine. On larger trawlers much greater technical problems had to be solved, such as the distribution of power for the ships propulsion and driving of the winch during shooting and hauling of the nets. In the first instance such vessels were fitted with two engines: a main motor for the propulsion of the ship and an auxiliary motor for driving the winch or, by choice, for strengthening the vessel's propulsion via a collective gear. Later, diesel-electric machines were increasingly installed.

With the use of motors instead of steam engines much space was saved on board. This permitted a substantial increase in the cruising radius of motor trawlers, particularly since oil is consumed more economical than coal. Also, calculations indicated that in the British and German trawl fisheries motor ships produced 40% larger catches per unit time than did steamers of a similar size and under comparable conditions. Other technical improvements were of benefit to both types of propulsion, though to a lesser extent for the steamers: reduction of vessel resistance by the introduction of the Maier stem in ships allowed for 30% higher catches, and increase of thrust by means of the Kort nozzle improved the towing power by 40%.

Motor ships also had the advantage of having larger capacities for fish holds but even more important was that space was gained on board to accommodate machines and additional personnel to process the catches to stable products; this formed a basis for the development of factory trawlers. Since the beginning of this century conventional steam trawlers had been fitted with simple boilers for the production of liver oil from cod and other lean fish and this provided a welcome source of additional income. Until the 1950s fish processing other than the traditional salting was limited to various experiments and on only a few ships. Some British, French and German ships had since the 1920s been equipped with installations for the production of fish meal and with deep-freezing plants, ice machines and filleting machinery.

The first vessels regularly engaged as factory trawlers for catching and fish processing operated from France, where the largest trawlers were

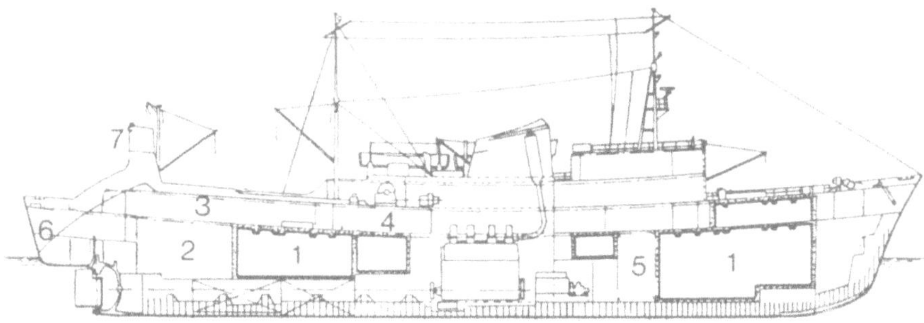

Fig. 72. First British factory stern trawler *Fairtry I*, 1954.
1, refrigerated fish holds; *2*, fish meal plant; *3*, processing room; *4*, packing room; *5*, bunker; *6*, slipway; *7*, fishing bridge

built for the fishery in Newfoundland waters and where, already by around 1930, steam engines had been abandoned completely in favour of motors. In 1934 the factory trawlers *Marcella* and *Jutland* were commissioned. The former was 70 m long, of 2340 GRT and driven by a 1000 hp diesel engine and the latter was 76 m long. Both ships were equipped with installations for deep-freezing, cold stores, a fish meal factory and boilers for train oil; curing of fish was also still carried out.

Important progress was then made during British experiments undertaken from 1947 with converted vessels, and led to the construction of the factory trawler *Fairtry* in 1954 and of two more ships of this type a few years later (Fig. 72). The *Fairtry* was 76 m long, of 2600 GRT and had a 2000 hp engine. Her processing capacity allowed for a daily catch of 20 - 25 t in two to three hauls and the storage space for various products was sufficient for runs of more than 2 months. The most important innovation was the ramp at the stern of the ship; this was formed like the slipway which had already been in use for a long time on whaling motherships (see Sect. 4.1).

This ramp not only simplified the operation of the trawl and facilitated the hauling of big catches but also allowed the use of much larger trawls than those which could be handled on conventional side trawlers. It was no longer necessary to keep the deck and bulwarks low for the operation of gear over the side and trawlers could be made considerably larger and with higher sides. This advantage promoted the accommodation of more personnel and of more extensive and sophisticated fish-processing installations.

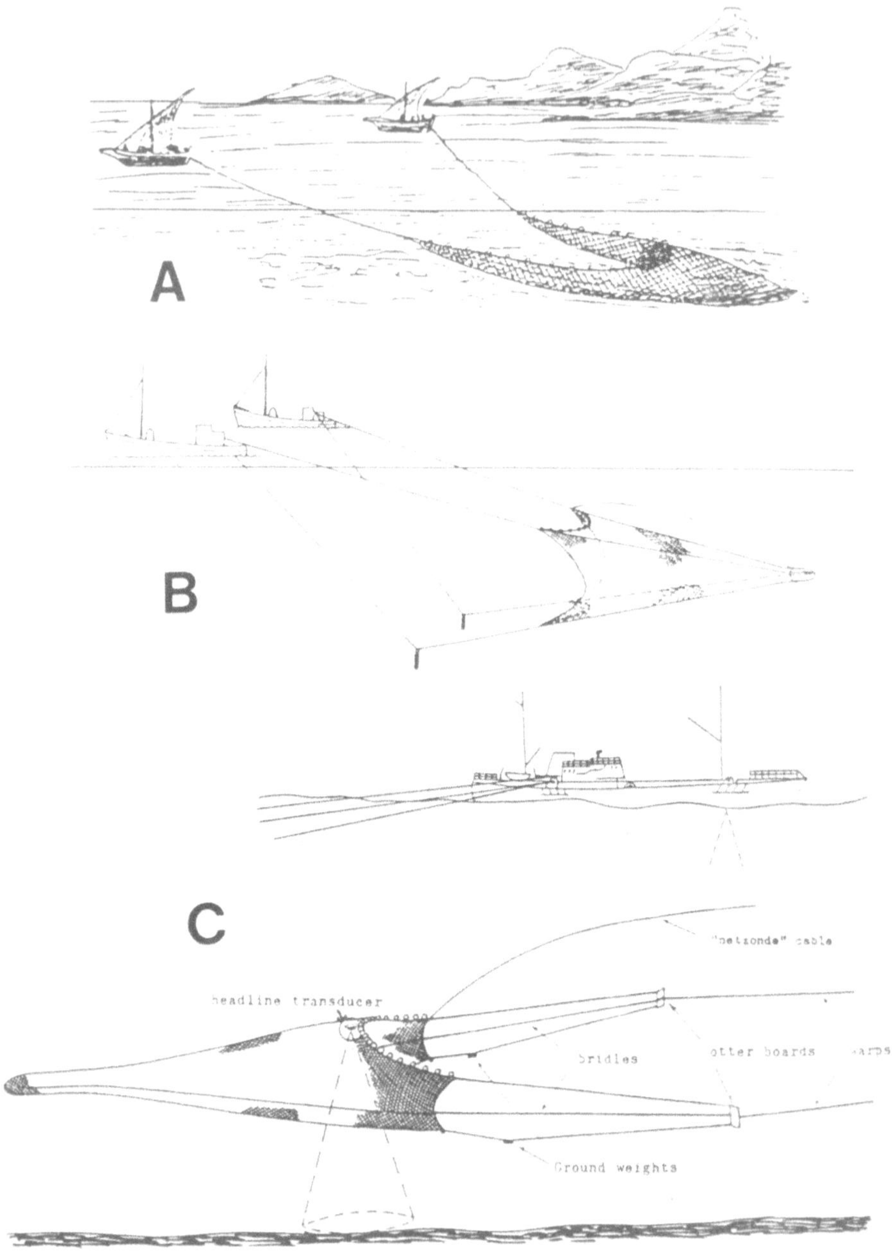

Fig. 73A-C. Pair trawling and midwater trawls: **A** French "gangui boef"; **B** Danish pelagic trawling for herring; **C** German one-boat midwater trawl. (**A** Le Danois 1937, **B,C** Schärfe 1966)

Fishing with stern-ramp factory trawlers proved so effective and profitable that within a few years many new ships of this type were built in a number of countries around the world. In the 1950s and 1960s most of these vessels were around 1000 GRT but new ships were made larger and larger and soon reached 3000 GRT and more. Particularly in Japan and the Soviet Union large fleets of factory trawlers developed in addition to motherships and floating factories which could not fish themselves but received fish from other vessels (see Sect. 5.2).

In order to maintain economic profits from the operation of such powerful and expensive vessels, the factory ships had to concentrate their activities on the most productive fishing grounds. In the search for such opportunities their fishing soon spread to all the world's oceans where fish were not only caught on or near the bottom at depths of more than 1000 m but also at all depths between the surface and the bottom.

The technique of trawling in midwater layers has been applied since the 1940s, especially in the herring fishery in the Skagerrak by small motor trawlers, fishing in pairs to keep the net open (Fig. 73). It developed from bottom pair trawling which had been used for a long time in the Baltic region, particularly in the "zeesen" fishery in the southern lagoons. Also, along the French Mediterranean coast, simple bottom trawls towed by two boats ("gangui boef") had been widely used for several centuries (Fig. 73). With the larger dimensions of vessels and equipment this technique was transferred to the Atlantic coasts and formed the basis for the development of the Spanish "pareja" fishery which is still successfully applied today. The further step of a pair of boats trawling in midwaters to catch pelagic fish became possible only when motors provided adequate power. Such fishing was quite feasible by small vessels but proved impossible by large, less manoeuvrable trawlers. Therefore, a one-boat midwater trawl was developed around 1960, mainly in Germany. The necessary opening of the net was accomplished by special hydrodynamically shaped otter boards. The first step was the development of high headline trawls which still fished close to the bottom but, during 1963 - 1965, the technique of three-dimensional trawling was perfected with the help of echosounding and netsondes to control the movement of the gear in relation to the fish shoals. Since then midwater trawls have been steered most efficiently by coordinating warp length and vessel speed.

The efficiency of trawling was increased enormously through mechanization and manifold technical improvements. It should not be forgotten, however, that a large amount of energy is required to tow the trawl gear

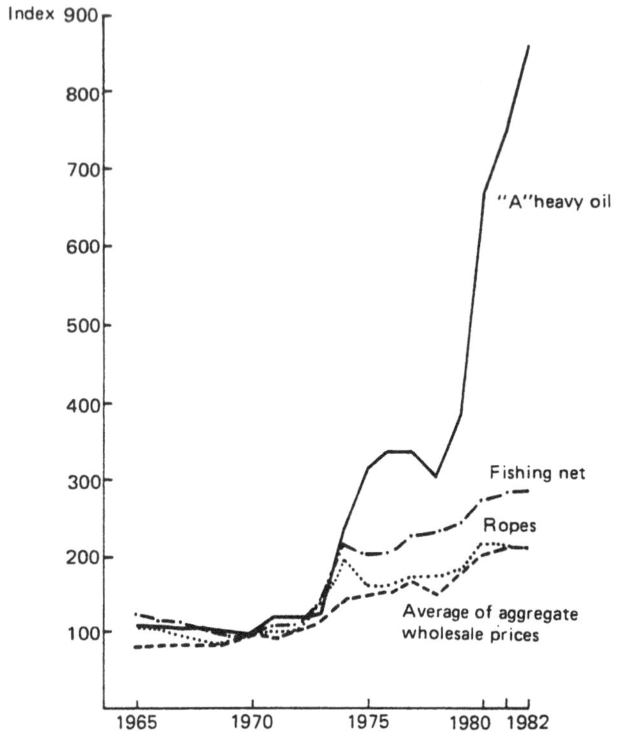

Source: Bank of Japan, "Annual Report of Consumer Price Indexes"

Fig. 74. Price indices for fuel and fishing materials, 1970 = 100

and that the consumption of fuel is high. Increases and fluctuations in oil prices on the world market reduce the profitableness of this fishery (Fig. 74). Therefore, in recent years attempts have been made to divert, where possible, fishing to other less energy-consuming methods.

3.3 Further Advances in Fisheries Technology

Mechanization and industrialization did not, of course, only affect trawling but also practically all sectors of fishing from that by large vessels to that by dugouts, which were fitted with outboard motors, and were used in artisanal fisheries.

In the field of fish processing major progress was made with the invention of canning. With the need to provide Napoleon's armies with fresh food, the French Revolutionary Directory offered, in 1795, a prize for the development of a new method to preserve food in good condition. The prize was won in 1809 by Nicholas Appert, who cooked the food in glass containers and sealed them while they were hot. At that time tin-plated iron containers were already used for packing and transporting salt fish and other food items. The utilization of such cans for heat-preserved food led to the rapid expansion of this preservation technique. After many experiments the first canning factories in Britain started to produce fish in tins on a commercial scale in around 1822. In France, sardines were canned from 1825. The development of the canning industry was greatest in the USA, where the first experiments date back to 1817.

One of the earliest steps in mechanization was the invention of a net-knitting machine in Scotland in 1820. From there this hand-driven machine was exported to the USA in 1858 where, since the 1840s, nets were made of cotton instead of flax or hemp. In Europe the Dutch drift net fishery changed from hemp to cotton nets in 1866 and fisheries in other countries soon followed this example. Increasing demand for nets stimulated the construction of a mechanical net-knitting machine in Paris in 1883. Since 1922 knotless nets have been fabricated in Japan. Another major step was the production of synthetic fibres in the 1930s. It was, however, not until the 1940s that such fibres were used for fishing nets, first in traps and later also in drift nets, trawls and purse seines. In 1950 the first gill nets of polyamide monofilaments were introduced and by around 1960 nets made of synthetic materials had replaced the old cotton nets everywhere.

The development of navigation aids, such as radio direction finders and radar, substantially improved the safe operation of fishing vessels. Of great importance was the invention of the echosounder. During the first decades of this century such instruments were installed on ships for navigational purposes but they also opened up the possibility of measuring water depths in the fishing grounds. In around 1930 experiments in Britain, Norway, Germany and Japan showed that fish concentrations could also be located by supersonics. An echo recorder was constructed in Britain and, in 1935, the Norwegians started to search for fish in the Lofoten waters by means of recording echosounders. By around 1950 most fishing vessels were equipped with vertical echosounders (Fig. 75). Many also had Sonar (sound navigation and ranging) instruments for horizontal detection of fish concentrations. These apparatus were developed during World War II and proved to be most useful in the midwater trawl and purse seine fisheries.

While the drift net fisheries lost their earlier importance, trawl and purse seine fisheries now produce the bulk of catches in the world. The purse

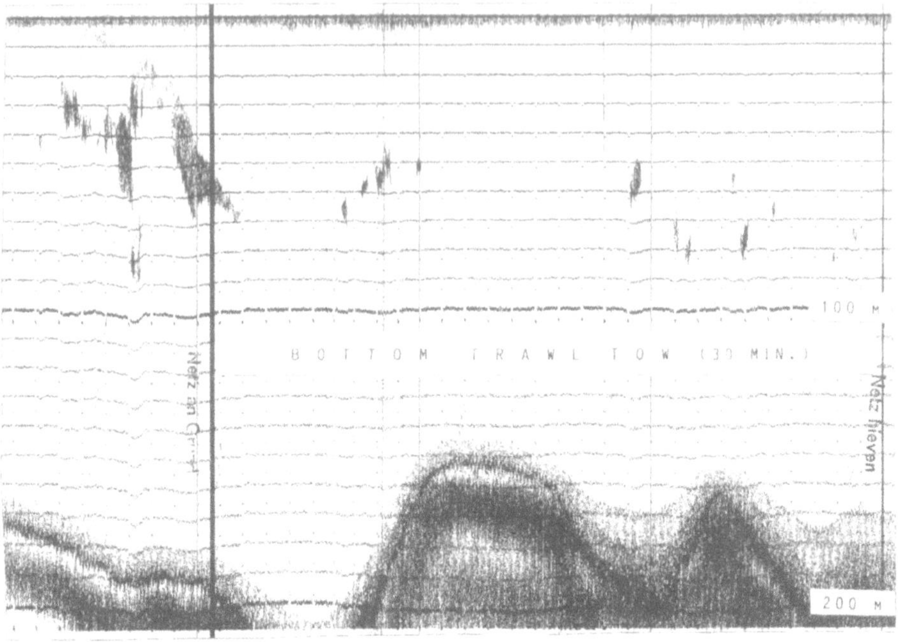

Fig. 75. Echogram showing krill concentrations at water depths of 20 - 80 m in the Antarctic Ocean; sea bottom indicated at 150 - 210 m depth

seine was invented in 1826 in Rhode Island, or perhaps in 1837 in Maine, in the American mackerel fishery. It is an encircling net which can be closed like a bag by a line running through rings on the lower edge (Fig. 76). After positive results the purse seine was soon also introduced to the menhaden fishery further south and to other American fisheries (see Sect. 5.4). From the USA the gear was transferred to Europe before 1880 for experiments in Norway and was first used commercially in the Swedish Bohuslän herring fishery from 1882. The application of this technique spread widely after 1903 when, as a result of motorization, the Scandinavian herring fisheries expanded from the fiords and skerries to the open sea and even to Icelandic waters. In 1882 purse seining for small pelagic fish was also transferred from America to Japan and, from 1913, the Japanese developed purse seines for skipjack fishing.

The use of purse seines was strongly enhanced with the growing demand for fish meal and oil. Originally masses of fish were used as fertilizer in agriculture. In 1875 the first experiments were made in Germany to use fish meal as feed for sheep. In 1892 these experiments were followed by

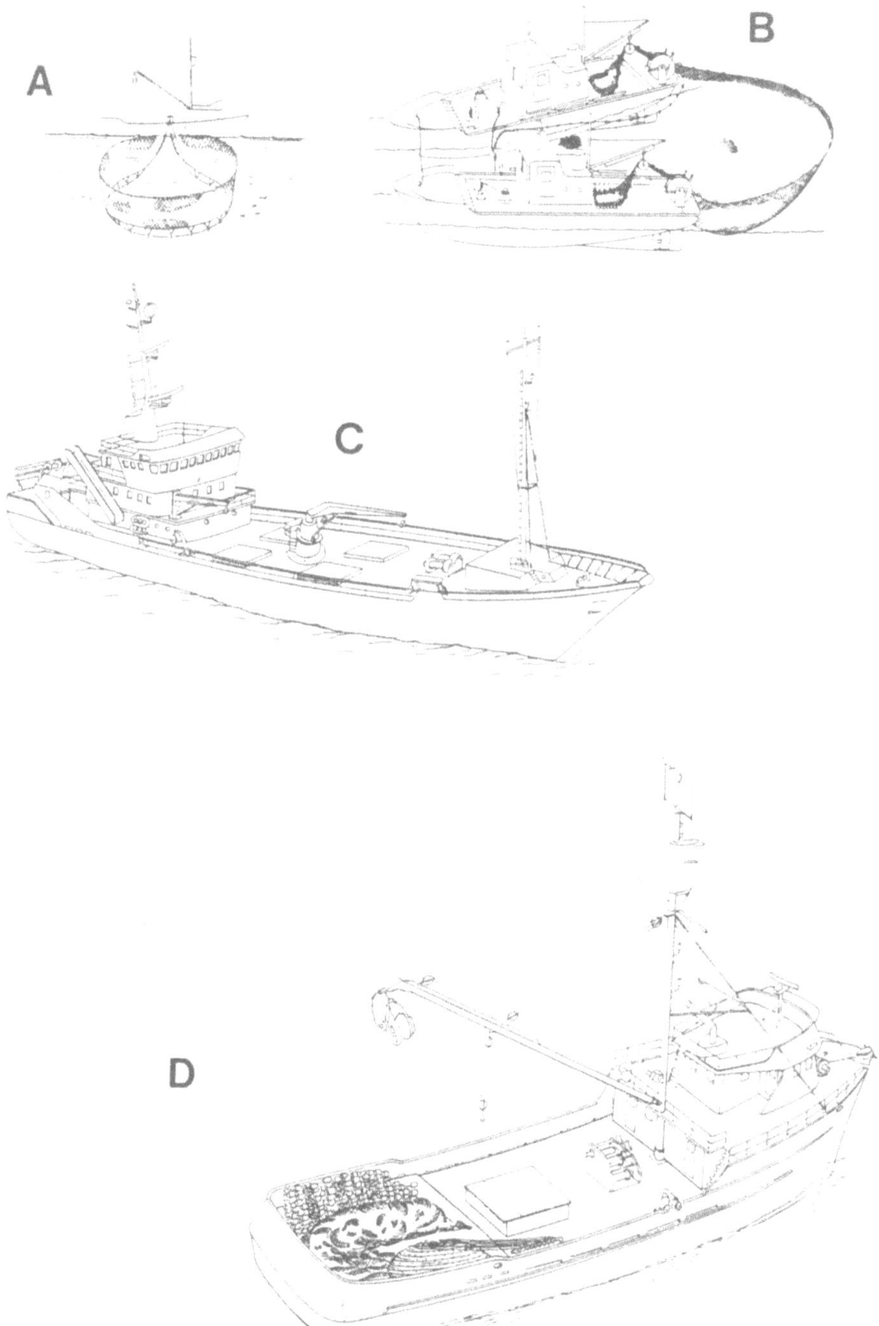

Fig. 76A-D. Purse seine fishing. **A** One-boat purse seining; **B** two-boat purse seining in Japan; **C** European-type purse seiner, 1100 GRT, 65 m; **D** North American-type purse seiner, 120 GRT, 29 m. (**A,C,D**, FAO 1985; **B**, Ministry of Agriculture, Forestry and Fisheries, Tokyo 1987)

large-scale trials in Norway and, in the first decade of this century, a market for fish meal as animal feed developed in Europe. Also, in America and in other areas of the world, the demand for fish meal and, consequently, for its industrial production grew very rapidly (see Fig. 97).

As with other gear, purse seines were improved with experience and with better technical possibilities. Their size increased considerably: in the 1950s many measured 100 - 600 m with depths of 15 - 90 m but much larger purse seines were constructed in the 1970s and these measured up to 1900 x 270 m. Originally, many people were needed to operate the large gear, particularly during hauling, but soon the nets became too large to be handled without mechanical power. In 1951 net drums were installed on purse seiners in the herring fishery of British Columbia and, a few years later, such equipment was also used in the salmon and many other fisheries. Considerable progress was achieved by the installation of hydraulic power-blocks which allowed the hauling of very large purse seines. These power-blocks were first used in the Californian tuna and sardine fisheries in 1953-1954 (see Fig. 122). As small auxiliary boats were then no longer needed, the purse seine fishery could expand to the high seas and even work under rough weather conditions like, for instance, in the Norwegian Sea. Motor pumps were constructed to suck the masses of small pelagic fish from the closed purse seines into the fish holds on board and to transport them from the ships to the factories on land. The manifold technical improvements allowed an enormous increase in the landings which are, today, mostly handled by a minimum of skilled personnel.

4 Hunting of Marine Mammals

4.1 Whaling

The utilization of whales is as old as the fisheries for herring and cod but its operation on a commercial scale started later. In contrast to the fisheries, whaling served, in the first instance, for technical purposes and only at a later stage was it used for the production of human food. As long as it was carried out by aboriginal people on a local scale to satisfy their daily needs for subsistence, there were no problems. However, after the inception of commercial whaling on an industrial level, the populations of these precious animals were rapidly reduced to very low numbers. With this sad event the general limitation of the living marine resources first became evident.

Since early times coastal people in many parts of the world have used stranded whales as a welcome source of food and materials. Only a few communities managed to enter into the active capture of larger whales by primitive means. Possibly the oldest pictures of whaling are Neolithic rock carvings, which are about 6000 years old, in several places along the Norwegian coast (Fig. 77). During pre-Columbian times Indians in America hunted whales from small canoes in a similar manner to that used by Eskimos with their kayaks (Fig. 8) to the present day. In Alaska, whaling can be traced back to 100 - 200 B.C.. Old reports from Japan describe the operation of numerous small boats to exhaust whales in the open sea so that they could be captured with harpoons and lances, and in certain districts also with the help of nets and poisons, and be towed back to the coast (Fig. 78). This method was used for right and gray whales, for humpbacks and even for blue and sperm whales. Well-known is the hunting of sperm whales with hand harpoons from small boats on the Portuguese coast and around the Azores: this was carried out until recent times. Such whaling around Madeira ceased in 1982 when the whaling grounds around the island were converted into a protected national park. At all times such hunting with simple gear and boats has been a most dangerous struggle, requiring much human skill and strength.

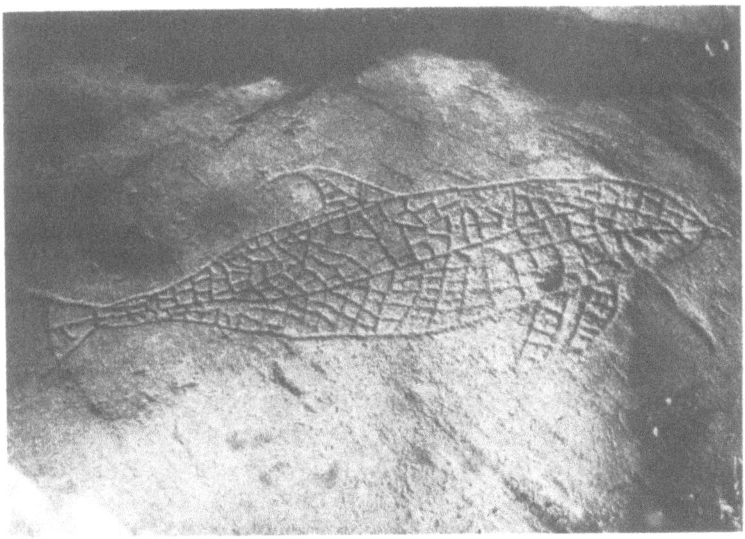

Fig. 77. Presentations of whales from the Stone Age. *Above*, whale and elks, rock carving in Leiknes, northern Norway; *below*, Neolithic rock carving of a toothed whale near Drammen, southwest of Oslo, Norway. Length 2.28 m. (Kühn 1952)

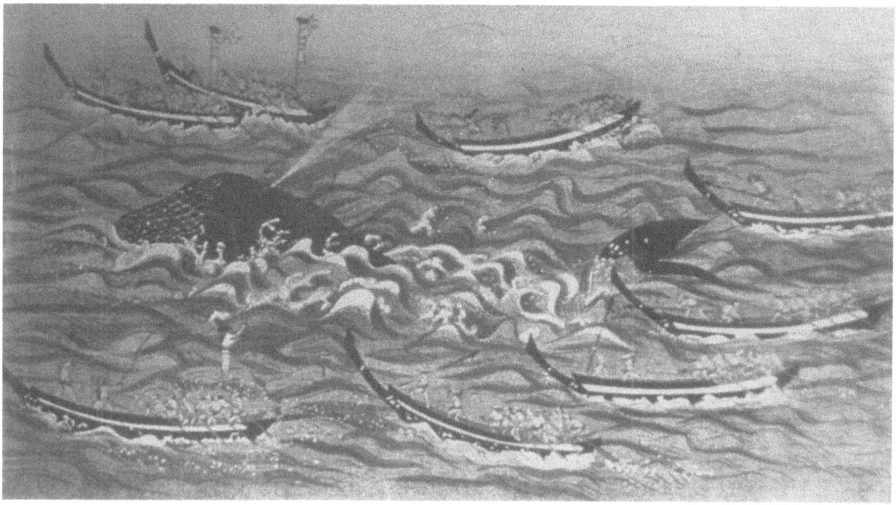

Fig. 78. Whaling in the province of Iki; Japanese colour block-print by Hiroshige III, ca. 1877

4.1.1 Northern Waters

The Norwegian Vikings are usually considered to be the discoverers of the whale resources of the North Atlantic. Stone Age pictures of whales have already been mentioned. In a report by Ottar of Helgeland to King Alfred the Great of England, dated 890, there was already reference to whaling along the coast of Tromsö. However, regular whaling on a larger scale was obviously implemented by the Basques in the coastal waters of the Bay of Biscay. There, the black right whale *(Balaena glacialis)* appeared in autumn and winter in large numbers during seasonal migration from the northern boreal waters. Already by the 10th century, the Basque "Confradias de Pescadores" had considerable experience in this field. The existence of suitable gear and organized associations of ships under the command of "maestros de chalupa", as well as of related trade connections and systems for the payment of fees, at this time seem to indicate a much longer history of this whaling. Biarritz and several cities in the Spanish Basque province bear a whale in their coats of arms (Fig. 79). Whales were caught not only caught for their blubber but also for

Fig. 79. Seal of the city of Biarritz from 1351, displaying a whale

their meat. The city of Bayonne introduced, in 1261, a customs fee for whale tongues.

It is assumed that Basque whalers reached North American waters around 1372, i.e. before Columbus and J. Cabot, during their search for new resources when the whales in the Bay of Biscay had become scarce. The first reliable reports of Basque whalers in the Gulf of St. Lawrence and in the Strait of Belle Isle date from 1526. Excavations on the southern coast of Labrador revealed the remains of a Basque whaling shore station from 1565. For about 200 years whaling ships from the region of Cape Breton and Saint-Jean-de-Luz operated in Newfoundland waters, where they also hunted the once abundant black right whale. Up to 60 vessels, each with a crew of 50 - 60 men, participated in this venture. Then, in the 17th century, political difficulties and war put an end to this flourishing activity. Basque whalers also operated along the coasts of Iceland, where reports for the year 1412 list 20 participating ships. The Basques sailed in caravels. The whales were hunted from small long boats, were flenched alongside the ships, and the blubber was then boiled on deck to extract the oil, which was stored in barrels.

A new phase began with the discovery of abundant whale resources in arctic waters, near Spitsbergen and Bear Island, during the exploratory voyages of Jonas Poole (1583) and Willem Barendsz (1596). Companies from England and the Netherlands, the leading countries in economy and shipbuilding, initiated the whaling there in 1611-1612. The Dutch hegemony was mainly based on the "Noordsche Compagnie", founded in 1614 and invested with a monopoly by the States General of Holland until 1642. Many other ventures followed during subsequent years. In the beginning, experienced Basque whalers participated as commanders, harpooneers and blubber cutters. Later, these functions were mostly taken over by people from the German Frisian islands who engaged in whaling after 1634, when a great tidal flood had ruined the basis of their agricultural living. In 1701 about 3600 men from the North Frisian islands, particularly from Föhr, worked on the whaling ships. Since Spitsbergen was erroneously considered to be a part of East Greenland, and the king of Denmark as the owner of Greenland levied royalties from the whalers, vessels participating in this new industry, were called the "Greenland farers".

During the first few decades whaling could be carried out very close to the shores. For this reason the Dutch, as the main group of participants, named it "Baien Visscherij". The animals, exclusively Greenland right whales or bowheads *(Balaena mysticetus)*, were hunted from boats with harpoons, killed with lances and towed ashore. On land they were processed by cutting up the blubber. Initially the whaling was most profitable and during the summer season real settlements developed around the blubber refineries. The largest was the Dutch settlement of Smeerenburg on Amsterdam Island and this reached the size of a little town around 1635 before it had to be abandoned, like many other localities, in the middle of the 17th century. In Smeerenburg the "Noordsche Compagnie" built warehouses where the barrels of whale oil were stored over winter for shipment during the following summer by specially chartered cargo vessels. English whalers worked in the waters of southern Spitsbergen and French and Spanish ships in the north. German whalers, mostly from Hamburg, appeared in Spitsbergen in 1617, 1625 and 1643, and after their shore station on Ulfeld Bay this water has since been called the Hamburgian Bay. The whale oil extracted from the blubber by heating was shipped to Europe and fetched high prices as fuel for lamps. Whalebone was taken as a most valuable elastic and durable material for corset-busks and for many other purposes. Furthermore, bones of the whales, notably the large lower jaws, were used as building materials, particularly on marsh islands where wood and stones were scarce. They can still be found in many such places. The remaining whale bodies were mostly left behind unused.

As a consequence of ruthless slaughtering coastal whale resources were rapidly depleted so that the Baien Visscherij ended in 1640. Already by 1626 whaling activities had shifted to the open waters, especially to the

Fig. 80. Hunting of the Greenland right whale or bowhead in Spitsbergen waters; etching by A. van der Laan (1730) from *Kleine Vissery*, Amsterdam

area along the ice edge between Spitsbergen, Jan Mayen and Greenland, in search of more productive grounds. The whaling ships were sailing ships, mostly with three masts, were 30 - 35 m in length and with a 7 - 9 m beam; they had a crew of 30 - 50 men (Fig. 80). They carried four to seven sloops, which were mobile rowing boats of light construction for hunting the whales with hand harpoons. After killing the whales with lances, the catch was towed alongside the whaler, the blubber cut off in pieces and packed into barrels for transport home. The average yield per ship during a season amounted to 7 - 11 whales. It was estimated that the total catch taken by the vessels of all participating nations during the productive period within a season might have been in the order of 1500 - 2000 whales. At the home ports oil was produced from the blubber in kilns (Fig. 81).

The traditional "Greenland" whaling reached its peak in 1650 - 1750. By around 1680 more than 300 ships were participating, of which 260 were

Fig. 81. Boiling of blubber for whale oil; etching by A. van der Laan (1730) from *Kleine Vissery*, Amsterdam

of Dutch nationality, carrying around 14 000 people. The work was very hard and dangerous. Many ships were lost during each season. In 1675 there were 83 vessels from Hamburg engaged in this whaling. In later years German ships also sailed from Bremen and from several other ports. Vessels from Altona, Glückstadt and Flensburg flew the Danish flag during those days.

The intensive exploitation of the right whale resources soon led to a serious decline in the catches. In the second half of the 18th century there were still about 135 - 170 Dutch, 34 - 75 English and around 80 German whaling ships engaged in the arctic seas but their seasonal catch hardly exceeded four whales per ship, while the cost of a voyage was covered only by the capture of the third whale. After the disturbances during the Napoleonic Wars, this whaling had only a short and rather modest revival and ceased at around the end of the 19th century. The last German whaling voyage of this kind was made from Bremen in 1872.

Table 3: Statistics of Dutch whaling enterprises (Lynge 1936)

	Spitsbergen		Davis Strait
	1669-1718	1719-1778	1719-1778
Number of voyages	6 883	7 284	3 161
Vessels lost	391	170	62
Successful voyages	6 492	7 114	3 099
Whales caught	34 479	23 111	6 986
Production			
Blubber (1000 barrels)	1 339	736	357
Oil (1000 barrels)	2 007	1 098	528
Whalebone (mill.pounds)	62	33	17
Balance (million hfl.)			
Proceeds for blubber	0.2	0.3	0.3
Proceeds for oil	81.7	53.9	26.2
Proceeds for whalebone	38.2	48.4	25.6
Total income	120.1	102.6	52.1
Minus expenses	92.8	85.1	37.7
Net profit	27.3	17.5	14.4
Means per year			
Successful voyages	130	118	52
Whales caught	690	385	116
Net profit (1000 hfl.)	546	292	243
Means per voyage			
Whales caught	5.3	3.2	2.3
Net profit (hfl.)	4 200	2 500	4 400

When whales became less and less abundant in the Spitsbergen region, part of the whaling fleets shifted their activities, in 1719, to the Davis Strait west of Greenland, where the Greenland right whales were found in large numbers. With increasing experience the ships followed the whales farther and farther to the north, finally entering Melville Bay when the ice conditions permitted. In the 1820s up to 90 British whaling ships worked in the Davis Strait while only less than 30 still went to the Spitsbergen area. The total number of British ships participating in northern whaling declined from 110 in 1825 to 18 in 1842.

Some economic results of whaling can be demonstrated by old Dutch statistics which are presented in Table 3. From these figures it appears that in the Spitsbergen area the average catch per voyage decreased from 5.3 whales in 1669 - 1718 to 3.2 whales in 1719 - 1778. During this later period the whales caught were obviously smaller, as indicated by the decrease in yields of blubber, oil and whalebone per whale. However, since the prices for these products had risen strongly, the net profit remained very high with 17% of the total income (1669 - 1718: 23%) in spite of much smaller catches. In the Davis Strait the catches per voyage were lower (2.3 whales) but the animals were larger and the yield in products higher. Here the net profit amounted to 28% of the total income. Even if the costs for the loss of ships were deducted from this sum, it is quite evident that whaling was a very profitable industry almost until the depletion of the resources.

Interesting long-term statistics, covering from 1709 until the present day, exist for the catch of pilot whales *(Globicephala melaena)* on the coast of the Faeroe Islands (Fig. 82). As from other islands in the North Atlantic, the hunting of this whale was carried out on the Faeroes each year by surrounding with boats the schools in bays and driving the whales towards the beach where the stranded animals were slaughtered. The oldest written records on the Faeroes are from 1584 but pilot whale hunting certainly goes back to much earlier times. In the past the catches provided the islanders with an important source of food which was distributed in accordance with long-established rules. Today, the old tradition is still continued and the Faeroese people are much criticised for this.

The level of catches is dependent on the weather conditions. In summers with good weather, when schools approaching the coast can be more easily sighted, the catches are larger than in bad summers. Scientific investigations have also shown a long-term correlation between warm periods and high pilot whale catches. An increase in the average number of whales landed on the Faeroes showed three peaks: in 1720-1740, 1830-

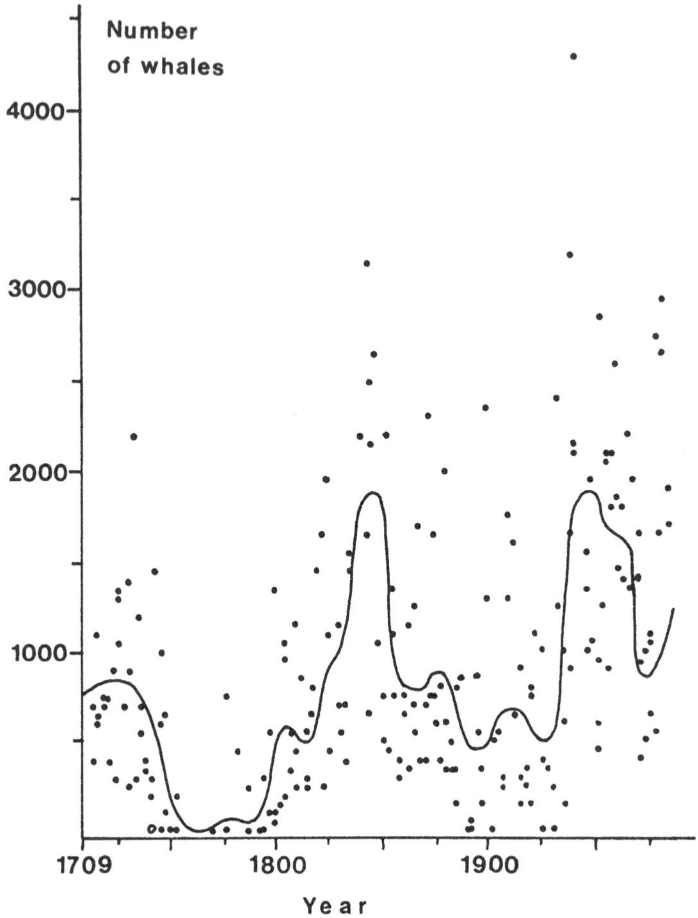

Fig. 82. Number of pilot whales landed on Faeroe Islands 1709 - 1984; curve indicates 10 years average. (Hoydal, in press)

1860 and 1930-1960. The latter period coincided with an increase in the sea surface temperatures. The good period of high pilot whale catches after 1830 was during a time when warmer conditions were also indicated by the abundance of cod off Greenland. Pilot whale catches in Greenland increased at the same time. The level of whale catches also seems to be dependent on the abundance of squid *(Todarodes sagittatus)*, prey which sometimes appear in masses in the coastal waters off the Faeroes, Norway and Iceland.

In the North Pacific rich resources of black right whales and bowheads were found in the Sea of Okhotsk, the Bering Sea, Bering Strait and Beaufort Sea. Commercial whaling by ships coming from New England started here in 1847 with very good catches during the first years. Heavy exploitation by up to 200 vessels had already resulted in a strong decline by the end of the 1850s and in around 1866 whalers changed to hunting for walruses and gray whales. Commercial whaling for bowheads ceased in the 1880s but subsistence whaling by Eskimos has remained up to the present day.

4.1.2 Sperm Whaling in the "Southern Seas"

The persistent demand for whale oil as fuel for lighting and related high prices, in particular when the arctic whaling declined, led to the concurrent development of another branch of whaling in the early 18th century, in which the Americans took a leading position from its inception. Encouraged by the profitable catches of French vessels and the rich resources of black right whales along the coasts of New England they began, in around 1650, to participate in this fishery. Originally, the whales were hunted by Indians from small boats close to the shores. In 1668 the first American whaling company was founded in New London. Long Island became the centre of this shore whaling. When the whales became more scarce near the coast, whaling on the high seas began in around 1715. Initially it expanded to the north: American whalers worked from 1732 in the Davis Strait, from 1751 in Baffin Bay and from 1761 in the Gulf of St. Lawrence and in Hudson Bay. Hunting was for both Greenland and black right whales.

With the decline of resources in the north, activities shifted increasingly to waters further south and concentrated more and more on the sperm whale *(Physeter catodon)* and partly also on the humpback *(Megaptera novaeangliae)*. A strong incentive for the capture of the former species was the production of spermaceti from the heads of sperm whales. This material was used for the preparation of candles, replacing the more expensive beeswax. In New England a flourishing candle industry developed, producing candles mainly for export.

The American vessels, mostly fully rigged sailing ships, brigs and schooners of around 300 GRT (Fig. 83), had their home ports along the Atlantic coast between Boston and Connecticut. New Bedford grew to become the most important port. With the expansion of whaling the ships were built bigger and bigger and shipyards in New England developed a special type of "blubber hunter" with low masts and transverse sails.

Whaling soon covered wide areas in all warm zones, mainly between 40°N and 40°S. In summer grounds nearer polar regions were also

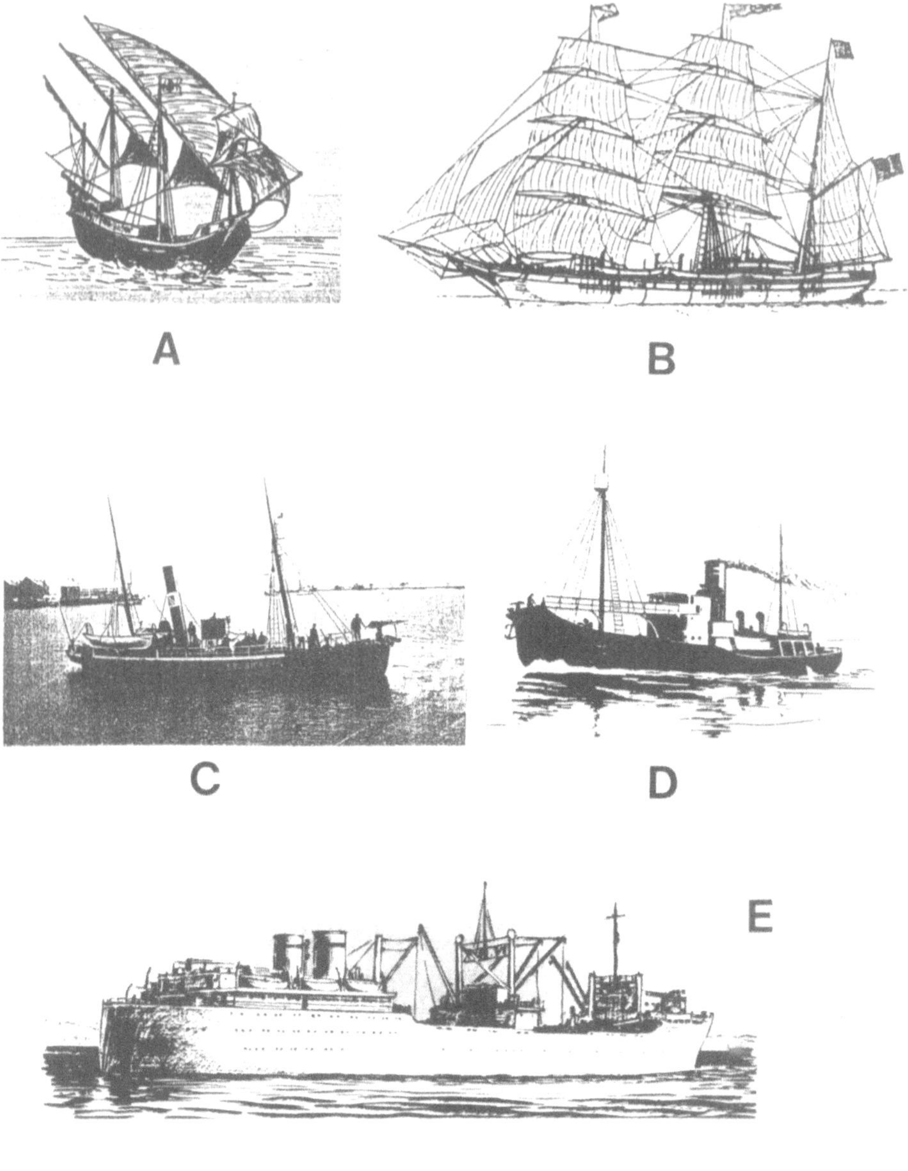

Fig. 83A-E. Whaling ships of various epochs. **A** Basque caravel of the 16th century; **B** American sperm whaling schooner; **C** Norwegian steam whaler *Svend Foyn* 1899; **D** Japanese catcher boat in the Antarctic; **E** modern whale factory ship

fished. Voyages became increasingly longer and finally lasted 2 - 4 years. In his book *Moby Dick* Herman Melville described most vividly the adventures of whale hunting during such a journey.

As whaling in the Arctic became unprofitable, so European ships, especially from Great Britain and France, also participated in this whaling. In 1775, immediately prior to the American War of Independence, around 350 American vessels were involved. Most of these ships were lost during the war but after its end the fleet was soon rebuilt. In 1846 a maximum number of 735 ships with more than 20 000 people were sperm whaling. A further 230 whalers were flying the flag of other states. From 1775 to 1844 participation from Britain consisted of a total of 2153 voyages with 860 ships. German whalers, mainly from Bremen, made around 60 voyages for sperm whales to the "Southern Seas" between 1836 and 1869.

The area of operations, originally ranging from the Bermudas and Bahamas to the Azores and Canary Islands, expanded in 1786 to include West Africa. In 1787 whaling spread into the Pacific Ocean around Cape Horn and soon extended to Australia, Japan and Kamchatka. The Pacific became now the most important region for this fishery. Depending on the sperm whale migrations, the fleets moved their activities between the Northern and Southern Hemisphere summers.

During the golden age of this whale hunting, i.e. the period from 1820 to 1850, the annual catch may have been in the order of 10 000 whales. The American catches in the period 1804 - 1876 were estimated at around 225 500 whales, mostly sperm whales. Like the earlier arctic whaling, this industry also became very profitable. Some ships returned from a single voyage with a loading that was of much higher value than the total costs for the construction of the vessel, its running expenses and the equipment. Since the duration of voyages was extremely long, the whale oil which had been extracted from the blubber by boiling in kilns on deck of the whalers was usually shipped to the home port by specially chartered cargo ships.

After 1850 an increasing number of voyages ended with financial losses. During the last quarter of the 19th century this whaling declined rapidly. The number of American whalers decreased from 336 in 1869 to 90 in 1889 and only 15 remained in 1898. The reasons for this decay were, on the one hand, the overall decrease in the abundance of sperm whales due to their heavy exploitation and, on the other hand, the fall in prices for the products. For lighting whale oil was increasingly replaced by petroleum which had been struck in 1859 in Pennsylvania. Furthermore, coal gas lighting was introduced in the cities from the early 19th century. Whalebone materials were replaced by those of steel. Nevertheless, the last whaling ventures of this kind were still being undertaken in the first half of the 20th century, forming a direct link with modern whaling.

4.1.3 Mechanization of Whaling

The common feature of all these old whaling ventures was that the whales were hunted with hand harpoons from small auxiliary boats. By these means it was only possible to catch the more ponderous whales, like the right and sperm whales, and the faster and often larger baleen whales remained virtually untouched. Their exploitation became possible after the introduction of shooting. The first trials in harpoon mechanization were made from 1731 but it was not successful until 1850 when a shoulder gun with a bomb-powered lance, and later a darting gun (hand harpoon with bomb), came into general use. A major breakthrough was the invention of the shell harpoon by Svend Foyn in Norway in 1868; this was shot from a gun and travelled a greater distance. The shells were constructed so that the wings of the harpoon stuck into the body of the whale. Around the same time steam ships were introduced to whaling (Fig. 83) which could cover a much wider area more easily and use more efficient methods for whale hunting.

The North Atlantic was, again, the first setting for modern whaling. From 1868 a number of land stations for the processing of whales developed along the coast of Finnmarken in northern Norway. In 1886 there were already 19 such ventures with 35 steam-driven whalers, mostly 30 m in length. The catches consisted exclusively of baleen whales, initially only blue whales but later all species and especially fin whales. In 1898 the catch off Finnmarken amounted to 547 sei whales, 448 fin whales, 43 humpbacks and 24 blue whales. It was estimated that during the period 1868 - 1904 almost 1500 humpback whales were taken in Norwegian waters. A disturbing decrease in the abundance of whales forced the transfer of land stations and extension of hunting areas. In 1903, for the first time, a floating whale factory was put into operation in Spitsbergen waters. As a result of quarrels between fishermen and whaling companies, whaling was banned along the Norwegian coast by the authorities in 1905 and was resumed under limited licence only after 1917 and in view of the critical food situation. In the meantime, baleen whaling had expanded to other areas with the establishment of land stations in Iceland (since 1883), the Faeroe Islands (since 1894) and later also on the Shetlands, Hebrides and in Ireland, all under Norwegian management.

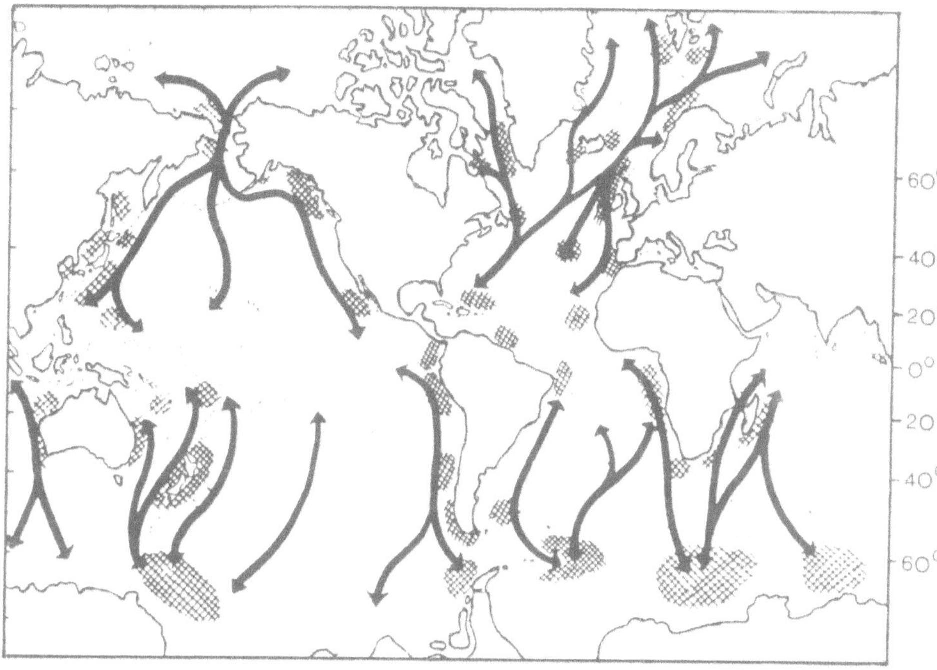

Fig. 84. Migratory routes of baleen whales between summer feeding grounds in polar regions and winter reproduction areas in subtropical and tropical waters (After Schubert 1955 and others)

By around 1905 up to 60 whaling ships landed a maximum of 2600 whales per annum from the Northeast Atlantic. From these catches more than 80 000 barrels of oil were produced per year. In contrast to earlier times, the whole body of the whale was now processed and the production of feed, bonemeal and manure increased the economy of the operation.

The heavy overexploitation of whale resources in the North Atlantic, the results of which were soon apparent, caused most of these ventures to vanish after several years of operation. Annual whale catches in the North Atlantic as a whole declined from about 3000 in the period 1905 - 1910 to 1500 during the 1920s and to 1000 during the 1930s. Only a few stations still exist at the present time.

A considerable number of whaling ventures from land stations also developed in other parts of the world, particularly where experience had shown that whales migrate from feeding grounds in polar waters along the coasts to subtropical and tropical regions for reproduction (Fig. 84). Most of these ventures were established by Norwegian companies. In

South Africa whaling started in 1908 and by 1913 26 stations were already operating along the west and east coasts with 89 catcher boats capturing more than 9000 whales. Later, other stations were built in Angola, the Congo and Madagascar. Similar developments took place from 1912 to 1916 on the coast of western Australia. In Latin America stations operated in Chile (from 1908), Peru (from 1925), Ecuador (1926) and Brazil (1911). Generally, results in tropical waters were not encouraging and even stations in more temperate zones were closed after a few years when whales became scarce in neighbouring waters.

Along the North American Atlantic coast shore stations concentrated on the area from Newfoundland to the Davis Strait. In the North Pacific region, which has rather similar natural conditions, the old traditions of whaling were further developed in Japan. From 1905 the European method was introduced and was accepted by the Japanese and Americans. Thus, a chain of whaling stations evolved from Japan via Siberia and Sakhalin to Alaska and California. Here the same events were observed as in the North Atlantic. After several years with high catches the whale resources, particularly those of the bowhead or Greenland right whale, declined and many of the stations had to be closed down.

Also, the resources of the gray whale *(Eschrichtius gibbosus)*, which migrates between the Arctic and lower California and in the western Pacific to Korea, were heavily exploited. Whaling off the North American coast, mainly from 1850 to 1890, reduced the stock from about 25 000 to only 250 whales in 1946. Fortunately, in the meantime, protection allowed a substantial recovery.

In 1940 Japan sent its first whaling fleet to the North Pacific, the Bering Sea and the Arctic Ocean. Since 1962 some factory ships with catchers from the Soviet Union have also participated in this pelagic whaling.

A small-whale fishery was developed by Norwegians in the early 1930s in North Atlantic waters between Norway, the Barents Sea, Spitsbergen, Iceland and Greenland. This whaling by small motor ships of up to 40 m length produced mainly meat from minke whales but also caught pilot, bottlenose and killer whales.

4.1.4 Antarctic Waters

Explorers, seal hunters and sperm whaling expeditions had reported rich whale concentrations in antarctic waters. With the decline in catches in traditional whaling grounds attention turned to these distant seas. The first trials by Scottish and German - Norwegian ventures in 1892-1895 to catch southern right whales in west antarctic waters were not very successful. However, in 1904, Norwegians, working with a company financed by Argentinian sources, established a shore station at Grytviken on South Georgia and began hunting of blue and fin whales. Already, 1 year later, the floating whale factory *Admiralen* and two catcher boats were transferred to the south from Spitsbergen. Encouraged by promising results, further stations were built on South Georgia, the South Shetland and Falkland Islands. During the season 1910-1911 the antarctic whaling was already being carried out by six shore stations and 14 floating factories with 48 catcher boats and a total catch of more than 10 000 whales.

Difficulties with the granting of concessions and payment of fees requested by the British authorities for whaling activities in coastal waters since 1906 resulted in 1923-1924 in a change to high-seas whaling in the Antarctic. In the same season the first whaling voyage to the Ross Sea was undertaken. Initially, the factory ships worked in or close to the drift ice as calm sea conditions were required to flench the whales alongside the vessels. From the 1924-1925 season more and more factory ships were equipped with a slipway in the stern to tow the whale carcasses on deck for processing. After this invention fleets became much more capable of operating in open waters. This promoted a further rapid increase in antarctic pelagic whaling. A further technical improvement was the breech-loading gun on the whale catchers. During the 1930-1931 season 41 factory ships with more than 200 catcher vessels were at work producing a catch of 37 500 whales. Together with catches from the land stations an irresponsible record of 40 201 whales was reached that year.

The shift to high-seas whaling required larger ships. The factory vessels were originally mostly converted tankers of 3000 - 4000 GRT but then their sizes increased to 6000 and finally to 10 000 - 20 000 GRT (Fig. 83E). Such sizes were necessary to accommodate not only the slipway and flenching platform but also the numerous personnel and extensive technical installations for the processing of whales. The technology advanced considerably. All parts of the whale were now used to produce not only oil for human nutrition (margarine) and technical purposes but also whale meat and meal for animal feeding, manure, whalebone and pharmaceutical substances. The replacement of steam

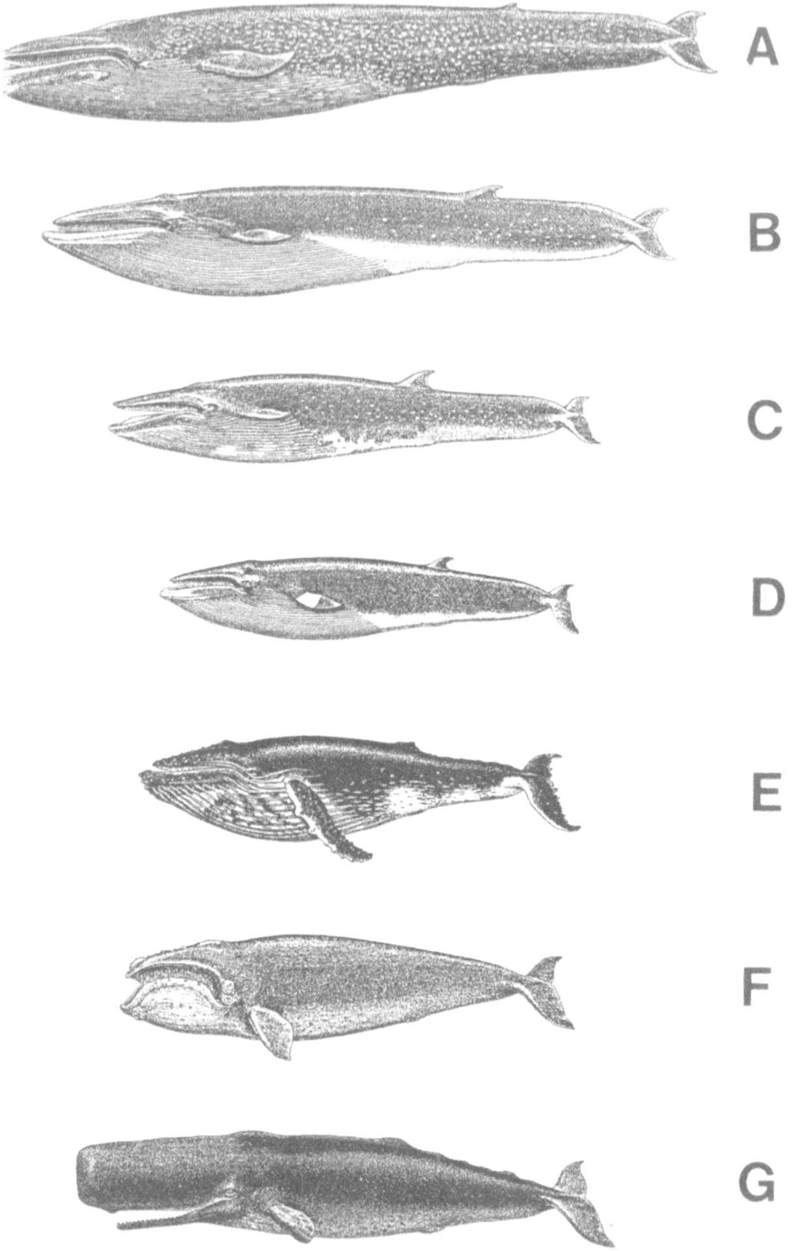

Fig. 85A-G. Main species used in commercial whaling (average lengths and weights in parentheses) **A** blue whale *(Balaenoptera musculus)* (24 m, 85 t); **B** fin whale *(Balaenoptera physalus)* (20 m, 50 t); **C** sei whale *(Balaenoptera borealis)* (16 m, 13t); **D** minke whale *(Balaenoptera acuto-rostrata)* (9 m, 7 t); **E** humpback whale *(Megaptera novaeangliae)* (18 m, 33 t); **F** black right whale *(Balaena glacialis)* (20 m); **G** sperm whale *(Physeter catodon)* (16 m, 35 t). (Peters 1938, drawings by F. Diehl)

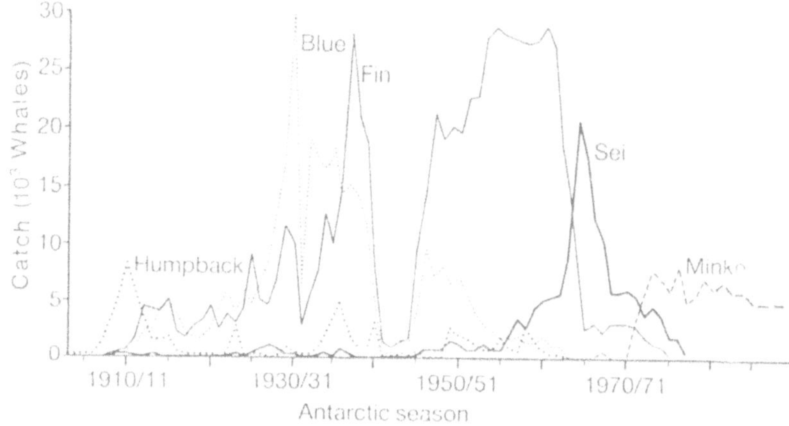

Fig. 86. Baleen whale catches in the Antarctic. (Walton 1987)

engines by less spacious motors provided additional loading capacities. Each mothership was accompanied by about six to eight catcher boats (Fig. 83D). While the fitting of harpoon guns remained the same, the efficiency of these ships was improved by increasing their size, engine power and speed and by installing equipment for Sonar detection of whales. At the end of the 1930s the catchers had an average size of 300 GRT with engines of 1100 hp, allowing a speed of 12 knots. In the early 1950s these figures had increased to 450 GRT, 1900 hp and 16 - 17 knots.

The areas for hunting pelagic whales extended to the waters around the Antarctic, roughly from the Antarctic Convergence at 50°S to the edge of the pack-ice in summer at about 65°S. Originally the whaling usually started in November and during the austral summer followed the retreating ice edge where the whales were most abundant in their search for krill as food. The season generally ended in March. Later, the hunting season was considerably curtailed as a part of the protective measures. The best catches were obtained in the Atlantic sector, with the exception of the Weddell Sea, and in the western Indian Ocean sector, while results were generally poorer in the Pacific sector.

The main species hunted were the humpback, the blue whale and the fin whale (Fig. 85); of these the blue whale was considered to be of highest economic value. Statistics show that after World War II, when catches of humpbacks and blue whales declined, fin whales made up the largest part of the production (Fig. 86). But by around 1960 fin whale catches had also diminished rapidly and whaling was then based mainly on sei whales and finally on minke whales.

While the land stations, spread over many parts of the world, belonged to a considerable number of countries, only a few nations participated in antarctic pelagic whaling. Until recent times this industry was entirely dominated by the Norwegians who served as experienced personnel, particularly fleet managers, harpooners, speckioneers, etc. not only on ships of their own nationality but also on those of foreign fleets. British companies were second in importance. However, since the 1931-1932 season, when the Norwegian ships were laid up as a result of low whale oil prices and the world economic crisis, British whalers caught more whales than the Norwegian vessels. In 1935-1936 Japanese companies started to participate in antarctic whaling followed, in 1936-1937, by German ventures. While most companies were mainly interested in the production of whale oil, the Japanese concentrated on the production of whale meat for human consumption.

The economic incentive for whaling on such a large scale was based on the demand for whale oil. When, in 1869, margarine was invented as a cheap food, there was a steep increase in the need for animal fat. Whale oil was used to fill the gap after it had become possible to convert it into edible fats by a hardening process (hydrogenation). Rising prices for whale oil led to the construction of additional whaling fleets. Then, as a result of overproduction, prices for this oil declined steeply and a hazardous competition arose between the producers, leading to economic losses and also to further attempts to improve productivity.

At the end of World War II animal fats were in great demand and whale oil prices boomed. Thus, whaling fleets were quickly rebuilt and in 1946 15 factory ships and 120 catcher vessels were already working in the Antarctic. In 1950-1951 the fleet of 19 factory ships came from Norway (10), Great Britain (4), Japan (2), Panama (1), the Netherlands (1) and the Soviet Union (1) and they were accompanied by 241 catcher boats. During the following years prices for whale oil declined as vegetable fats were increasingly used instead. Falling prices for the main product, combined with increasing running expenses and the decline in whale resources, forced most of the companies to terminate their whaling in the years between 1956 and 1968. Several of the fleets were bought by Japanese enterprises. In the early 1960s the Soviet Union built three very large new factory ships (25 000 and over 33 000 GRT) with large and powerful catcher vessels (843 GRT, 3600 hp, 19 knots). Only fleets from Japan and the USSR continued with antarctic whaling after 1968 until all such activities had to be stopped in the 1980s. The Japanese produced mainly fin, sei and minke whale meat, while the Soviets were primarily interested in sperm whales.

4.1.5 Regulation of Whaling

The history of the various whaling periods shows a common course of events: initially there was rapid development with good catches, followed by a decline in the abundance of whales and, subsequently, a shift of hunting to other areas or to other species. It turned out that whale resources are particularly vulnerable to overexploitation. Concerned with this problem and with the future of the whaling industry, the League of Nations endeavoured in 1924 and 1927 to draw up a convention for the regulation of whaling. It was, however, not until 1931 that a first convention was agreed upon in Geneva, based on Norwegian national law passed in 1929. Under this convention the capture of right and gray whales and female whales with calves was prohibited. Minimum lengths were introduced for the capture of blue, fin, humpback and sperm whales. The hunting seasons and areas were confined. A further conference in London in 1937 confirmed and tightened up the regulations. All these efforts, however, did not prevent the further fierce development of whaling.

In 1944 the nations participating in antarctic whaling were able to reach agreement on a seasonal catch limit of 16 000 blue-whale-units for all waters south of 40°S (based on oil production, 1 blue-whale-unit was equivalent to 2 fin, 2.5 humpback or 6 sei whales). This total catch limit became part of the International Convention for the Regulation of Whaling, signed in 1946 in Washington. Under this convention the International Whaling Commission (IWC) was established. Without the work of this commission whale stocks would certainly have been depleted much earlier but the IWC was not in a position to halt the decline. The main reasons were that management measures could effectively be introduced only after unanimous decisions and, under the pressure of economic interests, governments of whaling countries were unfortunately not prepared to accept the necessary substantial cuts in the allowed catches. Only after long and arduous negotiations and forced by low catches, was the catch limit for antarctic pelagic whaling gradually reduced to merely 2300 blue-whale-units in 1971-1972. Protection of endangered stocks would have been more effective with separate quota for each species. But catch limits for fin and sei whales in the North Pacific were only introduced in 1969 and it was not until the 1972-1973 season that counting in blue-whale-units was abandoned for the Antarctic. The agreements on the total protection of humpbacks in 1963 and of blue whales in 1967 came almost too late. In 1975 the IWC established a new management procedure based on the principle that all whale stocks should be stabilized at the level of maximum sustainable yield. In the meantime, public opinion, mobilized by environmental groups, particularly Greenpeace, had raised increasing opposition to commercial whaling. Finally, in 1982, the IWC with a membership now dominated by non-whaling nations decided to establish a zero catch quota, i.e. the suspension of all commercial whaling, from 1986. At the same time it was decided to undertake by 1990 a comprehensive assessment of the effect of the moratorium on whale stocks and to consider modification of the provision and the establishment of other catch limits. Proposals for limited catches of minke whales in certain areas were defeated during the 1990 meeting of IWC. The commission agreed, however, to continue the low catch quota allocated to aboriginal subsistence whaling for bowhead whales in the Bering Sea, Bering Strait and Beaufort Sea, for gray whales in the eastern North Pacific, and for fin and minke whales off West Greenland.

Protection of the large whales has been enhanced by the inclusion of all major species in Appendix I of the Convention for International Trade of Endangered Species (CITES), which was established in 1976 and came into force in 1981. This prohibits the international trade of eleven large whale species, with the exception of samples for scientific investigations, for which a special permit is required. However, in many countries the regulations of this convention are not yet strictly adhered to.

Nevertheless, there is now a real chance for whale stocks to recover. But, since the reproduction of these animals with one calf per female every 2 to 3 years is rather slow, this process will need considerable time. A comparison of old reports with modern accounts of this industry elucidates the fundamental change that has taken place in the attitude of most people towards the killing of marine mammals. Based on ethical principles, the slaughtering of these animals is, today, strongly opposed. Since most of the products of the whaling industry have meanwhile been replaced by materials from other sources, there is no justification for resuming whaling on a large scale, unless the food situation deteriorates to critical levels.

4.2 Sealing

Seals, a smaller target than whales and more accessible in coastal waters, have been hunted by man since the earliest times. Their bones were found in prehistoric middens in many places in the world, for instance in those of the Indians in California, Washington and Alaska and on the Aleutian Islands. Figure 87 shows engravings and a sculpture of seals from the Palaeolithic Age, discovered in southern France. For the subsistence of primitive man seals were, and still are in certain areas, a

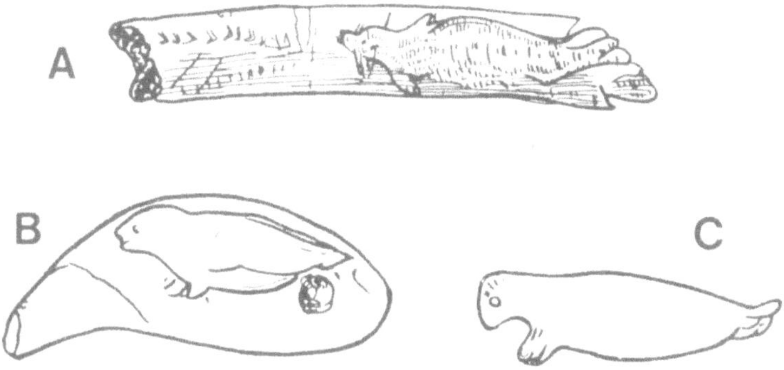

Fig. 87A-C. Palaeolithic presentations of seals discovered in France. **A** from Abri Mège; **B** from Brassempouy, Solutrean;
C from Sordes, Magdalenien. (Mohr 1952)

Fig. 88. Aino harpooning a seal; Japanese drawing

vital source of meat for food, of blubber for fuel, of bones for tools, and of skins for clothing, housing and boatbuilding. This holds true particularly for people in remote arctic and subarctic regions (Fig. 88).

4.2.1 Northern Hemisphere

The first records of sealing in the Arctic go back to 890, when Ottar of Helgeland mentioned the hunting of walruses *(Odobaenus rosmarus)* in the White Sea in a report to King Alfred of England. Shortly after the discovery of Spitsbergen and adjacent islands, the English made voyages, from 1603, to hunt for walruses, especially on Bear Island, where remains of blubber boilers can still be found today. Such voyages ceased when whaling started in 1611 but, as a by-product of whaling, seals were still taken. The animals were hunted from sloops with harpoons and lances. Hunting was quite dangerous as walruses go wild when attacked. They usually live on the pack-ice over shallow waters not further than 10 miles from land.

Russian sealers from the White Sea hunted in the Spitsbergen area and around Novaya Zemlya not only for walruses but also for belugas, polar bears and polar foxes. In 1799 about 2000 men worked from 237 ships. The level of this sealing fluctuated widely. From 1820 Norwegian sealers from Tromsö and Hammerfest also participated. By around 1840 walrus hunting had become uneconomic in Spitsbergen waters and, before the end of the century, also near Novaya Zemlya.

In the North Pacific the exploitation of walrus resources began much later, when whaling had reached the Bering Sea in 1847. Here, hunting for walruses was also a by-product of whaling, carried out mostly when whalers were waiting for the break-up of the ice and the arrival of whales. Between 1865 and 1870 about 60 000 walruses were killed with harpoons and rifles for the production of 50 000 barrels of blubber oil. It was estimated that in 1874 - 1898 around 345 000 pounds of walrus teeth from walruses in the Bering Sea were landed in San Francisco. Most of them were obtained from Eskimos in barter. The last large-scale hunting of walruses in the Bering Strait was by a Norwegian company in 1912-1914 before this sealing became uneconomic.

The blubber of walruses was boiled for the production of oil, which was often considered a welcome supplement to complete the cargo of the whalers. The hides were used for ship hawsers. The main goal of this hunt, though, was the ivory of the half-metre-long canine teeth of the males which fetched the highest prices. This ivory was already much in demand from the earliest times and was esteemed as a precious material not only for hilts of swords but also for a wide variety of pieces of art and tools. In particular ivory works from Arkhangelsk were widely known in Europe as highly desirable artistic objects.

The heavy exploitation of walruses in the Arctic, both in the Atlantic Ocean and the Pacific Ocean, resulted in the depletion of populations. Since 1952 this species has been protected under international agreement. No commercial harvesting is permitted, with the exception of local hunting for subsistence by Eskimos, Chukchis and some other indigenous people. As a result of this protection, the walrus population in the Bering Sea recovered and there is also hope for the population in the North Atlantic.

Fig. 89A-E. Some seals hunted commercially (maximum length and weight of males in parentheses). **A** Walrus *(Odobenus rosmarus)* (4.5 m, 1000 kg); **B** harp seal *(Pagophilus groenlandicus)* (2.2 m, 175 kg) **C** hooded seal *(Cystophora cristata)* (3.8 m, 400 kg); **D** Antarctic fur seal *(Arctocephalus gazella)* (2 m,160 kg); **E** elephant seal *(Mirounga leonina)* (6 m, 4000 kg). (*Kosmos-Lexikon der Naturwissenschaften* 1955, Kosmos, Stuttgart)

Commercial sealing on a much larger scale began in the early 18th century for the production of fur. For this purpose hunting concentrated on harp seals *(Pagophilus groenlandicus)* and hooded seals *(Cystophora cristata)* but bearded seals *(Erignathus barbatus)* and ringed seals *(Phoca hispida)* were also captured in considerable numbers (Fig. 89). Centres of this hunting were the breeding sites of the seals near Jan Mayen in March/April, in the mouth of the White Sea in February/March and in Newfoundland waters in February/March (Fig. 90).

Harp and hooded seals became the main target because they occur in much larger concentrations on the ice than other species and their young have pelts of the highest quality. The pups are born on the ice and cannot swim during the first few weeks of life and so they can easily be slain

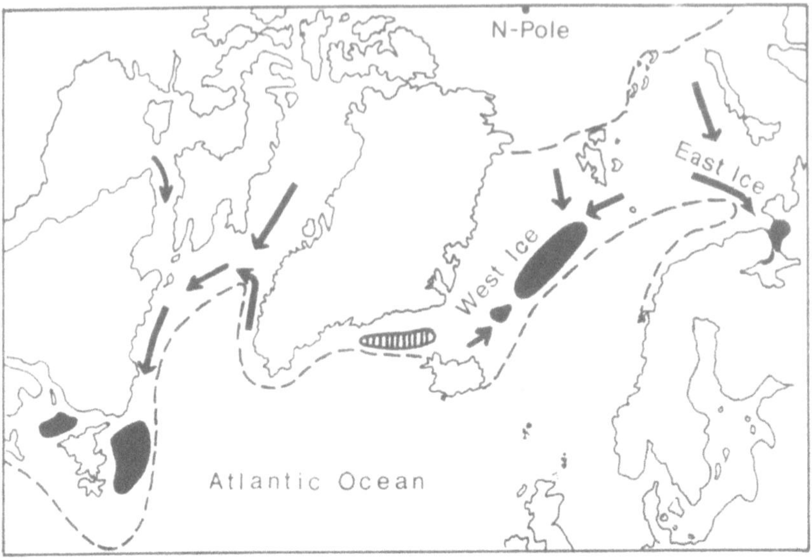

Fig. 90. Distribution and breeding sites of harp seals and hooded seals in the North Atlantic. *Dashed line*, boundary of usual distribution; *black*, breeding sites of both seal species; *striped*, breeding sites of hooded seals in Denmark Strait. (After Nansen from Mohr 1952; Iversen 1938)

with hooks and clubs. For about 2 weeks harp seal pups wear a white skin of special value and are called "whitecoats", while young hooded seals are named "bluebacks" for their dark grey-blue pelt. Besides the baby seals, 1-year old and larger seals were also killed, mostly by shooting, for their hides and blubber.

By around 1670, sealers from the Frisian islands of Heligoland and Römö had already visited Jan Mayen and produced blubber oil. With the decrease in whale stocks more and more seals were taken instead. From about 1720 ships sailed from Scotland, England, the Netherlands, Denmark and Germany. In 1760, 19 vessels from Hamburg landed 45 000 skins. The sealing voyages from Hamburg and other ports along the rivers Elbe and Weser were impeded by the English blockade during Napoleonic times. The last German sealing expedition in 1861 ended a tradition of "Greenland farers" which lasted for over 200 years and made about 6000 voyages. From 1847 onwards Norwegian sealers from Tönsberg on the east coast also participated in the hunting in these waters, now called the "west ice". The Scottish sealer fleet had the first steam-driven ships in 1860 and from 1865 the Norwegians followed with mechanized vessels (Fig. 91). In the 1880s Scottish and Norwegian steam-sail-

Fig. 91. Norwegian sealers. *Above*, old sailing vessel from eastern Norway with auxiliary engine, 1902; *below*, steam sealer from southern Möre, 133 GRT, 120 hp. (Iversen 1938)

ing ships dominated the sealing grounds and by the end of the century the Norwegians had ousted their competitors. The first Norwegian Sealing Act, enforcing an internationally agreed opening date of 1st April, was passed in 1876. The Norwegian vessels were barques of 200 - 300 tons, manned by about 45 men and operated with 5 - 10 long boats. From 1908 more and more sealing vessels were fitted with motors.

A new area for the capture of hooded seals was discovered by the Norwegians in 1874 in the Denmark Strait between East Greenland and Iceland (Fig. 90). Here, the peak season was in June/July so that vessels participating in spring in the sealing near Jan Mayen could undertake a second journey to these waters. The catches of the Norwegian sealers in the west ice (Jan Mayen and Denmark Strait) varied in 1900 - 1939 between 7000 and 50 000 seals per year, rose to a peak of 136 000 in 1951, and declined gradually until it fell to a few thousand animals in 1983, when import restrictions for seal skins were introduced by the European Community. The total harvest rose when the Soviet Union sent hunting expeditions to these grounds. But, since 1958, the catch of seals in both the west ice and east ice has been regulated by the quota established by the Northeast Atlantic Sealing Commission between Norway and the USSR.

Sealing in the "east ice", i.e. the region around Spitsbergen, Novaya Zemlya and the White Sea, increased after 1867, when the hunting of harp seals developed from Tromsö and Finnmarken. By the end of the century around 40 000 harp seals were killed each year by Norwegian sealers. Later, their catches increased steeply to reach a peak of 343 000 seals in 1925. The population could not bear such heavy exploitation and annual catches were down to an average of 37 000 in the late 1930s.

From 1888 Norwegian sealers also worked on the ice at the entrance to the White Sea, where the harp seal was abundant (see Fig. 109). This sealing was interrupted in 1921 when the Soviet Union excluded foreign ships from the White Sea, but in 1923 the sealers were again admitted under a licence system with payment of fees. At the same time Russian sealing was expanded and organized with the assistance of icebreakers and helicopters; their catches were considerably higher than those of the Norwegians.

In the Newfoundland region sealing developed independently from that in the Northeast Atlantic. Influenced by the hunting by indigenous Indians, sealing by white people began in the mid-17th century with the hunting of walruses. Only 100 years later these seal stocks were depleted.

The utilization of harp seals grew on a much larger scale. On the coasts of Newfoundland and Quebec these seals were driven into traps and on Magdalena Island they were caught by lines with fish-baited hooks. The most important sector developed with the killing of pups on the ice. For a long time sealing formed the main basis of living for many settlements in the region.

When sealing on land could no longer satisfy the growing demand for furs, hunting began from the sea in 1794. In this early period the ships used were sailing schooners of 30 - 50 GRT with a crew of 12 - 14. Good profits created incentives to build up a fleet and in the 1850s there were nearly 400 ships with about 15 000 men engaged in this activity. Between 1820 and 1860 about 300 000 to 500 000 harp seals were killed each year. After a peak in 1857 with around 700 000 skins obtained, the number of harp seals decreased and the fleet shrank. The introduction of steam-driven vessels could partly compensate for the decline in seal abundance as these ships were sufficiently mobile to also work in the outer edges of the ice. They could drop teams of sealers on the ice and later move around to collect the men and to pick up the marked piles of freshly stripped skins heaped up by the sealers ("panning"). The overexploitation of seal resources by the ships led to severe economic problems for the local settlers, the "landmen", who were dependent on the sealing from land.

In 1904 the first sealing vessel with a steel hull appeared on the grounds and in the following years many new ships of this type with an increasing capacity were put into operation. In spite of these efforts with short-term gains, production declined. Therefore, the fleet diminished again in the 1920s. A number of ships were sold to the Soviet Union to enlarge the sealer fleet in the Northeast Atlantic. In the 1930s the Newfoundland sealer fleet consisted of only 10 vessels with 1600 men.

During World War II the harp seal population recovered to nearly 3 million animals. This provided new impetus to resume sealing, not only by the Canadian fleet but also by Norwegian vessels. Sealing reached a very high intensity during the 1950s and early 1960s. Under the rising pressure of public opinion which objected to the cruel killing of seals, first protective measures were taken in 1969 and, from 1971, a catch quota system was implemented. The participation of Norwegian vessels ceased in 1982. Canadian production amounted to 100 000 - 140 000 harp seals annually during the period 1970 - 1982. It declined to 15 000 - 30 000 after 1983, when the European Community enforced a ban on the import of, and trade with, the skins of young harp and hooded seals. The capture of hooded seals, which was about 7000 per year in the early 1980s, fell to an insignificant number.

In the West Greenland area about 30 000 harp seals and 15 000 hooded seals were caught annually during the mid-19th century. In the period 1950 - 1985 catches fluctuated between 5000 and 20 000 harp seals and up to 5000 hooded seals. Ringed seals were a more important target with an average of 70 000 caught per year.

Sealing for skins in the North Pacific also has a long history. In 1786 a dense population of the Northern fur seal *(Callorhinus ursinus)* was discovered on the Pribilof Islands in the Bering Sea. Soon these resources of about 2.5 million seals were heavily exploited by the Russian-America Company, which employed Aleut in sealing on the island. The herds declined to a very low level until, in 1835, a regulation was enforced to permit only the killing of a limited number of juvenile males. After this protection the population increased again. When, in 1867, the Pribilof Islands together with Alaska and the Aleutian Islands were sold by Russia to the United States of America, a new phase of excessive sealing began. The American authorities soon introduced new restrictions. In the 1890s about 15 000 - 30 000 seals were still killed each year on the island. In the meantime, pelagic sealing from ships coming from Canada, Russia, Japan and the USA had developed outside the territorial waters not only of the Pribilof Islands but also in many other places in the North Pacific. This pelagic hunting, which reached a peak in 1894, was very harmful since it caught the pregnant females on their way to the islands and beaches to give birth. Until 1909 it produced more than 600 000 skins. The first conference in 1891 to end this overexploitation failed. In 1911 only 200 000 fur seals were left on the Pribilof Islands. In the same year a convention was at last agreed upon between the USA, Great Britain (for Canada), Russia and Japan to terminate pelagic sealing and to regulate land sealing on the Pribilof Islands (USA), the Commander Islands (Russia) and the Robben Islands or Tulyeni (Japan) by a quota system and by other management measures. The annual catch limits for the Pribilof Islands were gradually reduced from 60 000 to 20 000. When the population had largely recovered in the 1950s, a female harvest of 300 000 animals was permitted from 1956 to 1968 to reduce the female herd as overcrowding had caused mass mortality of the young. It was estimated that the population amounted to 1.5 million seals in 1975 but it seems that it is declining, possibly because many seals are getting entangled in nets from the trawl fisheries. Another possible explanation is that seals and fisheries compete for the same resource: Alaska pollock, herring and other fish. On the western side of the Pacific the recovery of fur seal herds progressed much more slowly.

4.2.2 Southern Hemisphere

In many places in the Southern Hemisphere seals were used by primitive communities for food, clothing, fuel and other purposes (see Sect. 1.1.3). The first record of commercial sealing dates from 1515, when the survivors of a Spanish expedition caught fur seals on the coast north of the river Plate and sold the skins on the market in Seville. Still today a sealing industry exists in Uruguay, producing 7000 skins per year. Sealing has been able to continue for almost 500 years because licences and catch limitations were introduced very early on and were retained over the years.

The occurrence of Cape fur seals *(Arctocephalus pusillus)* along the coasts of Namibia and South Africa during the austral winter was already known to early European expeditions, which often completed their forage from these resources. With the increasing market for blubber oil and skins in Europe, Dutch merchants established a sealing industry in the Cape Province in the early 17th century and this reached its peak in 1652-1657. The population of seals declined almost to extinction and the Dutch enterprises were discontinued in 1800. In the 19th century American sperm whalers visited the seal colonies from time to time to complete their oil cargo. Protective measures were introduced in 1893 and the seal stock started to recover. British and German companies were engaged in controlled sealing. In recent times South African enterprises caught 60 000 - 80 000 seals per year in the 1970s and around 30 000 animals in the mid-1980s. The present seal stock is estimated at 1.1 million animals.

From 1770 a new form of itinerant sealing developed with vessels searching wide areas for further productive sealing grounds. After 1775 they concentrated on the Falkland Islands, Patagonia and the region of Cape Horn. In 1792 they began to hunt fur seals *(Arctocephalus philippi)* on Juan Fernández Island and Isla St. Felix, west of Chile; by 1806 this species was almost extinct and still today there may be only 750 animals left.

In 1777 the publication of the results of Captain Cook's second voyage (1772-1775) indicated the wealth of seal resources on the subantarctic islands. Soon afterwards American, British and Russian sealers went south and stripped one breeding place after the other of almost all fur seals *(Arctocephalus gazella)* and many other animals (Fig. 92). Since the knowledge of hunting grounds was of high economic value, their location was kept secret for as long as possible. In 1790 hunting started on South Georgia and up to 1822 about 1.2 million seals were killed here. When the resources were depleted, the sealers extended their activities to other

Fig. 92. Killing of fur seals on subantarctic islands in the early 19th century. (Walton 1987)

islands, many of which were discovered during the search for seals. On the South Shetlands in 1819-1822 more than 320 000 skins were taken and 940 t of blubber oil produced. Populations on many other islands, also in the southern Indian Ocean (Kerguelen, Heard Island), met with the same fate.

The large incentive for sealing originated not only from the demand of the European market but also, to a great extent, from the trade with China, where furs were much liked by the wealthy people. Canton was the main port for this trade.

After an interruption of several decades sealing was revived for a few years in some places, particularly in 1870 on the South Shetland and South Sandwich Islands. However, hunting soon ceased from the lack of quarry. The destruction of seal resources was so great that still, on South Georgia in around 1930, only a few hundred fur seals were observed. The increase in population after the end of sealing was initially rather slow but since the 1950s it has speeded up to finally reach 17% annually. Today, the number of fur seals on South Georgia is estimated at more than 1 million and breeding sites on many other islands, e.g. South Shetlands and South Orkneys, have been repopulated. It may be assumed that the dramatic increase in the fur seal stock is the result of a larger food supply in the form of antarctic krill *(Euphausia superba)*, previously eaten by baleen whales which vanished during whaling.

The depletion of the fur seal populations made the hunting for skins unprofitable. Thus, in around 1810, the sealing fleets changed partly to the production of blubber oil from elephant seals *(Mirounga leonina)*. The main hunting grounds were on South Georgia and Macquarie Island, south of Australia. Here also the resources were quickly depleted. As it became increasingly difficult to find seal concentrations, groups of seal hunters went, sometimes even over winter, to search for seal in the remotest places and to boil the blubber for the production of oil. When the seal numbers were insufficient, even penguins, mostly king penguins, were taken. Under the rough climate living conditions for the sealers were horrible. By around 1890 elephant sealing became uneconomic and the ships turned to other regions. Hunting was resumed in 1910, when the populations had somewhat recovered. On South Georgia a whaling company was engaged in sealing under controlled conditions but this limitation did not halt a new decline in the number of elephant seals. The situation changed for the better when the British authorities introduced more rigid management measures in 1952. They restricted the catch to a maximum of 6000 large bulls to be taken from September to October and only from certain defined beaches, with alternation of the sites. In 1964 this sealing was discontinued and the elephant seal population has since fully recovered.

Today, seal populations in the antarctic region are generally in a good state. Species feeding on krill, such as the fur seal and possibly also the crabeater seal *(Lobodon carcinophagus)*, are more numerous than were their populations before sealing started. Most species are fully protected by the International Convention for the Conservation of Antarctic Seals, established in 1972. Catch quota were set at 175 000 crabeater seals, 12 000 leopard seals *(Hydrurga leptonix)* and 500 Weddell seals *(Leptonychotes weddellii)* but there is, at present, no economic incentive for sealing and in view of the strong public opinion against such commercial hunting it is not likely to recommence.

4.2.3 Sirenians and Sea Otters

Marine mammals of other zoological orders have also been used by man. An infamous example is that of the northern or Steller's sea cow *(Hydromalis stelleri)*, discovered on the Bering and Copper Islands, east of Kamchatka, in 1741 by Georg Wilhelm Steller during Vitus Bering's explorations in northeast Asia (Fig. 93). Steller, and other members of the expedition, were shipwrecked on the islands and survived only by living on the meat of these animals, which were up to 10 m long. Originally

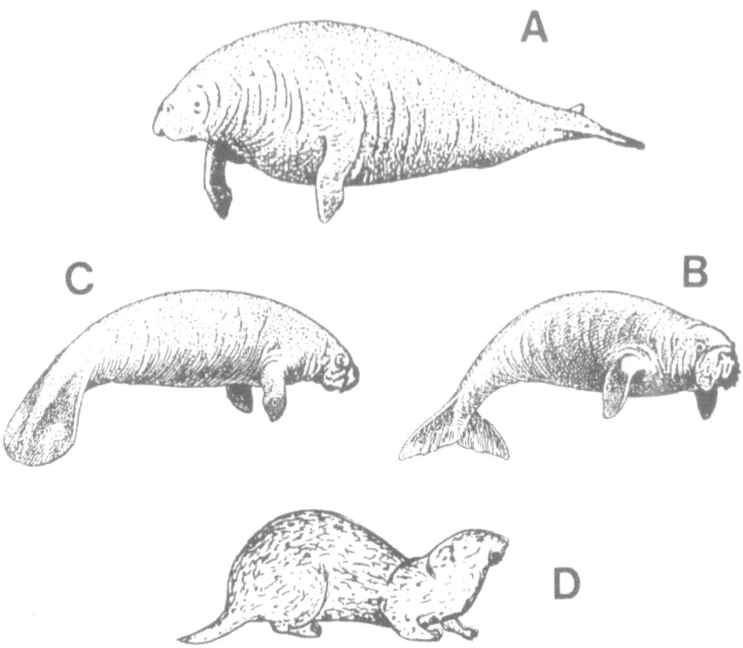

Fig. 93A-D. Sirenians and fish otter. **A** Steller's sea cow *(Hydromalis stelleri)* length 10 m; **B** dugong *(Dugong dugong)* length 3-4 m; **C** manatee *(Trichechus sp.)* length 3-5 m; **D** sea otter *(Enhydra lutris)* length 1.5 m. (**A-C** *Kosmos-Lexikon der Naturwissenschaften* 1955, Kosmos, Stuttgart)

there were about 2000 sea cows, feeding on seaweed close to the shore and therefore easy to catch. Already, by 1743, the first hunters and whalers had appeared. During the following 20 years hunters of fur seals and sea otters overwintered on the islands to slaughter sea cows for the production of salted meat as a food supply before they left for the hunting season on the Aleutian Islands and in Alaska. The barky skin of the sea cows was used as sheathing for the boats of the sealers. Killing was so ruthless that the last sea cow on Copper Island was killed in 1754 and the species became extinct with the slaughter of the last animal on Bering Island in 1768.

The dugong *(Dugong dugong)*, living in the Indo-Pacific region along the coasts from East Africa and the Red Sea to Australia, was, and still is, used in many places by local people for meat, oil and hides (Fig. 93). Parts of the animal are sometimes believed to have magic power. Conservation of this species is much needed but protective measures are very difficult to implement.

The same holds true for the manatees *(Trichechus* spp.), sea cows living along the shores and in the rivers of the Caribbean region and West Africa and in the Amazon River. Even until recent years manatees in Brazil were killed for the production of canned meat and hides for export. In the USA manatees are often endangered by high-speed boats.

Extensive commercial hunting and trading of sea otters *(Enhydra lutris)* developed after Steller had discovered this species in 1741 (Fig. 93). Very high prices were paid for the furs of these animals, which were abundant in rocky waters along the North Pacific coasts between northern Japan and lower California. In around 1850 the Russian-America Company traded about 118 000 sea otter skins, mostly for the Chinese market. In 1885 around 8000 furs were sold, in 1910 only 400 and, some years later, the species was almost extinct. Fortunately, as a result of protective measures determined under an international treaty and by national legislation the population has widely recovered, especially in Alaska and the Soviet Union. The total stock is now estimated at 35 000 - 50 000 sea otters.

4.2.4 Conflicts with Fisheries

As predators of fish and other animals of commercial value seals have always been chased by fishermen. The overall consumption of fish by seals is certainly substantial but, in most cases, relatively small in comparison to the resources and catches of commercial fisheries. Serious problems arise when seals prey on fish in nets, traps and other gear. Not only are considerable quantities of fish lost or mutilated but seals also scare fish from the nets. Furthermore, fishing gear, especially gill nets, are often severely damaged. The cost of such damage in recent times has been calculated for the fishery in the Gulf of St. Lawrence in the average at 300 dollars and off Norway at up to 2500 dollars per year per fisherman. In some areas damage was so severe that fishing had to be stopped altogether. The greatest problems arose near breeding sites of grey seals *(Halichoerus grypus)* in Norway, Scotland and Canada, especially where high-value fisheries or small communities were largely dependent on fishing.

The interaction between seals and fisheries became most evident during invasions of northern Norwegian coastal waters by large numbers of harp seals. Such invasions had already occurred in earlier times, e.g. at the beginning of the current century, but, since 1978, they were observed regularly every year between January and April. In 1987 and 1988 the invasions reached dramatic levels and harp seals were even found as far south as the Skagerrak coast. It was estimated that in 1987 more than 60 000 harp seals drowned in fishing nets. Fishing was greatly impeded and had to be stopped completely in some areas; loss of gear and catches was severe. Bounties were paid by the Norwegian authorities to compensate the fishermen for destroyed gill nets, lost catches and reduced fish availability in seal-invaded areas. Most of the harp seals immigrated from the Barents Sea, but tagging experiments showed that in 1987 and 1988 some seals also came from the west ice (Fig. 90). The Norwegian authorities blamed the conservationist groups, in particular Greenpeace, for the difficulties in the fishery as they had forced the termination of proper management of the seal populations in the early 1980s and this led to a strong increase in the harp seal stocks. Others explained the increase in seals as due to the contemporary low abundance of capelin, the main food of the seals, in the Barents Sea; this capelin shortage was thought to be the result of high fisheries exploitation and also ecological changes. Possibly both factors, seal population pressure and extremely low fish abundance, contributed to the cause of the seal invasions.

In a number of countries where damage to fishing became an important problem, like, for instance, in the Baltic, bounties were paid by the authorities for the hunting of seals. Such bounty schemes were introduced in several Scandinavian countries around 1890 but were later abolished when the seal stocks declined. Shooting is still used at some sites and so is poisoning with bait. Trials have been made with the application of bait with emetics, unpleasant substances to discourage seals from raiding nets. There is no doubt, though, that control of seals around fishing gear can be effective only with some killing of seals. It is thus necessary to come to an acceptable compromise between the care of seal resources and the demand of fisheries for food production.

Another reason to control seal populations is that certain species are intermediate hosts for nematode parasites of fish. The infestation of cod by the codworm *(Phocanema decipiens)* is a serious problem in fisheries in Canada, Great Britain and Norway. The most heavily infested fish must be discarded and the codworm must be removed by hand from the fillets of others. In Scottish waters, and elsewhere, infestation was found to be highest near breeding places of grey seals. The biological interactions are, however, rather complex and are not yet fully understood.

Sea otters along the coasts of the North Pacific also have a large impact on the yields of commercial fisheries for various shellfish species, like Dungeness crabs, lobsters, abalone, clams, scallops and mussels, because they feed extensively on these resources.

5 Development of Modern Fisheries

5.1 Statistical Account of World Fishing

One measure of the progress in fisheries development is the increase in production. Reliable catch statistics on a worldwide scale became available in 1947 when the Food and Agriculture Organization of the United Nations (FAO) started to publish regularly the *Yearbook of Fishery Statistics*. Data for the year 1938 were included in this series. Many improvements in the collection and presentation of statistics have been made since the first volume and some of the apparent increases in landings may in fact be a result of better statistics.

For the time before World War II such statistics are more scanty and are often difficult to attain. They are also limited to a small number of countries. From the few available sources it can be estimated that at around the end of the last century world fish production might have been in the order of 8 - 10 million t per year. The fisheries in Russia and Japan contributed about 1 million t each, and it is likely that the catches in China and in Southeast Asian countries, with old traditions and high fish consumption, were hardly lower than those indicated in the first statistics. In Europe at least 500 000 t came from the great cod fisheries, particularly catches off Newfoundland (400 000 t). The traditional herring fisheries produced at least 300 000 t, with Scotland providing half of this quantity. In the development of modern fishing England took the lead with 600 000 t, and about the same amount can be estimated for the United States of America.

More statistics are available for the 1920s and 1930s. They indicate a total annual fisheries production of 10 - 20 million t. A relatively complete set of catch statistics results in a figure of 17 million t for 1936. Most important were the landings in Japan (3.6 million t), USA (2 million t), Korea (1.7 million t), USSR (1.5 million t), Norway (1.2 million t) and Great Britain (1 million t). In the same year (1936) fishing fleets in the world included some 4300 ships larger than 100 GRT, 6100 steamers, 190 000 motorized vessels and 2900 trawlers. The

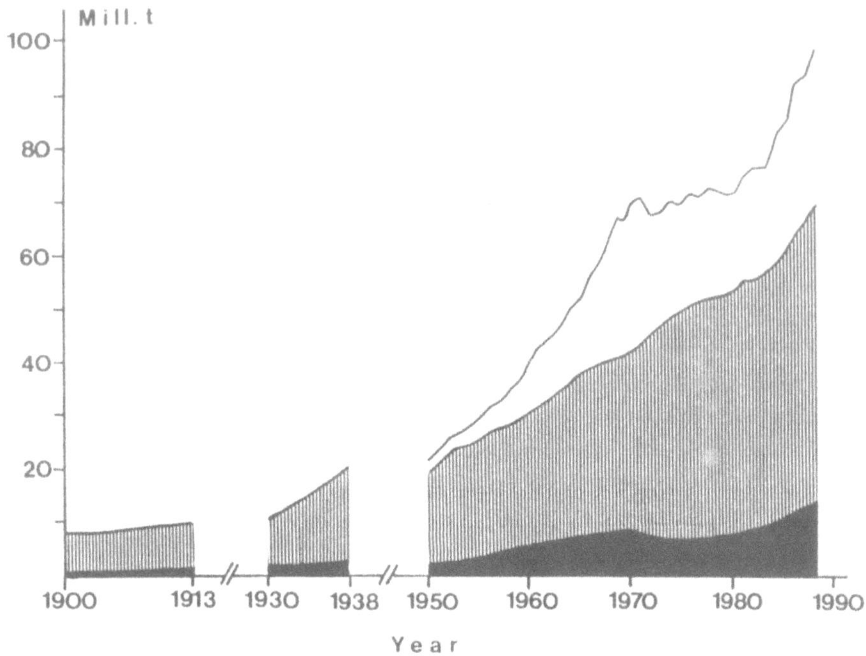

Fig. 94. Total landings of fish and other aquatic organisms, 1900 - 1988. *Dark*, inland waters and aquaculture; *grey*, marine waters, landings for human consumption; *white*, marine waters, landings for industrial purposes. (FAO 1947 - 1989)

number of motors installed in fishing vessels was 20 times higher than in 1913. The average tonnage of newly constructed ships over 100 GRT increased 2.5 times during the same period. The increase in catches effected by these improvements was estimated at 47% on average and at more than 200% in certain countries, like Germany.

The total landings of fish and other aquatic organisms increased enormously from around 8 million t in 1900 to 98 million t in 1988 (Fig. 94). From 1950 to 1970 the annual increment was 5 - 7%. During the 1970s growth slowed down as a result of the collapse in the anchoveta fishery off Peru and the decrease in distant-water fisheries excluded from the new Exclusive Economic Zones (EEZs). After 1983 the increment accelerated again. Between 25 and 30% of the total production is used for industrial purposes, mostly the manufacture of fish meal and oil.

Of the roughly 30 000 existing fish species only 1000 are utilized for consumption, and only a small number of species make up the bulk of the landings. Most important are the following (by landings in 1988):

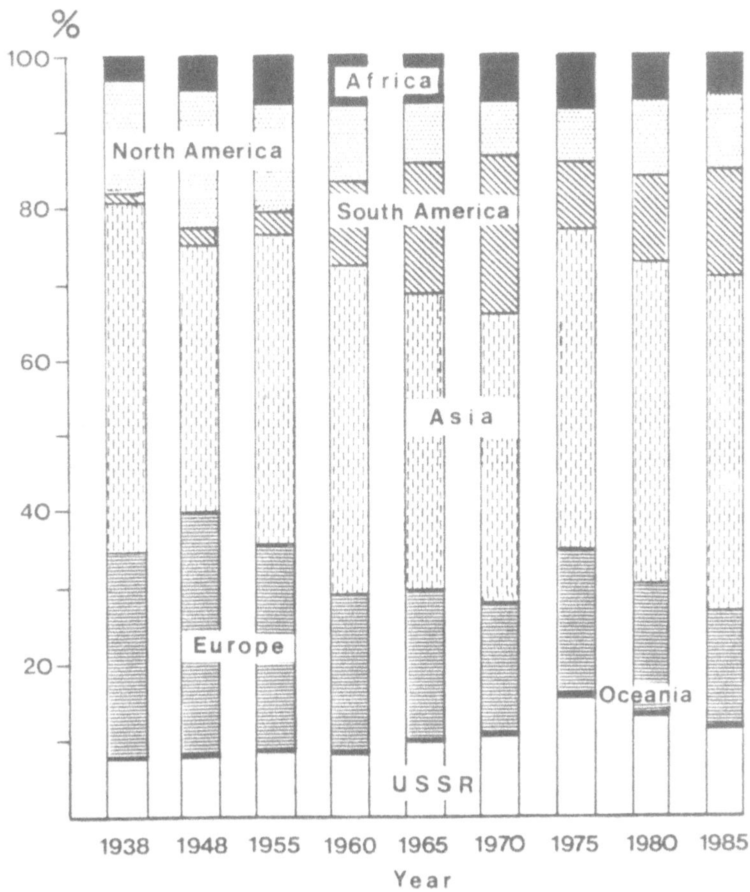

Fig. 95. Fisheries production by continents (%). (FAO 1947-1989)

Alaska pollock *(Theragra chalcogramma)*	6.6 million t
Japanese pilchard *(Sardinops melanostictus)*	5.4 million t
South American pilchard *(Sardinops sagax)*	5.0 million t
Anchoveta *(Engraulis ringens)*	3.6 million t
Chilean jack mackerel *(Trachurus murphyi)*	3.2 million t
Atlantic cod *(Gadus morhua)*	2.0 million t

Figure 95 shows the breakdown of landings by continents. With more than 40% Asia takes the largest share, followed by Europe. Over the years the percentage of European landings has decreased. South American landings were largest in the 1960s. Landings in the USSR amount to more than 10% of the total.

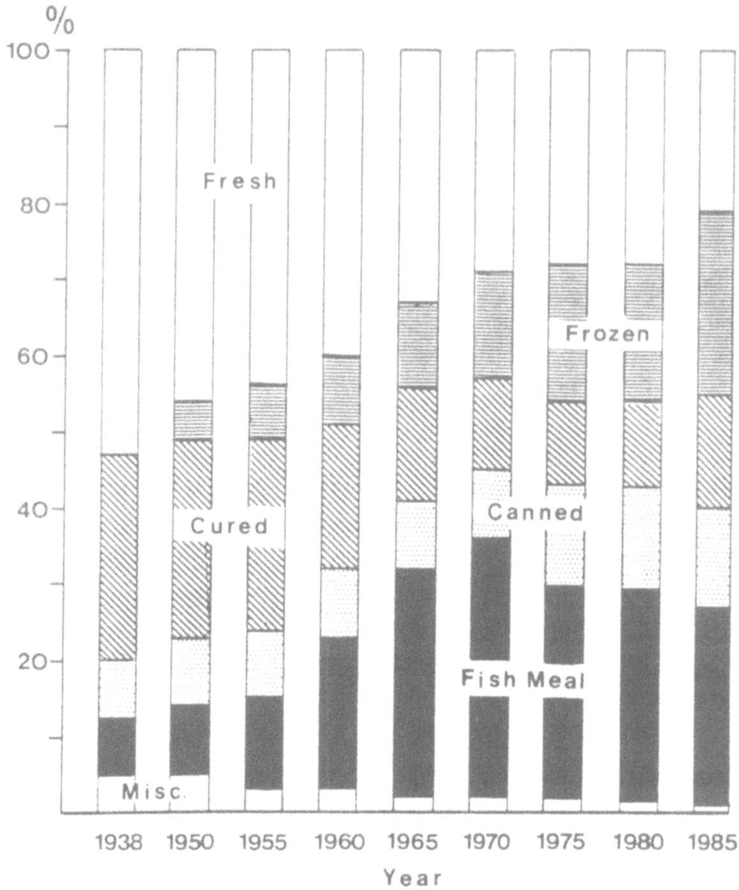

Fig. 96. Trends in the utilization of fish catches (%). (FAO 1947-1989)

Fig. 97. Production of fish meal (in million t). (FAO statistics)

Fig. 98. Increases in aquaculture production from 1975 to 1988. (FAO statistics)

Considerable changes have taken place during the last 50 years in the utilization of catches (Fig. 96). Most important is the consumption of fresh fish but, since 1950, preservation of fish by deep-freezing was increasingly employed. The demand for cured products, particularly for salted and dried fish, declined, while for canned products there was a slight increase. Most spectacular was the increase in the production of fish meal and oil (Fig. 97).

The increase in fish production during recent years has been partly the result of higher yields from aquaculture, which is now being developed most rapidly (Fig. 98).

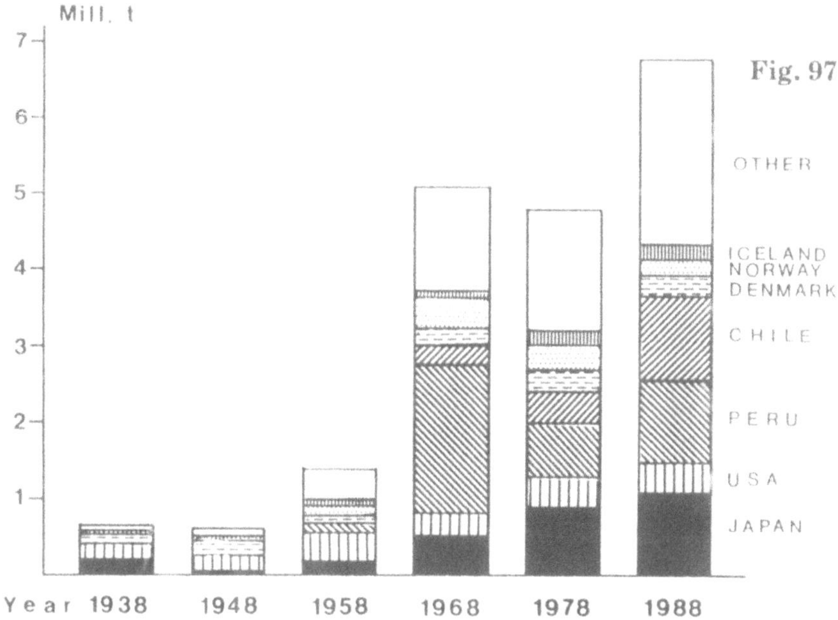

Mill. t
7
6
5
4
3
2
1
OTHER
ICELAND
NORWAY
DENMARK
CHILE
PERU
USA
JAPAN
Year 1938
1948
1958
1968
1978
1988

Fig. 97

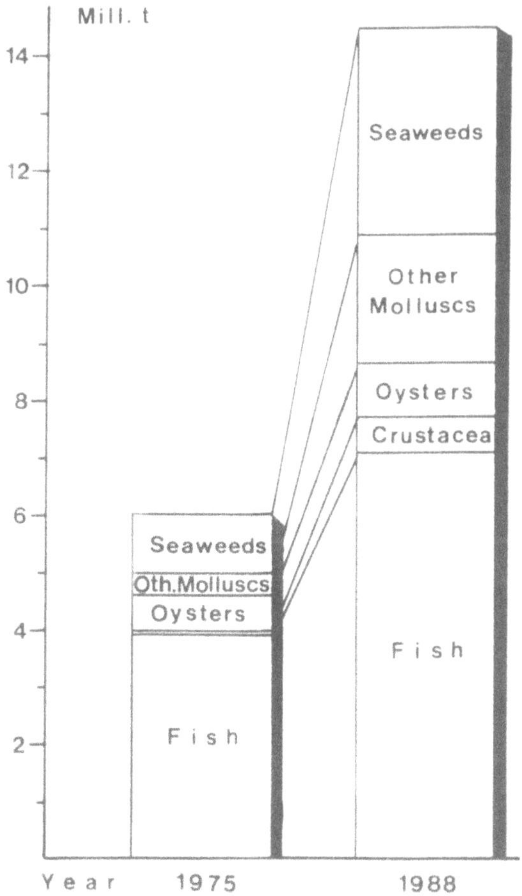

Mill. t
14
12
10
8
6
4
2
Seaweeds
Oth.Molluscs
Oysters
Fish
Seaweeds
Other Molluscs
Oysters
Crustacea
Fish
Year
1975
1988

Fig. 98

5.2 Overview on Major Fisheries

The development of fisheries in some of the most important fishing countries is described in the following sections. Landings from these countries amount to about 65% of the present world fish production. The development of these landings since World War II is shown in Fig. 99.

5.2.1 Japan

In an unprecedented development Japanese fisheries grew to be the most important and most diverse in the world. Already by the end of the last century fish production had reached 1 million t per year and at the time of World War I it surpassed the landings of Great Britain, which was, until then, the leading fishing nation. In the mid-1930s the total catch amounted to over 4 million t. During recent years the annual harvest of Japanese fisheries has increased to almost 12 million t, about 13% of the total world catch (Fig. 99).

The fishing fleet consists, at present, of about 420 000 vessels, of which roughly 380 000 are smaller than 5 GRT (30 000 are still not motorized). On the other hand, there are about 3000 ships of over 100 GRT, including 75 vessels larger than 1000 GRT and four even larger than 4000 GRT. While the number of small boats has remained constant for a long time, the number of fishermen has decreased considerably from nearly 1.5 million at the end of the last century to about 500 000 today as a result of industrialization and introduction of labour-saving equipment. Despite many technical improvements, the fishery still consists of two major components: numerous small-scale family enterprises in the coastal fisheries and aquaculture on the one hand, and some very large fishing companies on the other hand. Of the 200 000 fishing enterprises, about 95% belong to the former category, whereas the remaining 5% operate offshore and distant-water fisheries with large vessels.

Throughout history fish and other aquatic organisms have been a most important and indispensible source of food for the Japanese. At 86 kg per year the per caput supply is the highest in the world. Of the total domestic catch 70% is available to supply the Japanese market with fish and fish products consumed in an enormous number of varieties. Nearly 30% of the catch is used for industrial purposes, including the production of fish meal to meet the growing demand for animal and poultry feeds and for fish culture.

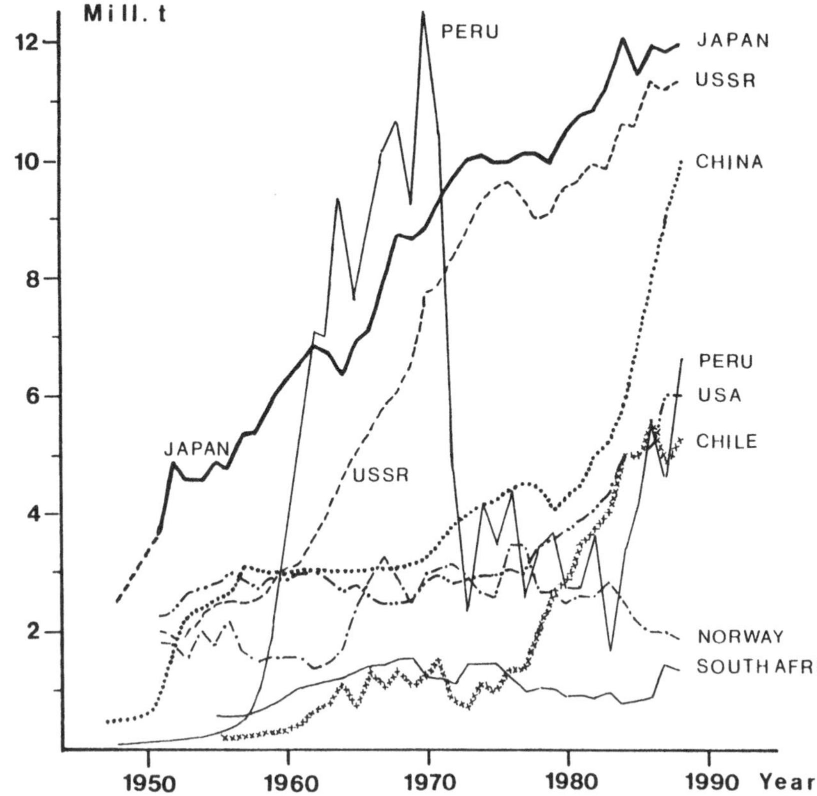

Fig. 99. Landings of fish and other aquatic organisms in major fishing countries, 1950 - 1988. (FAO 1947-1989)

Japan is also the most important country in fish trading. Domestic fish production is supplemented by large imports which have risen dramatically during recent years to reach 2 million t in 1987 with a value of 8.5 billion US dollars. Shrimps form the largest share in these imports at 30% by value. In contrast, the exports of fish and fish products amount to about 800 000 t per year with a value of only 20% of the imports. In value the largest components are frozen tuna and skipjack (25%), pearls (25%) and canned fish (12%). For both imports and exports the USA plays, by far, the largest role for Japan.

Basic Conditions for Fisheries Development

The Japanese islands, stretching over about 3000 km from subarctic waters in the north to subtropical regions in the south, are surrounded by productive fish-rich seas, influenced by interactions between the cold

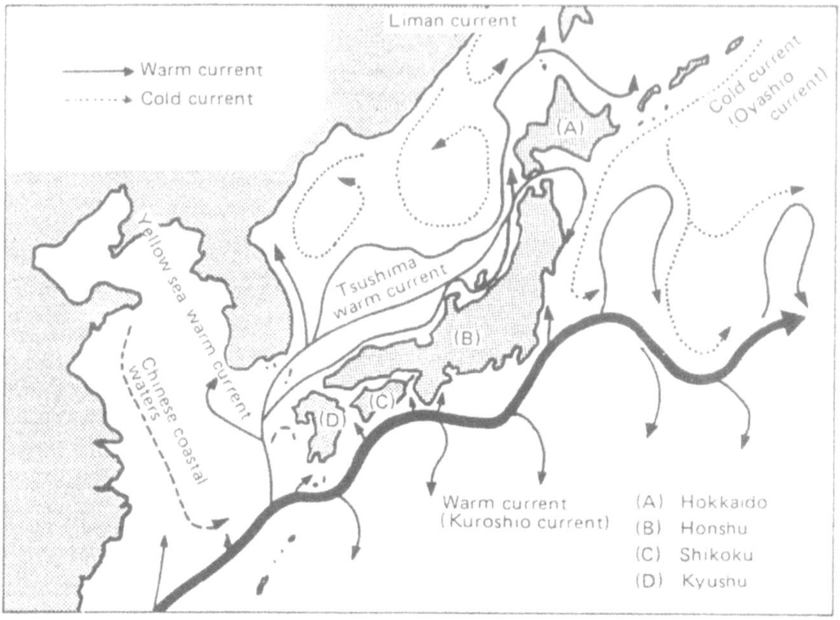

Fig. 100. Ocean currents around the Japanese Islands. (AICAF 1984)

Oyashio and the warm Kuroshio currents with intensive mixing of water and upwelling processes (Fig. 100). The coastline of almost 27 000 km around the more than 3000 islands is very irregular with many bays and relatively shallow shores, suitable for the development of coastal fishing. A central area is the Inland Sea, a depression between the islands of Honshu, Shikoku and Kyushu with an average water depth of 40 m, where a marine civilization developed in the 12th century. Still today this area with about 400 fishing villages is a centre of traditional fishing and cultivation of molluscs and seaweeds. The Inland Sea was the nucleus for the expansion of fishing along the Pacific side of the islands. This coast presented fairly rough conditions with high surf and only narrow shelf zones, but the oceanic waters proved much more productive than the seas along the western coasts of Japan. The catches on the Pacific side are nearly three times higher than in the Sea of Japan. For this reason the number of fishing villages along the Pacific coast reached almost 600 whilst on the west coast there are only 350. The lack of larger shelf zones on the Pacific side forced Japanese fishermen to develop pelagic and high-seas fishing at an early stage.

Although Japan is surrounded by very favourable fishing areas, there are also some hazards to fishing. Typhoons, most vigorous storms, occur each year in September/October and often destroy fishing vessels and installations. Furthermore, changes in the prevailing ocean currents often lead to sharp drops in the water temperature with detrimental effects on fishing.

Vast areas on the Japanese islands consist of mountains and only about 15% of the land is suitable for agricultural purposes. As long as the human population was relatively small, less than 30 million people in the early 19th century, the country was self-supporting on the basis of rice and fish. Meat production from livestock was always extremely limited so that fish played, and still maintains, the chief role in the supply of animal protein (in 1955 still 80%, now about 50%). When the population grew to 50 million in the beginning of this century, to 78 million in 1940, and to 123 million in 1990, much additional food was needed. Agricultural production was highly intensified often with two to three crops per year being planted. Fish and other marine organisms were widely used as fertilizer. Since no additional land could be cultivated and food imports had their limitations, this provided a strong and continuous pressure to develop food production from the sea, at first from nearby waters and later from all around the world. The success of fisheries development in Japan with consequent and well-considered efforts promoted by the government is really admirable.

Home Waters

As described in Section 1.1.4, fishing in Japan has a very long tradition. Excavations from a Stone Age village of rice-growing and fish-eating people near Shizuoka (Honshu) revealed several fish hooks. Later inhabitants of the islands were the Aino who were then dislodged by the Japanese Yamato culture to Yezo (Hokkaido) and further north. In early times the Aino lived along the coasts on shellfish and seaweeds. Large mounds of the shells of abalone *(Haliotis* sp.*)* are still found in many places. Later, the Aino also caught salmon, herrings and sardines from boats which were built by sewing together leaves of bamboo grass *(Arundaria* sp.*)*. Fish hooks of bone and harpoons were used regularly. Usually the catches, and also seaweeds, were dried in the sun. The importance of fish for the Aino is indicated by the fact that in their language they use the same word for fish and for food.

A large number of fish species, squid, molluscs and seaweeds were caught close to the shores by diving, traps, and with simple nets made of rice straw. Artificial planting of bushes to create shade to attract fish and poisoning of fish are very old catching methods. Also, whaling was carried out along the coasts from the earliest times (see Sect. 4.1). From the 12th century until the Tokugawa period in the 17th century, fishing methods, like fishing with seine nets for sea bream and various types of drag and surrounding nets, were developed, and more fish species such as cod *(Gadus macrocephalus)* and mackerel *(Scomber japonicus)* came into the reach of the fisheries. Freshwater fisheries were also important, as indicated by the existence of large traps with bamboo fences in Lake Biwa. In the Kamakura era in the 13th century special guilds had already been formed for fish trading.

During the Tokugawa period (1603 - 1867) Japan was, from 1639, completely secluded from other countries. The Japanese were forbidden on pain of death to leave their country, and no vessels larger than 50 tons were allowed. However, coastal fisheries from small boats developed further and became much more important than the catches in freshwaters. Beach seines for sardines *(Sardinops melanostictus)* were operated from two boats with about 100 people and gill nets were widely used. Very large traps for the capture of yellowtail *(Seriola quinqueradiata)*, bluefin tuna *(Thunnus thynnus)* and sardines already existed in around 1600 in the Toyama Bay north of central Honshu and on the northern coast of Kyushu. Traps were also built further north on the west and northeast coasts of Hokkaido for the fisheries on herring *(Clupea pallasii)* and salmon *(Oncorhynchus* spp.*)*. Much of this herring was transported to Osaka in southern Honshu as manure for agriculture and was bartered for rice needed in the north. In the south, originating from the Kagoshima region in Kyushu, pole-and-line fishing for skipjack *(Katsuwonus pelamis)* developed.

In spite of the seclusion, there was a considerable export of fisheries products, such as dried abalone, shark fins and trepang, to China. This export was routed via Nagasaki, one of the two ports open to foreign merchants and ships. At the same time Japanese fisheries expanded further north to Sakhalin, the Kurile Islands and even the coasts of Siberia.

The feudal system of the Tokugawa (Edo) period influenced greatly the social structure of fishing communities and the system of fishing rights with after-effects which even exist up to the present time. Increases in the human population were limited by birth control. People belonged to certain classes and changes in profession were prohibited. The lords decided on the establishment of fishing villages, on the main targets of fishing, and they also defined special districts in inshore waters with fishing rights for each village. Thus, the inshore areas were soon covered with a complicated system of reserved fishing districts; in the Inland Sea region these were mostly of small size and often jointly used by two or more communities. In later times these fishing rights were somewhat simplified and linked to fishing cooperatives and guilds. These cooperatives, at present about 4000 organized in a national federation, represent strongly the interests of coastal fishermen. When fisheries later developed further offshore, a rigid system of regulations and licences was introduced by the government for these new fishing grounds.

With the American expedition under Commodore Perry in 1853 and the conclusion of trade agreements with other states like England and Imperial Russia, Japan was opened up to the outside world. In 1867, in the first year of the new Meiji era, permission was granted to build larger seagoing vessels. However, until the 1880s, the Japanese fishery remained entirely at the level of coastal fishing with simple, though often sophisticated and skillfully applied, means. For fishing gear hemp materials increasingly replaced rice straw. With the end of birth control the population started to increase rapidly and all possibilities for fishing were used.

After the abolition of the class system in 1870 many fishermen engaged part-time in agriculture to improve their poor economic situation. Yet, their social standing was generally considered rather low. Since the times of the feudal system certain groups within the population were especially underprivileged, and some lived as nomads as they were not allowed to own land and were forced to separate from other people. One of these groups was the "Eta", mostly living on the south coast of Shikoku where they engaged in beach seining, long lining and diving for molluscs and algae. Other groups were the "Nori", boat nomads who fished for shrimp (ebi) with drag nets, and the "Yoshiwa", specialized in hand lining. The origin of the "Ama", who are also found in Korea and northern China, is uncertain. They are divers, both female and male, who collect molluscs, sea-urchins and algae in depths down to 20 m. They work without diving equipment from rocks and from boats. Most Ama live on the Shima Peninsula south of Nagoya. A relatively new activity is their diving for pearl oysters in the region of Toba, a tourist attraction.

Since early times catches from the sea were supplemented by production from aquaculture, both in freshwaters and in coastal bays and inlets. Culture of carp, imported from China in ancient times, was once rather important but decreased when many lakes were drained to provide additional land for agriculture. Eels *(Anguilla japonica)*, farmed since 1900 in central Japan, became more important. A third, more modern component is the culture of trout. Annual production from freshwater aquaculture reached 20 000 t in the late 1950s and amounts today to nearly 100 000 t, of which about 40% is eel.

The yields from mariculture are even 10 times higher after a steep increase during the last 30 years. They consist mainly of seaweeds (350 000 t), shellfish (750 000 t) and shrimps (5 000 t). The utilization of dried seaweed as a condiment, an ingredient in soups or a vegetable boiled with soy has a very long tradition. "Kombu" is prepared from the giant kelp *Laminaria japonica* which is abundant along the east coasts of Hokkaido. Highly prized "nori" or red laver from the red alga *Porphyra tenera* is cultivated in algal beds, particularly in the Bay of Tokyo since about 1700. The cultivation of clams in the Inland Sea dates back to 746. Today, the annual production of clams *(Venerupis japonica* and some other species) amounts to about 130 000 t. The farming of oysters *(Crassostrea gigas)* is probably also rather old, although written records date back to only 1600. It is mainly concentrated in the Bay of Hiroshima and on the coast of northeastern Honshu. About 230 000 t per year are presently produced in hanging cultures, with fairly stable yields over a longer period. Also important is the cultivation of scallops *(Patinopecten yessoensis)* on the north coast of Hokkaido with more than 200 000 t produced per year. Scallops harvested from natural beds first became important in the mid-Tokugawa period as a major export item to China via Nagasaki. The culture of pearl oysters *(Pinctada martensii)* has developed since 1864, when K. Mikimoto discovered the possibility of inducing the production of pearls by implanting foreign bodies in the oysters. In the 1920s the technique advanced to the level of industrial production of round pearls and, today, pearls with a value of 46 million yen are exported every year. The pearl oyster farms number about 2000 enterprises and are located in southern Japan, particularly in Ago Bay, Shima Peninsula. Mariculture of fish concentrates on the yellowtail and produces presently about 150 000 t per year. A large number of hatcheries have already been established since the 1880s for the artificial propagation of salmon and trout in Hokkaido, and for sea bream, mullet and other species further south.

At the end of the 19th century there were around 400 000 fishing boats (94% smaller than 10 m long) and more than 3 million fishermen, of whom half were engaged full-time in fishing. The total production in around 1880 was estimated at 200 000 t, consumed almost entirely as dried fish. In 1900 the catches had risen to more than 1 million t. Herring in the north and sardine further south made up more than half of this quantity whilst tuna and mackerel accounted for about 5%. More than 150 000 t of herring and sardine were used as manure for agriculture.

Offshore Pelagic Fisheries

Intensive fishing in the coastal waters soon led to a reduction in fish stocks and a decline in yields. Therefore, early protective measures were introduced and implemented through a system of corporate fishing rights and licences. Major expansions were only feasible in additional new fishing areas further offshore. Thus, in 1887, a law for the promotion of high-seas fishing was passed, and subsidies were provided for sailing ships larger than 30 tons, and later for steamers larger than 50 GRT. As a result, the number of vessels participating in high-seas fishing increased steeply. Already, by 1899, more than 3000 locally built vessels and 37 sailing and two steam-driven ships of European type (imported to Japan in 1897), with about 20 000 men, were engaged in this sector.

The offshore fisheries developed in the south mainly from Nagasaki and in the north around Hokkaido. They caught a large variety of demersal fish, including many flatfish in Wakama Bay, but the major production came from the pelagic resources, particularly herring, sardine, anchovy *(Engraulis japonicus)*, mackerel, saury *(Cololabis saira)* and squid *(Todarodes pacificus)*. Yields increased greatly with the application of improved gear, such as gill nets and surrounding nets, which were, from 1900, made of cotton. Purse seines, developed in America, were introduced to the Japanese fishery in 1882.

Under the influence of long- and short-term fluctuations in the ocean currents around Japan (Fig. 100), very large variations in the abundance of pelagic species occurred. In around 1900 herring, caught mainly west and northeast of Hokkaido, was the most important target with yields of up to 975 000 t (Fig. 101). Earlier records indicate that substantial herring fisheries existed in the south of Hokkaido in the mid-17th century and off northern Honshu in the 18th and 19th centuries. With rising sea-water temperatures since about 1910, followed by recruitment failures and decline in the herring stock, the catches fluctuated widely and decreased gradually until they became insignificant after 1960. Warm-water species progressed further north. The sardine resources around southern Japan increased and were more widely distributed farther north to Hokkaido and even Sakhalin. High sardine catches were obtained in the period 1925 - 1940 with a peak of 1.6 million t in 1936. This was followed by a sharp decline to 160 000 t in 1945 and, after some short-term increases, a low was reached in the late 1960s with less than 30 000 t. In 1973 catches began to rise quickly again, reaching 1 million t in 1976 and 4.4 million t in 1987. Thus, the sardine is now by far the most important species for the Japanese fishery. It is assumed that the large fluctuations in sardine

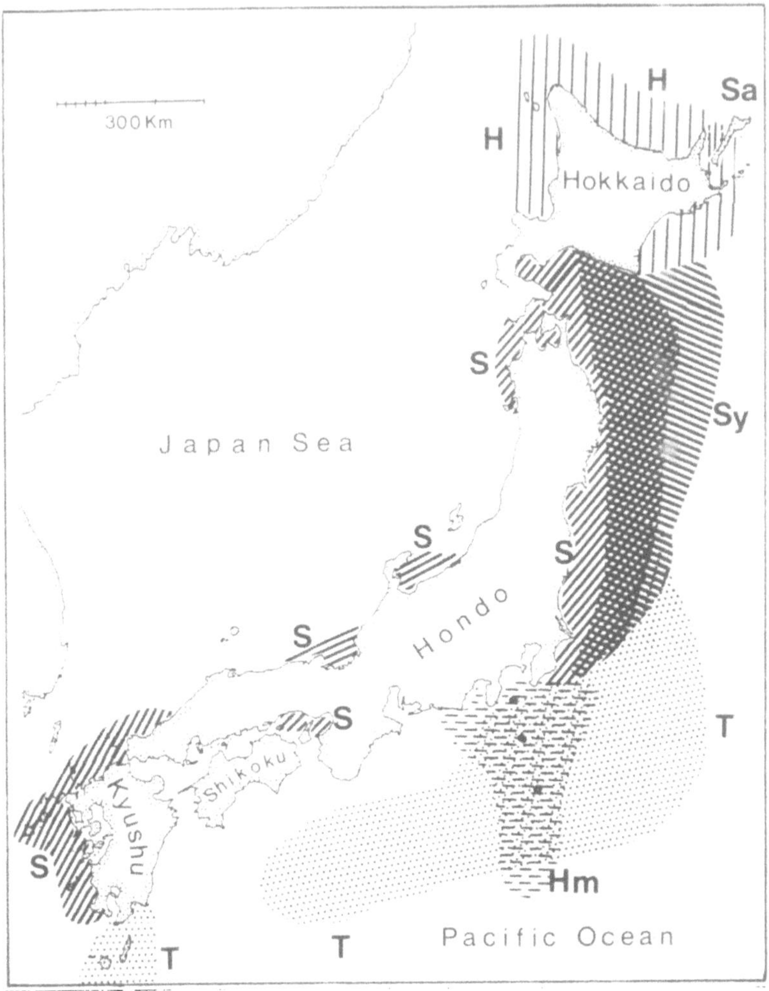

Fig. 101. Main fishing grounds for major commercial species in Japanese home waters. *S* sardine, *H* herring, *Sa* salmon, *Sy* saury, *Hm* horse mackerel and mackerel, *T* tuna

abundance are related to changes in environmental conditions rather than to the intensity of fishing. Sharp declines in the stock occur when cold water intrudes on the spawning areas and kills the young fish. Fragmentary records since the 17th century indicate that large variations in the yield of sardine fishing were also observed in olden times.

As sardines are not among the fish preferred by Japanese consumers, much of the catch is canned and exported or reduced to fish meal. The canning of sardines in oil was introduced to Japan from France in 1871.

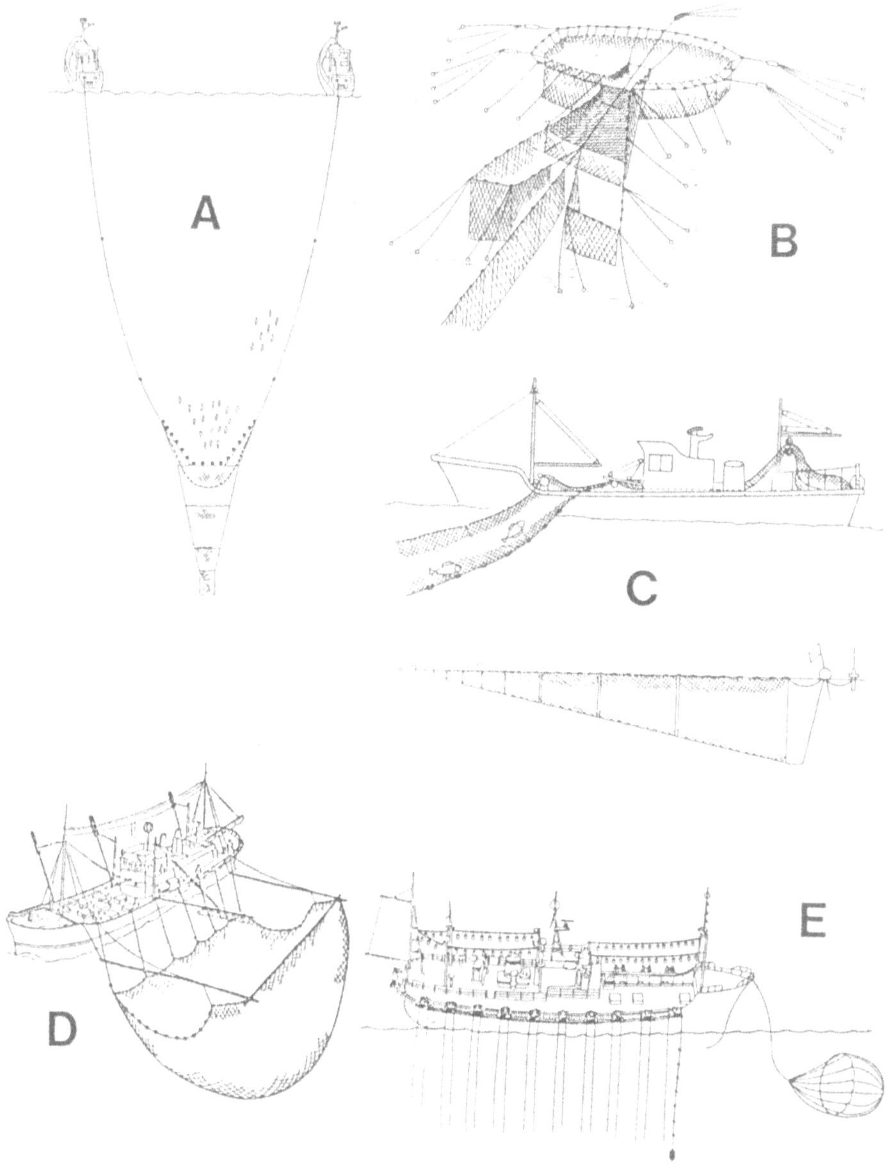

Fig. 102A-E. Some Japanese fishing techniques. **A** Pair trawler with bull trawl in East China Sea; **B** trap for sardine fishing; **C** drift net fishing for salmon; **D** saury stick-held dip net; **E** Jigging for squid. (Ministry of Agriculture, Forestry and Fisheries, Tokyo 1987)

As in Europe during Napoleonic times and in North America during the Civil War, the production of canned fish in Japan received a strong impetus from military needs during the Sino-Japanese War in 1894-1895 and during the Russo-Japanese War in 1904-1905. The large-scale development of the fish canning industry, though, began only in the 1930s with the enormous sardine catches.

It seems that stocks of other pelagic species also increased as a result of the widely observed warming in the North Pacific. The catches of mackerel in the waters off northeastern Honshu and Hokkaido, which were in the order of 40 000 - 150 000 t in 1910 - 1939, rose to more than 1 million t in the 1960s, due to the higher abundance of these resources. Catches of jack mackerel *(Trachurus japonicus)* started to increase sharply in around 1949 when there were some signs of increased abundance, and exceeded 500 000 t in 1960, before they levelled off to 100 000 - 200 000 t. The fishery for saury was carried out close to the shores of Shikoku with purse seines until drift nets replaced this gear in 1920 and the fishery expanded farther north and offshore to productive fishing grounds along the convergence of warm and cold waters. After World War II stick-held lift nets ("boke ami") were introduced and light was used to attract fish (Fig. 102). The landings of saury, which is important as bait in the tuna fisheries, reached 575 000 t in 1958 but declined to 200 000 - 300 000 t in later years.

Another major sector is squid fishing, which also has a long tradition. About 30 species of cephalopods are caught in the waters around Japan by angling and jigging, with boat seines, trawls, set nets and octopus pots. Fishing for flying squid, mainly in the seas south of Hokkaido, produced about 400 000 - 500 000 t per year in the 1950s and 1960s. Later, the catches declined to less than 200 000 t due to overexploitation, but expansion to other species and development of Japanese cephalopod fisheries around New Zealand, off Argentina, and in some other waters counterbalanced these losses. With a catch of about 800 000 t, roughly half of the total world production, Japan has the most intensive utilization of cephalopods.

Expansion into the North Pacific Region

Already in the 18th century Japanese fishermen had travelled north to Sakhalin and the Kurile Islands to fish for herring and salmon. Before 1750 they had even reached the coasts of Siberia in the Sea of Okhotsk. After the end of the Tokugawa period of seclusion this expansion intensified in around 1870, particularly in connection with the colonization of Hokkaido. The advance of Japanese fishing was facilitated by the fact that in eastern Siberia there were hardly any white settlers to use the fish resources and that the government of Imperial Russia was in a weak position in this distant region. The Russo-Japanese War in 1904-1905 strengthened Japan, and in the Treaty of Portsmouth (1905), which confirmed the Japanese right to the Kuriles and southern Sakhalin (Karafuto), the Japanese were conceded the right to fish along the Siberian coasts and to obtain fishing concessions in northern Sakhalin and on Kamchatka. In 1907 a Russo-Japanese fishing convention was concluded and this reserved fishing rights in certain areas, river estuaries and bays to Russian enterprises while the open-sea coasts were free to Japanese vessels, and the Japanese were allowed to establish land installations for the processing of their catches on grounds leased for a number of years. In the same year the first Japanese fishing base opened on Kamchatka.

In 1910 more than 7300 sailing ships and 49 steam-driven vessels were engaged in this salmon fishery with gill nets and traps. Japanese enterprises were especially concentrated in the Nikolaievsk district in Siberia close to northern Sakhalin where good catches of both chum salmon *(Oncorhynchus keta)* and pink salmon *(O. gorbusha)* were obtained. Later, when the catches declined, fishing in Kamchatka became more important; here other salmon species were also abundant. Japanese canneries on Russian territory produced between 60 and 90% of all tinned salmon, which was mostly exported and sold on the world market from this region. Dry-salted or pickled salmon from the salteries was consumed in large quantities in Japan.

Land-based fishing for king crab *(Paralithodes camtchatica)* for the manufacturing of canned crab-meat dates from 1905. This fishery expanded from Hokkaido to Sakhalin and the northern Kuriles.

With the expansion of Japanese fishing activities and the increasing settlement of Russian colonists more and more conflicts arose between the new settlers and the Japanese fishermen and processing personnel working there only during the short summer seasons. The situation resembled somewhat the struggles between European fishing interests and colonists on Newfoundland in earlier years (see Sect. 2.4).

The political uncertainties and the need to process quickly the perishable king crabs led the Japanese enterprises to divert their efforts partly from land installations on foreign territory to floating factories in open waters. This development started in 1920 and already by 1930 19 factory ships with capacities of between 1000 and 7800 GRT were participating. Each unit was accompanied by two to three motor ships of 80 GRT for the shooting of the tangle nets and carried on board up to 12 small motor boats for hauling the nets. In the 1930s this fleet, working mainly west of Kamchatka, together with 64 canneries on land, produced annually up to 600 000 cases of tins of king crab-meat from 30 million animals. However, the catches had declined to half this quantity within a few years.

In 1927 operation of motherships was also introduced for a pelagic drift net fishery on salmon outside the Russian waters. Already, by 1931, a fleet of 13 factory ships with more than 100 catcher boats worked mainly in the area near Kamchatka. Around 60% of the catches were salted or frozen on board, while the rest was transported to land factories for canning.

The initiation of offshore king crab and salmon fisheries affected the Russian resources and led to further tensions between the Soviet Union (founded in 1917) and Japan. A new fisheries convention was concluded in 1928 to restrict Japanese activities on USSR territories, but conflicts still persisted. In the 1930s the USSR authorities established a Far Eastern Fisheries Trust ("Dalrybtrest") and started to build up Soviet fishing centres and canning factories in Kamchatka and other places to compete with the Japanese (see Fig. 110).

The average catch of salmon by Japanese enterprises in the coastal and pelagic fisheries rose from 142 000 t per year in 1926-1929 to 317 000 t in 1936-1939. At the same time the USSR catches were further increased so that in the 1930s a peak in total production of 400 000 t per year was reached. Later, in the 1950s, the catches declined as a result of overexploitation, probably coinciding with the detrimental effects of climatic-hydrographic changes.

World War II terminated Japanese fishing in the waters close to the Soviet Union, and all land installations were lost in Kamchatka, on the Kuriles and in Sakhalin. Already by the early 1950s, however, Japanese fishing vessels had recommenced with their pelagic fisheries. After resumption of relations between both countries, the USSR used its increased power to repel the newly increasing pressure by Japanese fishing interests. A Japan-Soviet Northwest Pacific Fisheries Commission was established in 1956 and held annual meetings, during which catch quota were set for salmon, king crab and herring. Also, other management measures, including contributions to the artificial propagation of salmon, were negotiated, and certain areas were closed to Japanese fishing. In line with the overall decline in salmon and king crab resources the total allowable catches were rigidly reduced, and the USSR claimed an increasing share. In recent years the joint commission has been used as a forum for establishing annual catch quota on a reciprocal basis for a wider range of fish species in the respective Exclusive Fishing Zones of both countries.

Mothership operations in the Northwest Pacific from the 1920s formed the model for this type of fishing in later years on a much larger scale and in other parts of the world. This model was copied by the USSR and some other countries. The factory ships had the advantages ofhigh mobility and large loading capacities and furthermore, could supply their catcher vessels. Special transport ships with installations for freezing were put into service to land the fish products while the floating factories could continue to operate for longer at sea. The motherships in the salmon drift net fishery are vessels of 7 000 - 10 000 GRT, equipped with canning and deep-freezing machinery and a reduction plant for offal. Each mothership operates with a fleet of about 30 catcher boats of 80 GRT which fish with synthetic multifilament gill nets.

As the Japanese mothership fleets suffered more and more from restrictions in the Northwest Pacific, they soon expanded their activities into the Bering Sea and the Gulf of Alaska. Deep concern arose in America when the fleets appeared in these waters during the 1930s. In 1932 Japanese vessels with tangle nets fished for the first time for king crabs off the Alaskan coast, where Port Moller and Bristol Bay remained preferred fishing grounds for many years. The Americans became even more concerned when, in 1936, the Japanese also started pelagic fishing for salmon.

After World War II Japanese fishing fleets reappeared in the Northeast Pacific in 1952. The concern of the USA and Canada over this development led to the conclusion of a tripartite International Convention for the High Seas Fisheries of the North Pacific Ocean which became operational in 1953 with the formation of the International North Pacific Fisheries Commission (INPFC). Under this convention Japan agreed to abstain from fishing for salmon, halibut and herring east of longitude 175°W. Ten years later they obtained the right to fish for halibut with long lines in the eastern Bering Sea and for herring off the Canadian coast, near the Queen Charlotte Islands. Fishing for king crab in the eastern Bering Sea was resumed in 1953 and grew rapidly until 1964 when over 10 million crabs were caught. Under agreements with the USA the Japanese catch had then to be reduced continuously. The king crab stock began to show signs of depletion, and the fishery intensified the capture of tanner crabs *(Chionoecetes* sp.) which soon became the main target.

During the mid-1970s the total Japanese catches in the Northeast Pacific had reached 1.5 million t per year, of which about 800 000 t were Alaska pollock *(Theragra chalcogramma)* caught by trawling. After the establishment of 200-mile Exclusive Fishing Zones by the USA and Canada in 1977, the catch quota, which had to be negotiated yearly, was gradually reduced, and substantial licence fees had to be paid. Finally, in 1988, all fishing permits for Japanese vessels were withdrawn by the USA as a reaction to the continuation of Japanese whaling. Thus, the Japanese fishery was restricted to the rather limited international waters in the central Bering Sea between the fishing zones of the USA and USSR.

Trawl Fisheries

Various types of drag nets were used in Japan for several hundred years, but a real trawl fishery began in 1908 with the arrival of the first steam trawler from England. The fishery expanded rapidly. The number of otter trawlers had reached 136 by 1912. Conflicts with the inshore fishermen and the establishment of areas closed to trawling forced the trawlers to move into the East China Sea and the Yellow Sea, large plain waters with depths between 20 and 200 m. Here, good catches of bottom-dwelling fish were obtained, especially of highly valued sea bream and croakers. But soon the yields declined and other less valuable species dominated. In the interests of the inshore fisheries the trawlers were also excluded from the coastal waters of Korea which came under Japanese protectorate in 1905. As large vessels were required for other purposes during World War I, the number of trawlers decreased to only six by 1918. Through governmental regulation the number of otter trawlers was limited to 70 during the following years. The vessels had a size of around 200 GRT and engines of 250 - 400 hp. With the introduction of motor ships in the 1930s the sizes increased to 300 - 500 GRT and the power to 550 - 800 hp.

In 1920 trawling by two vessels was introduced with the bull trawl, a gear with long wings developed from small local drag nets (Fig. 102A). As the bull trawl caught a higher percentage of preferred species, sparids and sciaenids, than the otter trawl, this fishery expanded rapidly and there were many more pair trawlers than otter trawlers. The bull trawlers had sizes of 30 - 100 GRT. With almost 700 ships plus some 400 operating from Japanese bases in mainland China and Formosa, the pair trawl fishery contributed to overfishing in the East China Sea and the waters west of Japan. The ports of Shimonoseki and Nagasaki were centres of the trawl fishery.

A third category was that of the more than 2000 small boats of 20 - 40 GRT and 60 - 100 hp, most of which worked on the shelf around Hokkaido and northern Honshu with various types of drag nets and seines.

When the catches in the East China Sea declined, the trawlers extended their operations westwards to the South China Sea and the Gulf of Tonkin, and in the north to Kamchatka and the Bering Sea in 1931. Particularly productive was the trawling for Alaska pollock and other demersal fish in Bristol Bay in the eastern Bering Sea in 1933-1937. The trawlers delivered their catches to factory ships for the production of fish meal and later, also salted or frozen fish. During the second half of the 1930s fishing grounds in the Gulf of Thailand, the Arabian Sea and the waters north and west of Australia were explored. In several places good catches, particularly of sea bream, snappers and prawns, were obtained and either transported to Japan by special carriers or sold on the local markets. After exploratory surveys joint trawling enterprises were established in 1937 with local interest in Argentina for the fishery on hake and in Mexico from Mazatlan for the fishery on prawns in the Gulf of California. These prawns were frozen and packed for transport by refrigerated ships to Japan and the USA.

All offshore trawlers were lost during the war in 1941-1945. After the end of hostilities the US occupation authorities soon allowed the construction of new ships to help combatting the serious food crisis, but trawling was restricted to the eastern parts of the East China Sea. The rebuilding of the Japanese trawler fleet made fast progress after the suspension of restrictions. In 1959 it consisted of about 760 bull trawlers with an average size of 80 GRT, catching 338 000 t, 54 otter trawlers of around 400 GRT, catching 19 000 t, and more than 1300 draggers, catching 632 000 t. Catches in the East China Sea from resources which had made remarkable recovery during the war declined under heavy fishing pressure. However, the waters off Korea, China and Taiwan were now closed to Japanese vessels, and these countries expanded their own fisheries. Japanese ships entering such areas, in particular the 60-mile zone claimed by Korea, were frequently captured and released only after long negotiations.

Under these circumstances the Japan Fisheries Agency granted permissions in 1959 for the construction of larger trawlers for use in distant fishing grounds. Within a few years a fleet of factory stern trawlers existed with vessels of 1500 GRT and soon, also of other vessels up to 3500 GRT and beyond, fit for worldwide fishing operations. Technical improvements in gear design and the introduction of synthetic fibre nets and fish detection equipment increased the efficiency of fishing and modern processing machinery improved the quality of the products. With many exploratory surveys these vessels began intensive fisheries in all oceans.

In the first instance these large independent trawlers intensified the fishery in the Northwest Pacific, which had restarted in 1954 with the development of a fleet of medium-sized trawlers for fishing around Hokkaido, along Sakhalin, and off the Kuriles and Kamchatka, and the operation of mothership fleets with small trawlers in the Bering Sea. During the first years the mothership fleets produced frozen flounder, particularly yellowfin sole *(Limanda aspera)* in the Bristol Bay region. In 1961 the mothership activities reached a peak with five fish meal factory ships and 18 freezer motherships working with nearly 300 catcher vessels. Most of these catchers, otter and pair trawlers and seiners, were transferred from the coastal waters of Japan and from the China Sea. Three mothership fleets were specialized in fishing for shrimp *(Pandalus borealis)* in the waters near the Pribilof Islands. Around 1963 the fisheries changed. Catches of yellowfin sole and shrimps declined and the production of fish meal became unprofitable due to the situation of the world market. Emphasis shifted to the catching of Alaska pollock, which increased enormously within a few years and reached a peak of over 3 million t in 1972. From 1978, under quota agreements with the USA, the catches amounted to about 1.5 million t. The largest part of these catches was used for the production of minced pollock as the raw material for surimi. Most surimi products are consumed in Japan, but demand by the world market is increasing, especially for imitation shrimp-meat, which is exported to the USA and other countries. The large stern trawlers participated intensively in this Alaska pollock fishery in the Bering Sea and Kamchatka waters and in the fishery for ocean perch *(Sebastes alutus)* along the Aleutian Islands and in the Gulf of Alaska.

The Japanese trawl fishery also spread to other parts of the world. Already, by 1955, trawlers had reappeared in Argentina, and in Brazil joint companies were founded to exploit the rich shrimp resources in the northeast. The large stern trawlers went fishing off Northeast and Northwest Australia, New Zealand, in the Indian Ocean, off northwest Africa, in the Newfoundland waters and, lately, also in the Antarctic Ocean. Several operational bases were established, for example in Singapore, Cape Town, and Las Palmas on the Canary Islands. The ships are often away from their home ports for more than 1 year and their catches are shipped through foreign ports and transported to Japan or other countries by freezer-carriers. The preferred targets are highly prized food fish such as sea bream, croakers, shrimps, prawns and squid.

Tuna Fishing

Another branch of distant-water fisheries with worldwide activities developed with tuna fishing. In Japan the capture of tuna has a long tradition. Remains of bones of skipjack or bonito *(Katsuwonus pelamis)* were found in prehistoric shell mounds in Honshu. The oldest Japanese literature from the eighth century indicates that tuna meat was often eaten raw and usually dried on the roofs of fishermen's houses. Still today "katsuobushi", boiled and fermented skipjack meat, is a famous Japanese dish. Bonitos were most abundant along the coast of southeast and central Japan in the warm Kuroshio waters. Within a few miles from the shore the tuna were caught by pole-and-line from open sailing boats, manned by 25 - 30 people and equipped with wells for live bait. In 1913 motorization of these boats started and their range was extended to much further offshore to increase the efficiency. By around 1930 the vessels had sizes of 30 - 60 GRT and made trips of 4 - 7 days. The catches were cooled with ice. The skipjack boats sailed from almost 100 ports along the Pacific coast. A main centre was, and still is, Yaizu, which is now one of the largest fishing ports in the country (Fig. 103).

Bluefin tuna *(Thunnus thynnus)* have also been caught in Japan for a long time during summer and autumn as they migrate northwards along the coast. Large traps, similar to the madragues in the Mediterranean region, with fences of bamboo wickerwork were used from the 16th century. Later, tuna were also caught with hand lines in deeper waters. In around 1890 fishermen began to employ long lines which proved to be a most effective gear; long lining had already been invented in Japan in the 17th or 18th centuries.

From the 1920s larger vessels with sizes of 100 - 200 GRT and 200 - 400 hp engines were built for tuna fishing and were equipped with freezers and cold-storage units for fishing in distant waters with voyages lasting at least 1 month. It was now possible to fish during the whole year and to extend long lining to the most productive grounds lying between 30 and 40°N and stretching eastwards beyond the date line. In this fishery the albacore *(Thunnus alalunga)*, a high-seas species, gained in importance, particularly for the canning industry which preferred these fish with

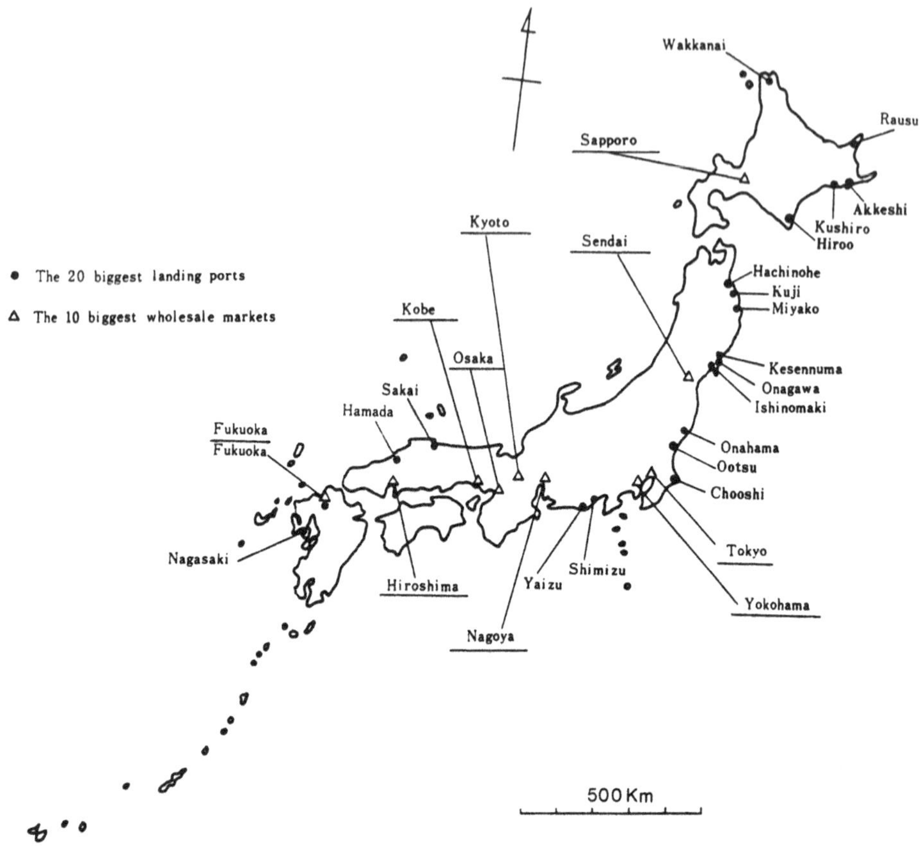

Fig. 103. Most important fishing ports and wholesale markets for fish in Japan. (Ministry of Agriculture, Forestry and Fisheries, Tokyo 1987)

their white meat. As the albacore is mainly concentrated in the upper 30 m of the ocean, pole-and-line fishing, using live bait such as anchovy, was frequently practised, as in the bonito fishery. For the tuna fishery in the distant waters of the central Pacific, especially in winter, it was economic to introduce motherships with freezers.

Mothership operations became increasingly important when, after 1930, a fishery on the hitherto unexploited equatorial tuna resources, mainly of yellowfin tuna *(T. albacares)*, was initiated. In certain areas bigeye tuna *(T. obesus)* made up 10% of the catches. As early as 1931-1933 motherships of the Japanese tuna fishery also visited the Indo-Australian region. The motherships carried small catcher boats on deck (Fig. 104C).

Immediately prior to World War II the Japanese tuna fishery, with about 2000 vessels, had reached an annual production of nearly 200 000 t, of

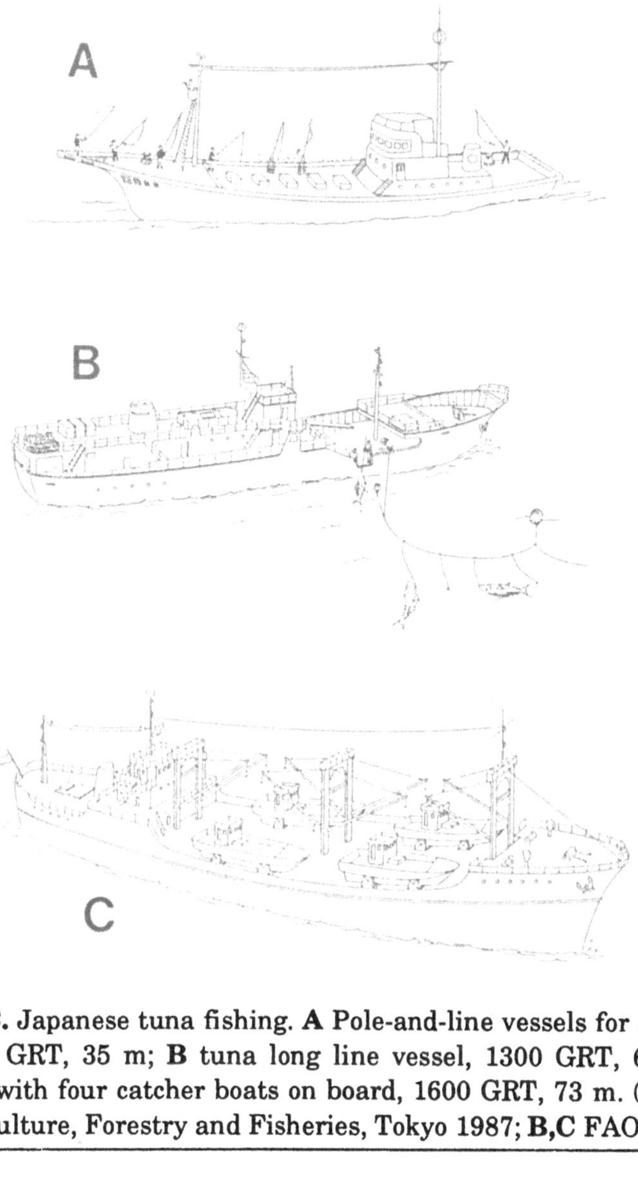

Fig. 104A-C. Japanese tuna fishing. **A** Pole-and-line vessels for skipjack fishing, 300 GRT, 35 m; **B** tuna long line vessel, 1300 GRT, 66 m; **C** mothership with four catcher boats on board, 1600 GRT, 73 m. (**A** Ministry of Agriculture, Forestry and Fisheries, Tokyo 1987; **B,C** FAO 1985)

which half consisted of skipjack and 15% of albacore. Of the catches 65% were made with pole-and-line, 25% with long lines and the rest with various other gear. Of the pole-and-line catches 95% consisted of skipjack and of the long line catches 70 - 90% were of yellowfin tuna.

After the war the tuna fishing fleet was rapidly rebuilt. In 1954 there were already more than 1200 ships of over 20 GRT. Through modifications in the licencing system the Japan Fisheries Agency encouraged, in the late 1950s, the construction of larger tuna fishing vessels, particularly for long lining in distant waters (Fig. 104). As a result, many new ships were built. The tuna landings, which were in the order of 250 000 - 300 000 t per year in the late 1950s, rose to more than 500 000 t in 1962 and amounted to 650 000 - 800 000 t in the 1980s with fairly constant proportions of skipjack (around 50%), bigeye (20%) and yellowfin tuna (15%).

Similar to the trawl fishery Japanese tuna fishing expanded gradually to all the world's oceans. With increasing distances from home ports the journeys became longer and more costly. Thus, some catches were sold in foreign ports and subsequently, bases were created for regular landings. The first of these bases was in American Samoa, where the US canneries were supplied with Japanese catches. In this region several Japanese motherships operated, each carrying up to eight fishing skiffs. From the Pacific tuna fishing spread into the Indian Ocean and reached the Atlantic in 1957. Penang and Singapore became major bases for the trans-shipment of tuna caught in the Indian Ocean. The first base in the Atlantic was Las Palmas. With the development of long lining for tuna in the tropical eastern Atlantic further centres were established in Dakar (Senegal), Abidjan (Ivory Coast), Tema (Ghana), Lagos (Nigeria) and other places. A large part of the tuna catch from the western Atlantic was landed in Puerto Rico for the US canneries.

Since the 1960s the Japanese have met with increasing competition from long line fisheries from Taiwan and Korea. These countries have the advantage of lower labour costs, an important factor in a fishery operating with lines that are often more than 150 km long and fitted with around 3000 hooks and thus involving extremely hard work during very long journeys. The hooks are equipped with bait of deep-frozen saury or squid, and the regular provision of bait is one of the problems of tuna fishing.

Recent Problems of Distant-Water Fisheries

In the interests of its vast distant-water fisheries Japan established the most extensive system of fisheries agreements with many other nations all over the world. A large number of joint ventures were formed for the production, processing and marketing of fish. Numerous Japanese vessels were registered to fly the flags of other states. Furthermore, Japan organized a substantial programme of technical and financial assistance for developing countries.

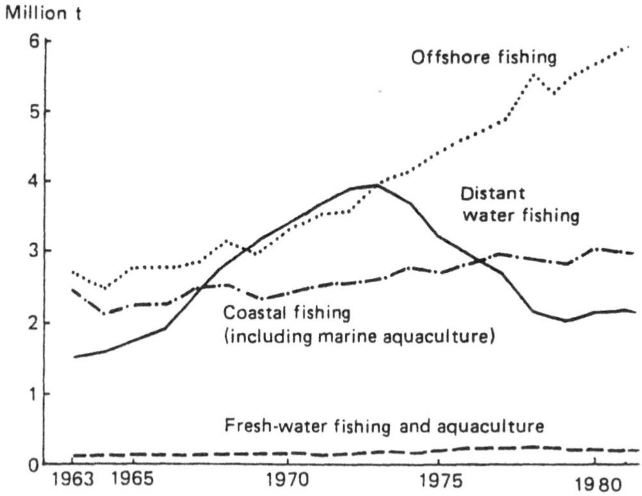

Source: Ministry of Agriculture, Forestry and Fisheries, "Annual Report Fisheries and Culture Prodction."

Fig. 105. Trends in the production of Japanese fishing sectors

When most coastal states established 200-mile Exclusive Economic Zones or Fishery Zones in the mid-1970s, the Japanese efforts to maintain fishing rights or to open new possibilities were intensified. However, this did not prevent serious problems for the distant-water fisheries. In 1974, Japanese ships caught about 4 million t of fish per year in areas now within the 200-mile zones of other states. Of this quantity around 1.8 million t were taken in USSR waters, mainly in the Sea of Okhotsk, and 1.6 million t in zones claimed by the USA in the Bering Sea and the Pacific. The Japanese authorities were now forced to negotiate contracts, mostly on an annual basis, with many countries for catch quota in exchange for returns, licence fees and reciprocal fishing rights. As a consequence, the production of the Japanese distant-water fisheries declined to about 2 million t in the early 1980s, half of the previous level (Fig. 105). Naturally, this system of annual negotiations makes these fisheries very susceptible to sudden reductions as a result of political or economic problems. This became evident when, in 1988, the USA excluded all foreign vessels from its fishing zone.

Nevertheless, as the total annual catch of Japan has increased to a level of almost 12 million t, this must be attributed to expansions in the pelagic offshore fisheries in the waters around Japan and in the coastal fisheries and aquaculture.

5.2.2 Russia - Soviet Union

With landings of around 11 million t per year the Soviet Union takes the second place in world fisheries after Japan. Fish production in Russia reached 1 million t in 1890 and in 1913 it surpassed 1.2 million t, a quantity which, after declines during World War I and the Revolution, was again reached in the mid-1920s. Immediately prior to World War II the annual fish production amounted to 1.5 million t. After heavy loss of vessels during the war, when one-third of the fishing fleet was destroyed, this level of production was again reached in 1950. The quantity doubled within the next decade. Then, with a major expansion of fisheries during the 1960s and early 1970s, catches reached an average of 9 million t by 1974 and more than 10 million t since 1984 (Fig. 99).

It is estimated that, at present, about 800 000 people are employed in the primary and secondary sectors of fishing. The USSR fishing fleet is the largest in the world. It comprises around 4000 vessels of over 100 GRT with an aggregate of 6.6 million GRT, about 50% of the world tonnage of fishing and related auxiliary ships. About 270 vessels are even larger than 4000 GRT. The fleet includes fish-collecting, transportation and processing ships to service the fishing fleets on the most distant fishing grounds around the world and to carry the catches back to the USSR. The largest is the floating processing plant *Vostock*, a ship of 35 000 GRT, 225 m long, carrying 14 catcher boats on deck and having a processing capacity of 300 t raw material per day. Many of the large factory ships were built in the German Democratic Republic and Poland.

A major change in emphasis was observed during the course of fisheries development. Until the early 1950s the largest part of fish production, between 600 000 and 900 000 t, came from inland waters. In the following decades the landings from marine waters increased dramatically, while the catches from inland fisheries rose only slightly to nearly 1 million t by the end of the 1980s. Now, roughly 50% of the total catch comes from the Pacific, 38% from the Atlantic, 4% from the Black Sea and Sea of Azov, and 8% from inland waters (3% from the Caspian Sea).

Basic Conditions for Fisheries Development

The Soviet Union covers a huge area and, with 22.4 million km^2, forms the largest country in the world. In an east - west direction it extends over some 9000 km and in the widest part from north to south is over 4000 km. Many rivers, streams, lakes and large landlocked seas create good conditions for inland fisheries. Possibilities for marine fisheries in home waters are less favourable. In fact, the coastline stretches over almost 43 000 km, but the greatest part of the shore faces the Arctic Sea in the north, with a rather low fish production due to cold climatic and oceanographic conditions. In the northwest, with the fish-rich waters of the Barents Sea, only the Murman coast, under the influence of the Gulf Stream, is accessible all year round, while the White Sea is blocked by ice during a major part of the year. Other coasts border on landlocked waters like the Baltic and the Black Sea with limited access to the open oceans. The coastal waters in the Far East are also covered with ice for several months. This is a very productive area, but it has the disadvantage of long distances to the main centres of consumption farther west. The Trans-Siberian railway with its branches is, therefore, since the end of the last century, a most important means for the distribution of fish products and many other goods in the country. As such, traffic lines are scarce in large regions, particularly Siberia; the large river systems also play a major role as transport routes.

A great demand for fish as a source of animal protein already existed in olden times. In Imperial Russia the main supply came from the Caspian Sea, from European freshwaters, and from the Murman coast and White Sea. Fish consumption was substantial, especially as the Lent rules of the Orthodox Church prohibited the consumption of meat on 132 days of the year. As in west European countries, monasteries were often engaged in fishing. Around the end of the 19th century Russia had a population of about 125 million. As the production of fish was not sufficient to satisfy the demand, large quantities of fish, mostly in the form of salted herring, were imported. In about 1910 Russia received about a quarter of all European herring catches and in 1913 almost 360 000 t of fish were imported, of which 285 000 t were salted herring. The imports were mainly routed via Königsberg, Libau and some other ports along the Baltic coast. Salted herring was even transported to the Asian regions. In comparison to these imports the exports were extremely low. They consisted mainly of luxury products such as caviar (1913: 600 t) which made up 50% of the value of all fish exports. Hamburg was the central place for this trade.

After World War I and the formation of the Soviet Union in 1917 the country suffered a serious crisis in its economy and in the food situation. In the following years it proved difficult to provide the essential supply of animal protein from livestock and fisheries in national waters. Therefore, within the framework of several 5-year-plans and as a part of integrated industrial development the government initiated a major scheme for the expansion of marine fisheries, starting in around 1930. The development was interrupted by World War II, but from the mid-1950s it was promoted with remarkable success, and soon Soviet vessels were fishing in most waters around the world.

Inland waters

The history of fishing in Russia dates back to rather early times; in the Ukraine it goes back to at least 3000 B.C. and near Lake Baikal even to 4000 B.C. (see Sect. 1.1.1). The numerous rivers, large streams and lakes played an important role in the colonization of the country, acting as the main lines of advance of man across the vast wilderness, where fish provided an essential food basis. In the Avarian Empire, which extended in the sixth century from the river Volga to the Baltic and to the river Elbe, salted fish, transported in wooden barrels and other containers, was already a very important commodity. Fisheries were also frequently mentioned in documents from the principalities of Novgorod and Kiev. While such activities in the former extended to the lakes Ladoga and Onega, and even to the White Sea, the development of fishing in Kiev progressed along the large streams to the shores of the Black Sea. Salting and drying of fish were the main preservation methods used, but the cold winters could be, and still are being, used for the distribution of naturally frozen fish. Moscow, as a large consumption centre, received much of the catches.

Further east, in Siberia, fisheries development was retarded not only by unfavourable climatic conditions but also by the lack of people and transport communications. The largest production (30 000 t in 1913) came from the Ob-Irtysh system, especially the Gulf of Ob which is an important fishing area for Arctic cod *(Boreogadus saida)*, navaga *(Eleginus navaga)* and various salmonids. At around the same time the catches in the Yenisey amounted to 6000 t and in the Lena to 4500 t per year. In the lower courses of the Siberian rivers fishing with traps and seines concentrated on highly valued large coregones like the njelma *(Stenodus leucichthys nelma)*, muksun *(Coregonus muksun)* and omul *(C.autumnalis)*. Fishing from sandbanks in the rivers and from dams in the narrower side-waters began with the breaking up of the ice, usually at the end of May, when the fish species, one after the other, started their upstream migration for spawning. Various species in the river estuaries also remained during winter and could be caught with harpoons, hand lines and nets through holes in the ice. Most rewarding was the fishing for sturgeon *(Acipenser baeri)* and stirlet *(A. ruthenus)*, carried out with hooks on long lines towed over the bottom. Cyprinids, like *Idus idus* and other freshwater fish, like pike *(Esox lucius)*, perch *(Perca fluviatilis)* and roach *(Rutilus rutilus)*, were caught further upstream. In the eastern Siberian rivers of Lena and Anadyr and their tributaries fishing for Pacific salmon *(Oncorhynchus* spp.*)* provided good catches. Beluga or white whales *(Delphinapterus beluga)* and seals were hunted or caught in large nets in the lower courses of the large rivers.

Originally, fishing and hunting were entirely in the hands of indigenous people like the Yenisei, Ostyaks, Koryaks, Samoyeds, Chukchi and others. These natives caught the fish with simple but effective gear like bow and arrow, spears and nets. For many of them fish was the main staple food, eaten raw or air-dried to supply them during the long winters; salt was rather scarce. When Russian traders and settlers advanced along the rivers, the natives bartered or sold part of their catches to them. Many natives became indebted to middlemen or were contracted by Russians as fishermen or workers. On the other hand, Russians started to participate directly in the fishing. From trade posts on the riverbanks catches were shipped to larger settlements. Since the 1960s fish production, particularly the fisheries for sturgeon and coregones, was affected by the construction of dams for hydroelectric power generation, e.g. in the river Yenisey.

Fishing in Landlocked Seas

Until the 1930s by far the largest part of the Russian fish production came from the **Caspian Sea** (1913: 65%, 1930: 50%). Prior to the Revolution the annual fish catches in this region were in the order of 600 00 t, as compared to 220 000 t in the 1870s and 350 000 t in the 1980s.

The Caspian Sea, with an area of about 370 000 km^2 the largest inland water in the world, is rather shallow in its northern part where silting is caused by the rivers Volga and Ural and produces salt marshes along the coast (Fig. 106). In the south the sea is up to 980 m deep. During winter the northern part is covered with ice. Large variations in water temperature and salinity can be observed. Below 600 - 700 m depth the water is low in dissolved oxygen and hydrogen sulphide limits the penetration of life. The water level varies widely under the influence of river floods in spring and desiccation in summer. Furthermore, a serious fall in sea level has taken place due to climatic changes and was much accelerated since the 1930s by anthropogenic influences.

Russian colonists and fishermen had already advanced along the river Volga to the Caspian Sea long before the city of Astrakhan was conquered

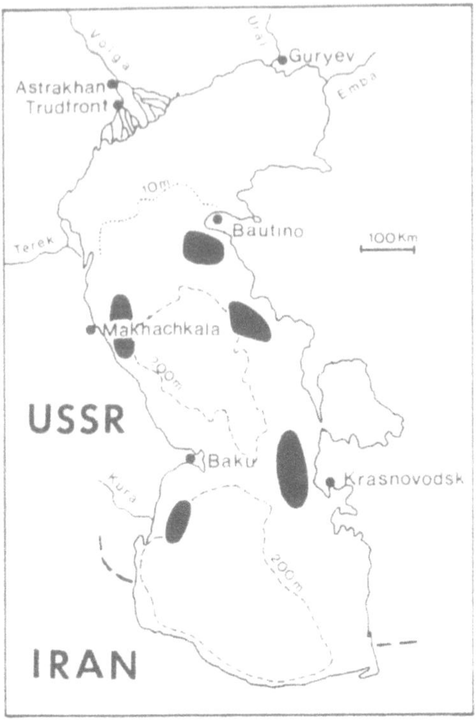

Fig. 106. Caspian Sea with main Soviet fishing ports. *Black*, main fishing areas for kilka

by Russian troops in 1557. Rich fish resources in the Volga delta, along the shores and in other rivers provided a major incentive for the Russian colonization around the Caspian Sea. Fisheries extended along the west coast to Daghestan and Azerbaijan where the estuaries of the rivers Terek and Kura proved to be very productive fishing areas. Already, by around 1800, hundreds of boats sailed from Astrakhan in spring to the entrance of the Emba River in Kazakhstan for sturgeon fishing. On the Ural and neighbouring rivers fishing was mainly in the hands of the Ural Cossacks who settled in the region around Guryev from the end of the 16th century. During the 19th century the Russians gained control of all coastal areas. Even on the Persian coast in the south a long-term fishing concession was granted by the Russian authorities to a private enterprise. In 1928 a joint Russian-Iranian company was formed, and it was not until 1953 that Iran obtained full authority over the fisheries on its territory.

The fishery in rivers, in the first place in the Volga delta, and in coastal waters concentrated initially on a number of highly valued fish species, in

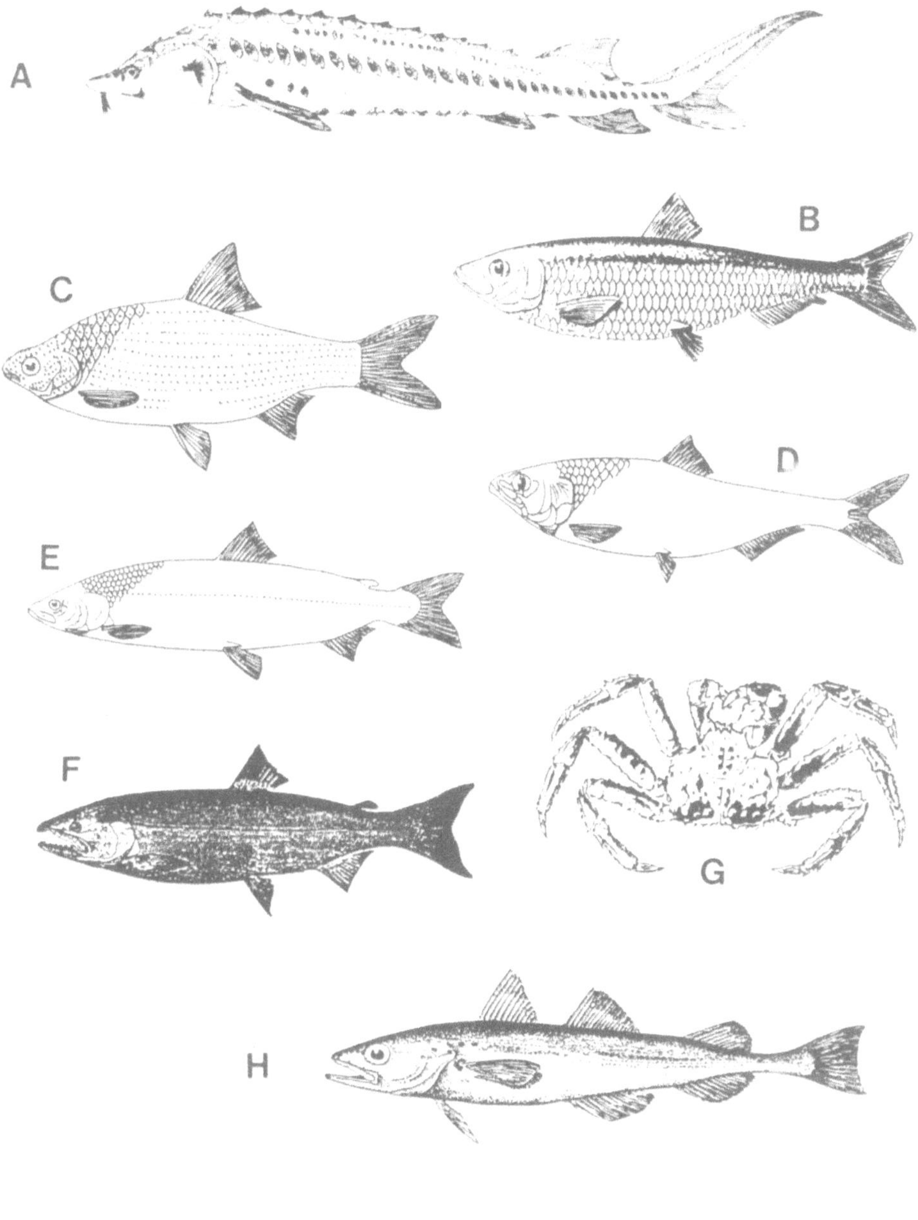

Fig. 107A-H. Some commercial species of importance in USSR fisheries. **A** Russian sturgeon *(Acipenser güldenstädti)*, max. 210 cm; **B** kilka or tyulka *(Clupeonella deliculata)*, 20 cm; **C** vobla *(Leuciscus rutilus caspicus)*, 20 cm; **D** shad *(Caspialosa pontica)*, 60 cm; **E** njelma *(Stenodus leucichthys nelma)*, 70 cm; **F** Pacific salmon *(Oncorhynchus* sp.*)*, 100 cm; **G** king crab *(Paralithodes camtschatica)*, carapace >20 cm; **H** Alaska pollock *(Theragra chalcogramma)*, 60 cm

particular sturgeon, shads (called "herring") and Caspian roach, which entered the rivers and migrated upstream, mainly in spring but also in autumn for spawning (Fig. 107). In winter catches were also made under the ice. Other species sought included carp *(Cyprinus carpio)*, bream *(Abramis brama)*, pike *(Esox lucius)*, perch-pike *(Lucioperca sandra)* and silurids. In earlier times traps and set nets were the main gear. Later, drift nets and seines became the most frequently used catching methods.

With annual catches of about 40 000 t in the first half of the 19th century and 28 000 t in 1913, sturgeons were the most economically important species for production of caviar from the eggs and glue from the swim-bladders. Up to 6000 t of caviar, over 80% of the world production, were manufactured yearly and exported to all parts of the world. The meat from the backs of sturgeon was dried and salted as "balyk", a product highly esteemed in Russia and abroad. The catches consisted mainly of Caspian sturgeon *(Acipenser stellatus*, Russian, sevruga), Russian sturgeon *(A. güldenstädti*, Russian, osetr) and great white sturgeon or beluga *(Huso huso)*, which provides the best quality caviar. The yields of the fishery varied rather widely and declined specifically during the Revolution and World War II. In the 1930s and 1950s production amounted to about 13 000 t per year. The general decrease was partly the result of overexploitation of the stocks, accelerated by the development of a proper sturgeon fishery in the Caspian Sea at the beginning of this century. Another reason was the construction of large dams and reservoirs for hydroelectric power schemes at Volgograd and Kuybishev on the Volga and in other rivers in the 1950s; these blocked the upstream migration of fish with rather detrimental effects on their reproduction. Also, the drainage of freshwater for irrigation contributed much to ecological changes in the river deltas and to the increasing desiccation of the Caspian Sea. Only the establishment of numerous stations for the artificial rearing of sturgeon, particularly in the region of the Volga delta, could remedy the situation. Also, offshore fishing for sturgeon was forbidden in 1962. With these ameliorations the improved sturgeon stocks have provided again, in recent years, annual catches of about 20 000 t with a production of more than 10 000 t of caviar, a most valuable source of revenue in foreign currency.

Another major sector was the catching of shads, of which 17 different species are found. The most important are *Caspialosa caspia* and *C. volgensis* which, together with several other species, provided, in 1912-1917, annual catches of more than 300 000 t. Under the influence of excessive fishing the catches declined to around 130 000 t in the 1930s and 1940s. Also, these resources were much affected by the construction of dams in the rivers and by subsequent environmental changes, so that the catches decreased further to 50 000 t per year in the 1950s and to a few thousand tons at the present time.

The vobla or Caspian roach *(Rutilus rutilus caspicus)* fishery in the coastal waters and rivers at the northern end of the Caspian Sea showed the same sequence of events. Its production reached 140 000 t in 1913 and a maximum of 260 000 t in the 1930s but fell to less than 100 000 t in the 1950s and around 10 000 t in the 1980s.

It is difficult to assess to what extent pollution has contributed to the decline of these resources. With the underwater oil exploitation near Baku and off Turkmenistan, chemical industries and the intense traffic of ships, water pollution presents serious problems in this landlocked sea without effluents.

While catches of these valuable anadromous species declined strongly, a marine fishery on kilka *(Clupeonella deliculata, C. engrauliformis, C. grimmi)* developed in the Caspian Sea from 1925, moving progressively from the north to central and southern waters. Originally, these small clupeoids were caught in traps and large beach seines and were salted for transport in barrels. After the war fishing was intensified with the use of conical lift nets during the night, when fish were attracted by light. Kilka catches reached 20 000 t by 1950 and increased sharply to more than 400 000 t in the early 1970s after the introduction of industrial fishing with the advanced technique of pumping light-attracted and electrically stupefied fish on board the vessels. A fleet of medium-sized freezer-seiners was built and several factory ships of the *Dnepr* type (14 000 GRT) were put into operation for freezing the catches in blocks, preserving the fish in tins and producing fish meal and oil. The factory ships were supplied with catches from the seiners and processing was increasingly carried out at sea.

In contrast to the river and coastal fisheries for anadromous species and fish-culture activities which are in the hands of fishermen's kolkhozes, the kilka fleet is operated by a single state enterprise administered by "Kaspryba" at Astrakhan. As the site of a huge combine with large processing facilities Astrakhan is still by far the most important fishing centre of the region. However, because of rapid silting the large fishing vessels had to be stationed at the port of Trudfront near Astrakhan and at other harbours closer to the fishing grounds further south (Fig. 106).

Fishing in **Lake Aral** began later than in the Caspian Sea region but, as with the exploitation of sturgeon, here only shyp *(Acipenser nudiventris)* were caught, in the delta of the Amu Darya and the mouth of the Syr Darya. A larger scale commercial fishery developed after the construction of the Orenburg-Tashkent railroad in 1905, which allowed the transport of catches out of the deserted area. A maximum in fish production of 40 000 t was reached in around 1915 but at that time sturgeon catches had already declined. Instead, the catches consisted mainly of freshwater bream, vobla, carps and other cyprinids adapted to low salinities and to large variations in water temperature. Conditions for fishing are seriously deteriorating because the lake, which had a surface area of 64 000 km^2 and an average depth of only 14 m, is drying up rapidly, a development accelerated by drainage of water for irrigation.

Many other adjacent shallow lakes with links to Lake Aral are also fished during the flood season, mainly for carp. This species also provides more than 75% of the catches in **Lake Balkash**, a shallow water where fishermen from Bessarabia and the Caucasus settled. Here, the total catches amounted to 13 000 t in the 1970s. In contrast, **Lake Baikal**, further east, is, with a depth of 1700 m, one of the deepest lakes in the world. The catches of 12 000 t (in 1940) consisted to 75% of highly valued omul *(Coregonus autumnalis)* but declined in later years due to overexploitation and possibly in connection with pollution by effluents from woodpulp factories.

In the **Black Sea** and **Sea of Azov** fisheries have a much longer tradition than in the Caspian Sea area although production was smaller (Fig. 108). During ancient times the wealth of many Greek colonies along the northern shores of the Black Sea was based on fishing and fish trade. Strabo (63 B.C. - 20 A.D.) described this fishery. Istrios on the Danube estuary, Olbia on the Bug and Tanai on the river Don were famous places for fishing, mainly for sturgeon, and so were the shallow waters of the Sea of Azov, called the "Maeotic Swamp". Demosthenes (384 - 322 B.C.) referred to the trade of salted fish in earthenware jars from Feodosiva and picked up at Pantikapeja in the Strait of Kerch. The salt was procured from Krymsk in northern Crimea.

Later, fish exports to Greece, Italy and other places suffered from the advance of the Turks who conquered Constantinople in 1453. The fisheries were revived with the immigration of Russian fishermen along the main rivers. New trade links were then established to the north. The Russian fishermen concentrated on the fishery for sturgeon in the Sea of Azov, in the river Kuban and the Danube delta. The Caspian sturgeon *(Acipenser stellatus)* was the most important species. Sturgeon catches in the Sea of Azov alone amounted to nearly 2000 t per year in around 1850.

Fig. 108. Major USSR fishing ports on the Black Sea and Sea of Azov

Another important fishery exploited the abundant resources of various species of shad *(Caspialosa pontica, C. maeotica, C. tanaica)*, particularly in the narrow Strait of Kerch, when, in spring and autumn, these fish migrated from the Black Sea into the Sea of Azov and vice versa. In earlier times this fishery was carried out mainly with seines and traps but since 1840 set gill nets were mostly used.

Since 1900 a fishery on anchovy *(Engraulis encrasicolus)* was introduced, which soon became more important than the one on *Caspialosa*. After their summer feeding in the Sea of Azov the anchovy could be caught in prime condition during autumn in the Strait of Kerch on their migration back to the overwintering grounds in the Black Sea. Also, the stocks of tyulka *Clupeonella deliculata, C. cultriventris)*, which even enter the lower courses of rivers, like the river Don, provided good catches. In the Gulf of Taganrog and other waters with very low salinities perch-pike, bream and roach prevailed.

A large part of the catches of small fish were sun-dried or salted and widely distributed in Russia as a popular ingredient for fish soups and other dishes. Kerch and Azov were very important fishing and trading centres. During certain periods up to 70 000 fishermen were working in the Sea of Azov and large seines were even towed with the help of horses or oxen.

Overexploitation of the resources led to a serious decrease in fish production from the Sea of Azov and its tributaries; this was estimated at 88 000 t in 1893. In particular the percentage of highly valued food fish declined while the percentage of less-esteemed fish increased. As a result of forced protection during the war years fish stocks increased again somewhat, but soon fish living conditions deteriorated in connection with the diversion of freshwater from the rivers for irrigation and industrial purposes, leading to higher salinity in the Sea of Azov. However, this did not prevent the further development of modern anchovy and tyulka fisheries. Eggs of Baltic herring were successfully transplanted to the Sea of Azov and also to the Aral Sea in the mid-1950s.

Generally, the Sea of Azov, a rather shallow inlet of the Black Sea (maximum depth 13 m) with extreme variations in temperature and salinity, is much more productive than the Black Sea (Fig. 108). It serves as a feeding ground for many Black Sea fish during summer. As in the Caspian Sea and Lake Aral, the fish fauna is rather unique with many endemic species of the Pontic-Sarmatic region which was a large brackish-water lake during the Tertiary period. With major geological changes in the early Quarternary period and the formation of the Bosphorus, Mediterranean species migrated into the Black Sea and some even into the Sea of Azov. Today, about 60% of the fish species in the Black Sea are Mediterranean, like bonito *(Sarda sarda)*, bluefin tuna *(Thunnus thynnus)*, horse mackerel *(Trachurus trachurus, T. mediterraneus)* and mackerel *(Scomber scombrus)*, and about 40% are Sarmatic, like sturgeons, *Caspialosa* spp. and *Clupeonella* spp..

Also, in the Black Sea fish are subject to special living conditions. With a surface area of some 462 000 km^2 this sea is rather large; it is 2200 m deep, but the zone of fish life is limited to the upper water layers since below 125 m depth (200 m in coastal waters) the sea contains high levels of dissolved hydrogen sulphide and little or no oxygen. Large quantities of freshwater and organic materials are carried into the sea by the rivers Danube, Dniester, Dnieper and other streams. Sediments sink to greater depths where the highly saline water is poor in oxygen as the inflow from the Mediterranean through the Bosphorus is very limited and no convection occurs with the low salinity waters near the surface. Since no oxidation is possible, degradation of organic matter is effected by bacteria producing hydrogen sulphide. Thus, the demersal fish resources in the Black Sea are restricted to coastal waters and the pelagic fish to the upper water layers.

Exploitation by commercial fisheries began along the northwestern coasts of the Black Sea where the river deposits formed wide shallow areas. Odessa developed to become a centre of this industry; and the first canneries were built here in 1870. Fishing for sturgeon, mainly *Acipenser güldenstädti* and *A. stellatus*, concentrated on the waters between the rivers Danube and Dnieper. Several species of shad *(Caspialosa* spp.*)*

were caught in beach seines and traps, especially in spring when they entered the rivers to spawn. Many other species like mackerel, horse mackerel and mullet *(Mugil cephalus, M. chelo)* also formed part of the catches. Altogether the landings remained small in comparison to those from the Sea of Azov and amounted to nearly 29 000 t in 1910. Of this quantity 16 000 t came from the northwestern Black Sea and the coasts off Bessarabia and 13 000 t from the Crimean waters. Other areas were of no importance.

With the construction of larger vessels and advances in gear technology the fisheries expanded further offshore and to other parts of the Black Sea. The total annual landings from the Sea of Azov and Black Sea increased to more than 200 000 t in the early 1960s and reached an average of 365 000 t during the last 10 years. This large increase was mainly due to the introduction of purse seining and midwater trawling and to the more intensive utilization of the anchovy, mackerel and horse mackerel resources near the Strait of Kerch, along the Caucasian coast and off the Turkish shores in the south. In particular, anchovy is caught in quantities of over 200 000 t per year using modern methods, including the locating of shoals from aircraft.

Since the 1960s the ports of Odessa, Sevastopol and Batumi developed to become major bases for large trawler fleets employed in distant fisheries in many parts of the world, particularly in the South Atlantic and Antarctic seas.

Another landlocked sea, the **Baltic**, was always of minor importance to Russian fisheries. Total annual landings amounted to more than 40 000 t in 1913, around 30 000 t in 1926 and, in the 1970s, reached a maximum of over 300 000 t. They consisted mostly of herring *(Clupea harengus)*, cod *(Gadus morhua)* and sprat *(Sprattus sprattus)*. Before World War I about 18 000 fishermen from the Baltic provinces were engaged in spring to catch the herring in large traps. At least 20 000 t per year were caught. In the Finnish Gulf fishing was carried out under the ice and, in inshore waters, often with the help of horses to tow the nets. In recent years the eastern parts of the Baltic, up to the island of Gotland, have been intensively fished, but catches have declined to a little more than 200 000 t owing to the poor state of the cod stock.

Around 1955 a fleet of herring drifters was transferred from Murmansk to Kaliningrad (formerly Königsberg) and Klaipeda (Memel) as these ports were situated closer to the fishing grounds in the northern North Sea and around Ireland. For similar reasons Kaliningrad was chosen, in the 1960s, as a second centre for the operation of long-distance trawlers fishing in the North and Central Atlantic.

Northeast Atlantic and White Sea

Abundant resources of salmon *(Salmo salar)*, cod, herring and other species along the Murman Coast and in the White Sea were already being fished for from early times (see Sect. 1.1.1 and Fig. 109). In the 10th century Novgorod, originating from a Viking settlement, concluded agreements with the Norwegians for fishing in the north. Some reports speak of perhaps several thousand boats and 30 000 fishermen who were engaged, during the 16th century, in hand-line fisheries for cod, haddock, halibut and plaice along the Murman coast and in Finnmarken. A large part of the catches was salted or dried to stockfish. Salted salmon was exported to the Netherlands, Flanders and France. Most of the fishermen were Pomores, a people, who lived on the shores of the White Sea and on the Kola Peninsula. Pomores travelled each summer to Norway for bartering. From the 16th century they also went in boats of an old Viking design to Spitsbergen and Novaya Zemlya to hunt for seals and white whales (belugas).

At the end of the 18th century the fisheries began to decline for economic reasons and by around 1900 there were only 960 boats with 4000 fishermen left on the Murman coast, producing a catch of about 10 000 t. Many fish were imported from Norway. Arkhangelsk was the centre for the fish trade and here herring from the White Sea, particularly the Kandalaksha Bay, navaga cod, and Atlantic salmon caught with drift nets and seines in the rivers Mezen, Dvina and Onega were marketed. Harp seal skins from the sealing grounds at the entrance to the White Sea were also traded in Arkhangelsk.

So far the fisheries had been limited to coastal waters and produced relatively small quantities. In 1924 a commercial fishery in the Barents Sea began to operate from Murmansk with 15 trawlers landing 6000 t. This fishery grew steadily during the following years and, by the end of the 1930s, the fleet consisted of some 90 trawlers, catching more than 200 000 t per year. Until 1950 the trawlers fished mostly in the southern Barents Sea for cod, haddock and flatfish, but then the fishery expanded to Bear Island and Spitsbergen, and deeper grounds were exploited for redfish *(Sebastes* spp.*)*.

In the mid-1950s the port of Murmansk, which was not blocked by ice in winter, was developed to a very large industrial centre with extensive

Fig. 109. Fishing harbours and fishing grounds of the White Sea. *Hatched*, main herring fishing areas; *dotted*, sealing grounds

installations for fish processing and as a base for the vessels fishing in the North Atlantic. Many new factory trawlers were commissioned and fishing operations were improved by the implementation of scouting vessels and close cooperation within flotillas. By 1960 the catches of demersal fish had increased to 650 000 t.

Since the early 1950s fleets of small drifters, accompanied by motherships and auxiliary vessels, operated from Murmansk to participate in the herring fishery in the North Atlantic. The herrings were salted and barrelled at sea and transported home by carriers. Large Soviet fleets fished in the North Sea and the waters west of Scotland. In around 1955 these ships were transferred to Kaliningrad and other Baltic ports. Most herring catches of the Soviet fishery were, however, obtained further north in the Norwegian Sea and off the east coast of Iceland and were landed in Murmansk. Herring catches in these waters were largest in 1965 with 500 000 t and the total herring landings peaked in the same year with 660 000 t. But, as a result of the vast overexploitation of the herring resources in the Norwegian Sea and in the North Sea, by fleets of many nations, the Soviet herring landings from the Northeast Atlantic declined in around 1969 to insignificant quantities. Only in the Baltic could a catch of slightly more than 100 000 t be maintained.

The total annual landings of all fish species in the USSR from the Northeast Atlantic varied between 1.4 and 2.6 million t in the 1970s and early 1980s and, since 1986, have decreased to 950 000 t.

Originally, fishing in eastern Siberia concentrated almost entirely on the rich salmon resources (Fig. 110). Chum salmon *(Oncorhynchus keta)*, pink salmon *(O. gorbusha)* and sockeye *(O. nerka)*, entering the streams and rivers in May to October from the North Pacific in enormous numbers on their spawning migrations, mainly provided the essential food basis for the indigenous population in a hostile environment. Salmon were so abundant that usually the whole supply for the long winter period, 500 fish for a family and their dogs, could be obtained within 1 week. In shallow river waters the fish could even be caught by hand or in seines operated from the banks or from canoes. In the upper courses of rivers with fast running waters, like on Kamchatka, village communities erected timber dams across the streams and each family had a trap. Later in the year the salmon migration became insignificant but some waters remained open all year round and also provided fish, mainly coho salmon *(O. kisutch)*, in winter. People were absolutely dependent on the fish. Lack of catches, for instance through loss of fishing installations, often caused hunger and death in winter. Originally, fish supplies for the cold season were buried in holes in the soil where they decomposed slowly. Later, this method was used only for dog feed, always needed in large quantities, while, for human consumption, the fish were dried in the open air since salt was hardly available.

When the Russians penetrated the region during the 18th century, the early settlers lived, like the indigenous people, mostly on fish. They went into trading with the natives, copied the endemic catching methods and developed their own installations with indigenous workers. The salmon fishery reached a commercial scale in around 1885, when it first became possible to preserve the fish for storage and transport with salt, and from 1900 also by canning in newly built factories. The salmon fishery expanded, both in time and space, as it was extended from the rivers into the estuaries and adjacent coastal waters, where fish at the beginning of their upstream migration were of the highest quality. Very large traps were erected in these waters to supplement catches made with beach seines.

The construction of the Trans-Siberian railway contributed greatly to the economic development of the region. In 1912 about 12 000 t of salmon were transported by rail to Europe. It was estimated that in around 1914 the total Russian salmon production amounted to about 150 000 t per year, of which 65 000 t came from Kamchatka and 50 000 t from the Amur River district. This level was also maintained during later decades

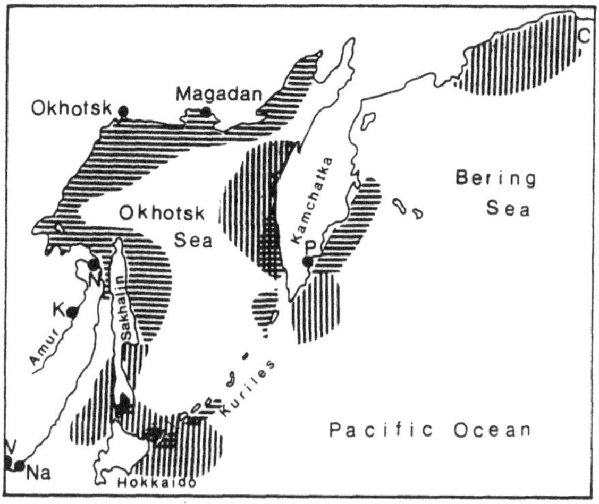

Fig. 110. USSR fishing areas for Pacific salmon *(horizontal hatching)* and king crab *(vertical hatching)* in the Okhotsk Sea and Bering Sea: *C*, Cape Navarin, *K*, Komsomolsk, *N*, Nikolaievsk, *Na*, Nachodka, *P*, Petropawlovsk, *V*, Vladivostok

until the 1950s with the help of conservation measures and artificial fish propagation. Later, the annual catches decreased to 70 000 - 90 000 t. The main centres of this industry, with numerous canning factories, salteries and refrigeration plants, are Petropavlovsk on Kamchatka, and Nikolaievsk and Komsomolsk on the river Amur (Fig. 110). At present there are 17 salmon farms in operation at sea near Sakhalin and the Kuriles for the production of pink and chum salmon.

As described in Section 5.2.1, the Japanese increasingly penetrated Russian territories with their fishing activities, until in around 1928, up to 80% of the fishing lots were in their hands. Japanese companies also bought a large part of the Russian production of salted salmon and salmon caviar. After the establishment of state enterprises and the systematic promotion of Russian settlements the Japanese influence was repressed. Finally, in 1945, Japanese fisheries were expelled from southern Sakhalin and the Kuriles, former Japanese territories.

In the Amur area rich resources of two sturgeon species *(Huso dauricus,* and *Acipenser schrenki)* were used for the production of black caviar, but this fishery had already decreased by the 1920s.

With the development of salmon fishing in coastal waters off Kamchatka, the Kuriles and Sakhalin, a fishery was also started to exploit the king crab *(Paralithodes camtchatica)*, caught in entangling set nets and processed in canning factories on land (Figs. 107 and 110). For better mobility catches were mainly processed on board large factory vessels from the early 1930s. The crab fishery reached a production of 25 000 t in the 1930s and annual catches of between 20 000 and 45 000 t were obtained in later years, although it appeared that local stocks were rather vulnerable to heavy exploitation and regulatory measures had to be introduced. In 1959 the Soviet crab fishery extended to the eastern Bering Sea, particularly the Bristol Bay area and the waters near Kodiak Island in the Gulf of Alaska. This fishery ended with the introduction of the EEZ by the USA.

As the salmon fishery was limited to the season from May to October, there was also an incentive to develop the exploitation of other marine fish species. The first was the herring *(Clupea pallasii)*, which was abundant along the shores of the Okhotsk Sea, Kamchatka and Sakhalin. The landings of this fishery increased from 7500 t in around 1910 to 50 000 t in 1940 and reached a peak in the mid-1950s with more than 300 000 t. After 1964 pair trawling and purse seining largely replaced drift netting. In the 1970s and 1980s the annual production mostly varied between 100 000 and 200 000 t. Magadan and Okhotsk are major ports for this fishery, which supplies Siberian cities with salted herring. Other species fished increasingly since the 1930s with seines and trawls were several flatfish, particularly the yellowfin sole *(Limanda aspera)* and the Pacific cod *(Gadus macrocephalus)*.

Vladivostok, a port which can be kept open by ice breakers during winter, developed to become the main fisheries centre of this region. The building up of a trawler fleet started at the end of the 1920s, and by 1935 the port hosted already some 30 trawlers and a few freezer-factory vessels. A major expansion of the fleet began in the early 1950s, and in around 1955 the fishery spread from the coastal waters into the open Bering Sea. In 1959 fleets of Soviet factory ships with trawlers began to fish around the Pribilof Islands and in Bristol Bay off Alaska. Initially, the trawlers concentrated on the fishery for flatfish, especially yellowfin sole, but when catches declined they shifted to ocean perch *(Sebastes alutus)*. Dramatic increases in the number of ships were achieved in the early 1960s with the introduction of large stern trawlers, motherships, factory vessels, freezer-transporters and other auxiliary ships. At the end of the 1970s the Far East fishing fleet comprised more than 1000 high-seas vessels. Besides Vladivostok, Nachodka had developed to become a major fishing port.

Fish production increased to high levels mainly through fishing for Alaska pollock *(Theragra chalcogramma)*, which was initially used for the production of fish meal and oil and later especially for fish paste (Fig. 107). This fishery grew very rapidly producing insignificant quantities during the 1950s but more than 100 000 t in 1960 and 2 million t by the mid-1970s. In recent years the annual catch was even over 3 million t. During the early years this fishery concentrated on the eastern Bering Sea. When, after the establishment of fisheries conservation zones by the USA in 1977, considerable fees were requested for access to these waters, the Soviet fishing activities soon reverted more and more to the EEZ claimed by the USSR in the Northwest Pacific. The bulk of the catches was made by small trawlers, operating for 3-4 months at sea with the help of motherships, and by large stern trawlers.

Distant-Water Fisheries

After the end of World War II the Soviet fishing fleet was rebuilt. From 1947 until 1957 the reconstruction programme concentrated on small drifters, trawlers and seiners for coastal and medium-distance fisheries. Most of these ships were built in the German Democratic Republic as war indemnity. Many of them were multipurpose vessels which often operated in flotillas with motherships and with carriers to land the catches. Since 1954, following the construction of the British *Fairtry*, larger factory trawlers for fishing in distant waters were ordered (Fig. 111). The first were ships of the *Pushkin* class (2480 GRT, 85 m), built in the Federal Republic of Germany. From 1960 bigger stern trawlers of the *Leskov* and *Mayakowskij* classes (2500 - 3200 GRT), built in Poland and the USSR, and, later, ships of the *Kosmos* class, were added to the fleet. Further factory trawlers, the *Tropik* class (2435 GRT, 79 m), built in the former German Democratic Republic since 1962, were specially designed for operation in tropical waters. An improved version was introduced with the *Atlantik* class (2760 GRT, 82 m). Also, many other types of vessels were built in considerable numbers, like motherships (some up to 20 000 GRT), freezer- and refrigerator-carriers, floating factories, supply ships, scouting vessels, research ships, etc.

The fleet not only increased enormously in the number of vessels and in tonnage but also the production of fish per man and per year tripled from 1960 to 1975. In 1978 the fishing fleet, without auxiliary ships, comprised 2558 high-seas vessels. Of these 1088 operated in the Far East, 707 from Baltic ports, 443 from Murmansk in the north, 229 in the Black Sea and 91 in the Caspian Sea. Of the 531 large factory trawlers 143 were sta-

Fig. 111. Soviet stern trawler

tioned in the Far East, 201 in Baltic ports, 156 in Murmansk and 31 in the Black Sea. With a share of more than 80% trawling is by far the most important fishing method used by the USSR fisheries.

With the expansion of the fleet Soviet ships began to fish on many distant fishing grounds until a worldwide operation was achieved (Fig. 112). Already, in 1954, USSR vessels were making their first exploratory voyages to the waters around Newfoundland. From 1956 they fished on the Grand Banks and off Greenland mainly for cod and redfish. In 1961 they started an extensive herring fishery on the Georges Bank off Maine; this ended in around 1970 because of the depletion of the herring stock, which was being fished by several nations. Soviet fisheries also expanded to the South Atlantic and into the Indian Ocean, and from the Far East to fishing grounds in the Pacific, particularly off Alaska and Peru. In 1969-1970 USSR trawlers began to work in Antarctic waters, mostly the Atlantic sector, fishing for fin fish and, since the early 1970s, also for krill.

The Soviet Union entered into a large number of agreements with other countries to establish operational bases for the expansion of fisheries and for the formation of joint ventures. More than 20 million US dollars were spent per year in over 40 developing countries to ensure access to produc-

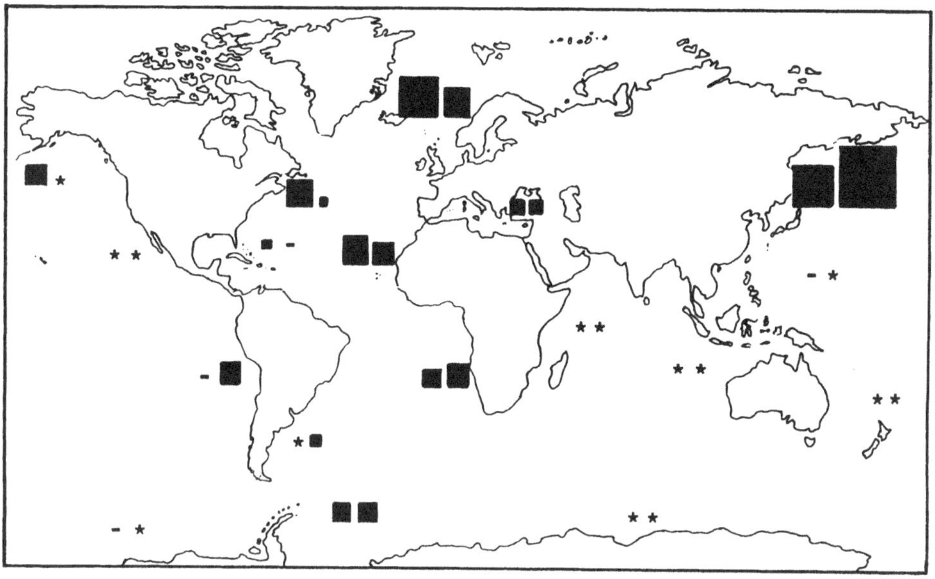

Fig. 112. Landings by USSR ships from various regions of the world's oceans in 1975 *(left column)* and 1985 *(right column)*. (FAO 1947-1989)

tive fishing grounds overseas. It was estimated that the USSR obtained more than 2 million t annually from such waters. Most of the catches were, and still are, transported to the Soviet Union, either via shore bases or directly reloaded on to carriers at sea. Some catches, mostly deep-frozen, were also sold in foreign countries.

In 1975 the total landings in the USSR from worldwide fishing in marine waters amounted to 9.45 million t. The distant water fisheries met with serious difficulties when, in 1977, many countries introduced 200-mile EEZs and the Soviet fishery lost access to many fishing grounds. Consequently, the landings decreased temporarily to less than 9 million t in 1978 (Fig. 99). The greatest declines in catches were observed in the Northeast Pacific, the Northwest Atlantic and the Northeast Atlantic (Fig. 112). However, the difficulties were overcome and the landings rose further to more than 10 million t in 1985. This was mainly achieved by intensification of the fishery in the Northwest Pacific and off Peru and Chile. Since access to fishing grounds on the shelf was now much more limited, the USSR fleets expanded their midwater trawling in offshore waters and their demersal trawling to depths below 1000 m on the continental slopes.

Under the influence of these changes, and also as a result of the depletion of a number of commercial fish stocks in the North Atlantic, a substantial alteration in the composition of the landings took place. Landings of valuable species like cod, haddock, saithe, hake, herring and redfish declined, while, often, less valuable species like jack and horse mackerel, blue whiting and Alaska pollock made up for the losses.

5.2.3 China

China, the third largest country in the world, with more than 1 billion people and very densily populated in its central and eastern provinces, has, at present, a fish production of over 10 million t per year. Fisheries have developed enormously during the last 40 years, with an increase in the total production from 500 000 t at the end of the 1940s to 10.4 million t in 1988 (Figs. 99 and 113).

It is possible that the catches before 1958 were higher than indicated. Early statistics were rather inaccurate and were certainly greatly improved when fishermen, who had been working in cooperatives, were amalgamated in 1958 with people's communes and organized into production brigades. About 4.5 million fishermen, many also working in agriculture during certain periods, belonged to such collectives in the 1970s. Around 860 000 of them fished in marine waters. In addition, only 20 000 people were employed by 17 state-run fishing enterprises.

It is difficult to estimate the number of fishing vessels. A very large number of small sampans and sailing junks, possibly more than 400 000, are used for this purpose. Motorization started during the 1920s, and today the number of motorized ships is estimated at 157 000. Most of these vessels are, however, small and only suitable for operation in inshore or inland waters. A high-seas fishery began only in 1985 and is now carried out with more than 50 larger stern trawlers.

Almost all of the fish production in China is consumed in the country. Fish and other animals from freshwaters are usually preferred by the consumer but a wide range of marine organisms, including sea-cucumber (trepang), jellyfish, copepods and seaweeds, are eaten as well. Fish are preferred fresh or, nowadays, frozen, while many other organisms are dried and/or salted in the traditional way. Shrimp paste and other fermented products have also always played an important role.

Accordingly, exports of aquatic products were never very large. In the 1970s they amounted to around 100 000 t/year and rose to 250 000 t in

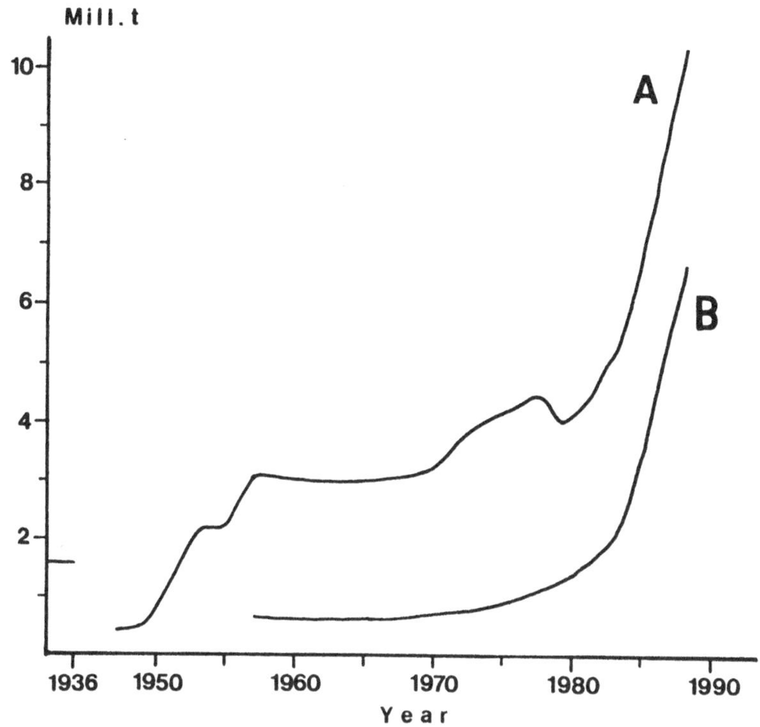

Fig. 113. Increase in fish production in the People's Republic of China from 1948 to 1988. **A** Total production, **B** production from aquaculture. (FAO 1947-1989; Bureau of Aquatic Products, Beijing 1985; Zhu de-Shan 1980)

the 1980s. Fresh carp and frozen shrimps were the most valuable items. In comparison, imports were negligible, amounting to about 20 000 t.

Marine Fisheries

The coast of China extends from north to south for some 16 000 km (Fig. 114). In the Pohai Sea, a shallow water semi-enclosed by the Shandong and Liaodong peninsulas and receiving large quantities of silt from the Yellow River, and in the Yellow Sea cold- and temperate-water species like cod *(Gadus macrocephalus)*, Pacific herring *(Clupea pallasii)* and flounder *(Paralichthys olivaceus)* prevail. Further to the south, in the East China Sea and the South China Sea, a large number of subtropical and tropical species are abundant. Here, a large part of the demersal catches consist of hairtail *(Trichiurus haumela)* and large yellow croaker

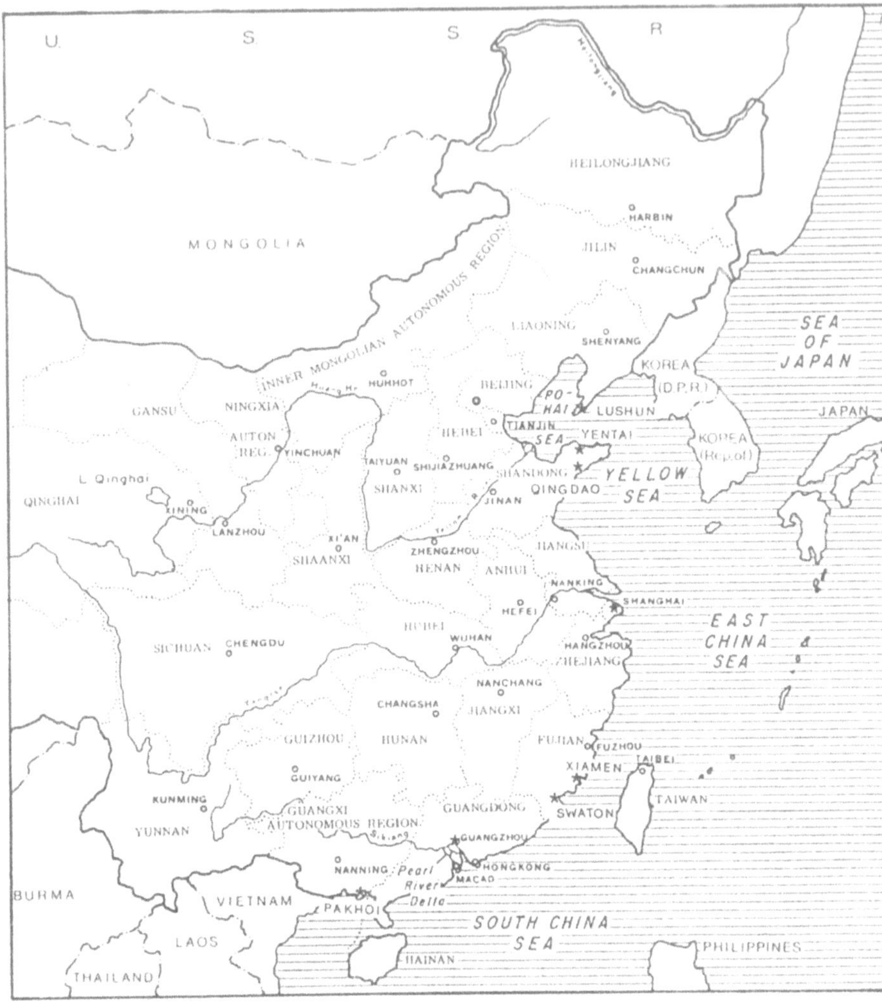

Fig. 114. China and adjacent seas. *Asterisks*, main fishing ports

(Pseudosciaena crocea). The shelf areas are quite extensive, particularly in the seas north of Taiwan.

In the traditional inshore fisheries a large variety of fishing gear was, and still is, used, ranging from hand harpoons and fish weirs to cast nets, stake nets, dip and lift nets, set gill nets and hand and long lines. These gear are operated from the shore, from bamboo floats and from junks up to 30 m in length. A junk of 40 - 65 tons often has 20 - 45 people on board. Several junks, assisted by sampans, collaborate to fish with large seines for yellow croaker and other species. Light to attract fish and noise to scare fish into the nets play a large role in fishing.

Coastal fisheries are most extensive in the Guangdong, Fujian and Zhejiang provinces but also off the Shandong and Liaodong peninsulas. In southern China they were hindered by piracy for a long time. For this reason fishing vessels operated mostly in groups for protection. Another obstacle is frequent typhoons. There are many small fishing localities along the whole coast with numerous places for salting and drying and for the production of shrimp paste, trepang, shark fins and many other items.

A substantial catch is harvested by trawling. This technique was introduced to China by Japanese trawlers sailing to the East China Sea in 1912 when they were expelled from home waters. Japanese fishermen eventually settled in certain places, especially in Shanghai. Motorization of Chinese fishing vessels began in the 1920s. In 1933 nine steam trawlers operated from Shanghai, which developed gradually to become the main fishing and fish trading centre, favoured by the vicinity of productive fishing grounds in the Yangtze estuary. Other places of major fish landings are Lushun (Port Arthur), Qingdao and Yentai (Cheefoo) in the north and Xiamen (Amoy), Swaton and Pakhoi in the south (Fig. 114). At the end of the Civil War only 600 small wooden trawlers were available, left by the Japanese. Some European type trawlers were donated by the USA. But in 1949 several ships left for Taiwan and there formed a core for further fisheries development. The construction of modern trawlers started in 1955, and soon frozen fish were exported to Europe and shrimps to Japan. In 1959 the number of coastal trawlers was estimated at 400, with four larger ships of 33 m length and 260 GRT. Following the Japanese example, most of the trawling is done by pairs of vessels jointly towing the gear. Shrimps are caught by beam trawlers with up to 22 small nets on long bars.

With the intensification of fishing on the shelf demersal fish stocks were overexploited and yields decreased in the Pohai Bay, Yellow Sea and East China Sea. The total production from these waters could only be maintained and increased by a shift in emphasis to file fishes (cantherines, *Navodon* spp.). With the expansion of line fishing and trawling in the South China Sea, where 300 - 400 different fish species can be observed in trawl catches, the demersal fish resources here also became fully exploited. For this reason, in the late 1960s, the government promoted the introduction of drift gill net fishing and purse seining with use of light to attract fish and to make use to of the abundant pelagic fish resources. Drift nets are mainly used for the capture of Spanish mackerel *(Scomberomorus niphonicus)*, Chinese herring *(Ilisha elongata)* and mackerel *(Scomber japonicus)* and the purse seines for round scad *(Decapterus maruadsi)*, small sardines *(Sardinella* spp.*)*, Pacific herring *(Clupea pallasii)* and many other species.

Through planned development the total annual landings of sea fisheries rose from about 1.5 million t in the mid-1950s to around 3.5 million t in the 1980s. About 90% of this quantity is caught by small vessels fishing within 20 miles of the coast. The growth in landings from marine waters has been reduced during recent years. Most of the demersal resources in Chinese home waters are already overexploited and the pelagic stocks fully utilized. High-seas fishing initiated in 1985, partly in the form of joint ventures with other countries like Guinea, Iran and the USA, might lead to some additional catches and income in foreign currency. But the real possibilities for substantial further increases in fish production lie in aquaculture.

Inland Waters

With a territory of 9.6 million km^2 China is a huge country and correspondingly large are the inland waters, estimated at 17 million hectares. They comprise rivers (about 50%), natural lakes (30%), reservoirs (11.5%) and ponds (8.5%). Most important are the large rivers, the Huang-Ho (Yellow River), Yangtze-Kiang, Sikiang and their affluents, and the great lakes, the Taihu Lake west of Shanghai, Lake Poyang in Jiangxi and Lake Tung-King in Hunan.

As described in Section 1.1.4, these waters have been used intensively since early times for fishing with a large variety of gear, used by Chinese farmers and fishermen with ingenuity and skill. Many of the traditional fishing methods are still employed, including fishing with tamed cormorants (Fig. 16) and with fish otters driving the fish into nets. The number of sampans and junks used in these fisheries was estimated at more than 300 000 and there may be more than 2.5 million people working in this sector.

Since ancient times farmers in the central plains of China have fished in summer when their fields, and even their reed houses, were under water during the high floods. With their families they used to live on boats for half a year and to eat mostly fish before they could return to agriculture.

Life became much easier when the water run-off was regulated by the construction of high dykes and by using the great lakes as reservoirs.

While Chinese farmers were forced to work as fishermen during the summer flood season, there were other groups of the population who were fully engaged in fisheries. These folk were of a different, obscure origin and were considered as an inferior class of people, often excluded from rural societies. The most important of these groups are the Tangka, who lived almost entirely on house and fishing boats, mainly in Guangdong, and particularly in Guangzhou (Canton) and the estuaries of the Sikiang and Perl rivers, and on the river Min in Fujian. Still, in the 1930s, around 1 million of these people lived on boats in water settlements. After the Revolution many settled in fishing villages on land and their living conditions improved. Numerous boat people also lived along the lower Yangtze-Kiang. They led a vagrant life, fishing and trading their catches in various places. Another group in southern China are the Hoklos, of whom many left the People's Republic of China after the Revolution to settle in Hongkong, where they form part of the large population living permanently on boats.

It was estimated that catches in natural freshwaters were at a level of 500 000 - 600 000 t per year during the second half of the 1950s, after a peak of 683 000 t in 1953. In later years catches declined and, since the 1970s, have remained at an average of 300 000 t. Various reasons were responsible for this decrease: large areas of land were reclaimed from lakes and other waters for agricultural purposes, for instance in the Yangtze valley more than 1 million hectares were reclaimed. With the construction of dams and water-gates migration of many fish was inhibited and spawning conditions for such species like hilsa herring *(Hilsa reevesi)*, long-tailed anchovy *(Coilia mystus)*, eel *(Anguilla japonica)*, various carps and river crabs *(Eriocheir sinensis)* deteriorated widely. Furthermore, water pollution and overexploitation of the resources contributed to the decline in production. Under these circumstances the government decided to promote the development of fish production by aquaculture.

Aquaculture

The rearing of fish and other aquatic organisms has a long tradition in China. Written records of fish culture date back as far as the era of the Yin Shang and early Zhou dynasties (12th-11th centuries B.C.). At that time Chiang Tzu Ya organized the fisheries with licence systems, and

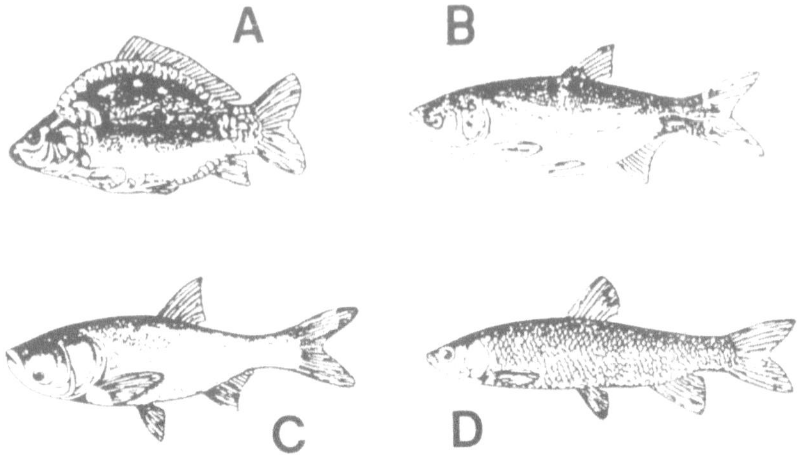

Fig. 115A-D. Some fish species used in Chinese aquaculture. **A** Common carp *(Cyprinus carpio)*, max. 90 cm; **B** silver carp *(Hypophthalmichthys molitrix)*, > 1 m; **C** bighead carp *(Hypophthalmichthys nobilis)*; **D** grass carp *(Ctenopharyngodon idella)*, 1 m

protective measures like closed seasons in summer were already being enforced to secure fish recruitment. In the book *Zhou Rites* various fish species and their use as royal food, sacrificial offerings and gifts to guests are described. Famous is the *Treatise on Fish Culture* by the fish culturist Fanli in Wuxi; it describes common carp rearing in ponds and was published in 473 B.C.. The philosopher Mencius (357 - 300 B.C.) warned against overfishing by casting too many nets in small ponds.

Until early in the seventh century A.D., the common carp *(Cyprinus carpio)* was the only food fish raised in ponds and open waters. This tradition ended abruptly when, during the Tang dynasty (618 - 906), the name of the Emperor was similar to "li", the common name of the carp. Fishing, sales and consumption of this species were banned. For this reason fish culturists turned their activities to other species, in particular the family of carp, the grass carp *(Ctenopharyngodon idella)*, black carp *(Mylopharyngodon piceus)*, silver carp *(Hypophthalmichthys molitrix)*, bighead carp *(Hypophthalmichthys nobilis)* and mud carp *(Cirrhina molitorella)*, and to other fish like the white Amur bream *(Parabramis pekinensis)* which could easily be caught in the large rivers and reared together in polyculture ponds (Fig. 115).

In contrast to the common carp, these species do not normally breed in ponds and prefer natural waters. This gave rise, all over China, to the

Fig. 116. Chinese fish-ponds surrounded by mulberry trees. (Bureau of Aquatic Products, Beijing 1985)

development of a substantial fishery for, and trade in, fish fry, which can be traced back to the times of the Song dynasty (960 - 1276). The trade covered wide areas from the river Amur in the north to Thailand and Malaya in the south. The fish brood was caught in rivers, like the Yangtze-Kiang, in spring with conical small-meshed nets moored on piles in the strong current. The eggs were kept in barrels or small ponds until they hatched and the larvae were transported in wells on boats or in bags on land; they were fed and handled with great care. Fish farming was thus almost entirely dependent on the provision of fry from natural waters. This situation lasted until artificial spawning was invented in the early 1960s.

In 1937 the area under fish culture was estimated at 166 000 hectares and fish production reached 382 000 t. After a serious drop during the war years, the same level was regained by the early 1950s. At that time the fish culture areas were mostly concentrated in the Pearl River delta in the Guangdong province and the Yangtze valley, particularly in Lake Taihu, in the provinces of Zhejiang and Jiangsu. Large-scale construction works with the building of dams for flow control, hydroelectric power generation and irrigation strongly reduced the fish brood in the rivers. It therefore became urgent to search for methods of producing the fry from pond-reared fish. Crucial progress was made when, in 1958, silver carp were raised to sexual maturity in ponds and spawning was artificially induced. In the following years artificial breeding was also brought about for other species of the carp family and fish culture in ponds spread all over China (Fig. 116). The production of cultivated freshwater fish rose to 800 000 t in the 1970s. The old methods of stocking, feeding and harvesting the fish were further developed to improve the production. Not only the number of ponds increased but also lakes and reservoirs were regularly stocked with carp fingerlings. In 1978 about 2.7 million hectares were used for fish culture (26.5% in ponds, 44% in reservoirs and nearly 10% in enclosures in rivers and lakes).

As marine resources in coastal waters were overexploited, so that catches from natural stocks could not be increased, the government started a special programme in 1979 for the further development of aquaculture. This included the promotion of mariculture; China was one of the first countries to develop this technique. Already, before the Ming dynasty (1368 - 1644), oysters *(Crassostrea gigas)*, mother-of-pearl *(Pteria martenssi)*, mullet *(Mugil cephalus)*, laver *(Porphyra* sp.*)* and other organisms were being cultivated along the coast. In 1978, 148 000 hectares were used for the cultivation of marine organisms, of which 65 000 hectares were under intensive culture. Annual production reached 450 000 t, consisting mostly of seaweeds, molluscs and shrimps. Artificial culture of kelp *(Laminaria japonica)*, by planting seeds on floating rafts, was developed from 1950, mainly in the coastal waters from the Liaoning province in the north to the Fujian province in the south. In 1978, 250 000 t of dried kelp were produced. Also, laver, red algae of the species *Porphyra haitanensis* and *P. yezoensis*, are cultivated in a similar way along the coast of the East China Sea and the Yellow Sea. Cultivation of molluscs concentrates on oysters, grown in traditional cultures by the vertical hanging method, and on razor clams *(Sinonovacula constricta)*, the beds of which are ploughed and levelled. These shellfish are cultivated mainly along the coasts of Guangdong, Fujian and Zhejiang in southern China. In the north the Liaoning province is the centre of mussel *(Mytilus edulis)* cultivation. Together with harvests of *M. smaragdinus* further south the production of mussels reached more than 400 000 t in 1988. The main species of crustaceans cultivated are Chinese shrimp *(Penaeus chinensis)*, reared in northern China, and several other *Penaeus* species in the south. The cultivation of warm-water shrimps yielded 85 000 t in 1987 and was the largest in the world. Other marine animals intensively cultivated are sea-cucumbers *(Stichopus* sp.*)*, scallops *(Pecten* sp.*)* and abalone *(Haliotis* sp.*)*. Altogether, mariculture production was 540 000 t in 1983 and had risen to more than 1 million t by 1988.

The development programme for aquaculture was an unprecedented success. Total production rose sharply within a decade to 6.6 million t in 1988 (Fig. 113). Thus, aquaculture production from both freshwaters and coastal areas made up 63% of the total fish production in China and 46% of the world aquaculture production. It is symptomatic that the spectacular increase coincided with the time when farmers were again permitted to use some fields and ponds privately. With the present increment it is likely that Chinese fish production will very soon be the largest in the world, surpassing the catches of Japan and the USSR.

5.2.4 North America

With landings of almost 6 million t in 1988 the USA ranks as number 5 in the list of fishing nations. At the same time 1.6 million t were caught by Canadian vessels. In both countries a most remarkable increase in the landings, almost a duplication, has occurred since the introduction of EEZs in 1977, which led, subsequently, to the gradual suppression of large foreign fishing fleets in American waters. In the USA this enormous development was the result of the 1976 Magnuson Fishery Conservation and Management Act (MFCMA) by which not only an EEZ was established but also Regional Fishery Management Councils were formed for the elaboration of Fishery Management Plans. Reconstruction of the fishing fleet was promoted through financial and tax facilitations. The MFCMA led to the formation of joint ventures under which fish caught by US vessels were directly sold to foreign ships for processing (1986: 1.3 million t, mostly Alaska pollock). Similar measures were taken in Canada.

From early statistics the total fish catches in the USA in around 1880 were estimated at 774 000 t and for 1920 at 1 million t (Canada 344 000 t). After World War II landings in the USA amounted to 2 million t per year and in Canada to 850 000 t. From this time until 1977 fisheries in both countries stagnated, largely for socio-economic reasons (Fig. 99).

The natural conditions for fisheries development are quite favourable with rich resources along the east and west coasts and in vast inland waters.

With the exception of the old sperm whaling (see Sect. 4.1.2) and the tuna fishery, North American fisheries were always concentrated on waters near to the shore and carried out with relatively small ships. Of about 107 000 commercial fishing vessels registered in 1977 in the USA only 247 were larger than 200 GRT and up to 1600 GRT (mostly large tuna purse seiners). Only 240 of the Canadian fishing vessels were longer than 30 m. The various sectors of the fishing industry were mostly fairly isolated but there is now a trend to build ships suitable for participation in more than one fishery.

The development of fisheries in North America was greatly influenced by immigrants from Europe, especially from Portugal, Italy and Greece as well as from Scandinavian countries, and also from China, who introduced their fishing gear and methods and adapted them to local conditions. However, new endemic techniques were also developed. The streamlined schooners and efficient purse seines are American inventions which were transferred to many other parts of the world.

In recent years the number of full-time fishermen was around 220 000 in the USA and 80 000 in Canada. About 17 million people in the USA engage in recreational fishing, catching more than 300 000 t of mostly high-value fish each year. Often conflicts of interests arise between commercial and sport fishing.

In contrast to Russia, fish did not play a major role during the colonization of the North American continent as meat was sufficiently available in the interior. Only in the northeast, especially in New England and in the region of the Great Lakes, was fish used to a greater extent as food and as manure for agricultural production. During early times fish were mostly salted to preserve them for transport and as food for the winter period. Since the mid-19th century canning became more and more important and, at the same time, a market for fresh fish developed, starting with the landing of live fish or fish on ice in Boston in around 1850. As in Europe and other places, the construction of railways facilitated fish distribution, and in America this was to New York and other centres of consumption. A further major step was the introduction of freezing in the 1930s. Now, cured products like salted, smoked or pickled fish are only of minor importance. Today, with a population of 240 million, the USA has a huge market with high demand for fresh, frozen and canned fish, crustaceans and molluscs. The major fisheries, by value, are those for salmon, shrimps, crabs, tuna and scallops, making up 58% of the total value of all US landings.

Roughly 50% of the US fish production is used for human consumption, while the other half, consisting mainly of menhaden, is used for bait or processed to fish meal, oil and solubles for animal feed and industrial purposes. As the demand is not satisfied by home production, about 2 million t of fish and fishery products are imported to the USA every year, making the country, together with Japan, the greatest fish importer in the world. Exports are also sizeable at around 450 000 t per year, now steeply increasing. Canada, with a population of only 25 million, on the other hand, exports roughly 70% of its fish production to the USA and many other countries.

East Coast Fisheries

The development of North American commercial fisheries began in the waters around Newfoundland and in the Gulf of St. Lawrence where European fishing fleets had already operated since long time (see Sect. 2.4). During the 18th century more and more English colonists settled in Newfoundland and along adjacent coasts, like those of Nova Scotia and Labrador, and many of them fished for cod *(Gadus morhua)* in inshore waters. The fish were caught with seines and hand lines, and from 1875, also in a great number of cod traps; they were gutted and split, salted and dried in the open air on flakes along the beach. Initially, companies in the west of England strived to impede the development of this independent fishery until the English fishery suffered from economic depression after the Napoleonic wars. Later, a large number of ships from the southern coast of Newfoundland, in 1911 2200 boats, participated in the offshore cod fishery on the Grand Bank and on other banks. From around 1860 hand lining from schooners was replaced by long lining from little dories, as in the European fishing fleets.

A cod fishery also developed on the east coast of Labrador. Each year a fleet of "floaters" sailed in summer to these waters from Newfoundland and New England to join the local inhabitants in fishing for cod with traps, operated close to the shore from small boats. This seasonal fishery was still carried out until the 1950s.

Further south, along the coast of New England, fishing grounds were already being explored early in the 17th century. With colonization from 1620 a fishery, mainly on cod, developed to satisfy local needs. In contrast to the fisheries based in European ports which were limited to the summer, this cod fishery could also be carried out in winter. With this advantage a substantial export industry for dried salted cod (klipfish) developed with Boston as its centre, particularly when, in the mid-17th century, the Civil War in Britain displaced the English fishery to operate from New England instead of Europe. Klipfish of good quality were exported primarily to Portugal, Spain and France, while products of lower quality were transported to Virginia and the West Indies as cheap food for the slaves on the plantations. Increasing demand for this food provided incentives to expand the American cod fishery to more northern waters, first off Nova Scotia, then to the Gulf of St. Lawrence, Newfoundland and Labrador. Originally, this fishery operated inshore but from around 1700 larger schooners participated in the long line fishery on offshore banks, like the Grand Banks, Western Bank and Banquereau (Fig. 59). Gloucester, Plymouth and Marblehead near Boston developed to become the main ports for this fishery.

In the United States the market for dried salted cod declined after the 1860s, while, in Newfoundland, Nova Scotia and on the Gaspé Peninsula in Canada, this industry has maintained its importance until the present time. For a long time the economy of this region was entirely dependent on klipfish production and on the price this staple food fetched on the world market. Still, in the mid-1960s, there were about 20 000 fishermen in Newfoundland and East Labrador, and 100 000 t of cod were processed to klipfish, mostly for export to southern Europe, the Caribbean region and Latin America. In recent years the number of fishermen has declined and klipfish production fell to slightly less than 20 000 t per year.

Large by-catches of halibut *(Hippoglossus hippoglossus)* in the cod fishery off the New England coast gave rise to a special line fishery on this highly valued species. The heyday of this fishery was from 1840 to 1900 and Gloucester was the central port. Most of the halibut were salted and cured in smokehouses. When the resources in home waters were exhausted, the fishery expanded in around 1870 to the Grand Banks and even to West Greenland and Iceland.

Another important fishery in the northeast exploited the mackerel *(Scomber scombrus)* resources between Chesapeake Bay and the Gulf of St. Lawrence. Mackerel were already caught in beach seines by the first colonists. From the end of the 17th century a fishery with long lines developed which was at its peak between 1820 and 1870, when 600 - 900 vessels fished around Cape Cod and further north. In around 1850 purse seines were introduced to this fishery and operations extended further offshore and concentrated more on the Gulf of Maine. During the 1880s about 450 ships (400 purse seiners) participated in this fishery, sailing mostly from Gloucester and Provincetown. The annual production reached 100 000 t (65 000 t in the USA). Two-thirds of the catch was salted and barrelled and the rest landed fresh on ice. After 1882 the mackerel catches declined steeply. The reasons for this are not quite known. It seems that there was a natural decrease in the mackerel stock with a rise in the herring stock at the same time. Variations in fish abundance have always been very large. Possibly the production of salted mackerel declined when the market for fresh fish expanded. Mackerel landings at the end of the 1930s amounted to 40 000 t per year and in the 1980s to about 30 000 t.

Fishing for herring *(Clupea harengus)* also started early. In 1885 the herring catches surpassed the rapidly declining mackerel landings and fluctuated, during the following 45 years, around an average of 60 000 t. Main fishing grounds were the southern parts of the Gulf of St. Lawrence, especially around Magdalen Island and off New Brunswick with Fundy Bay, along the Nova Scotia coast and in the Gulf of Maine. Herrings were caught in traps and with gill nets; later also with purse seines. Many of them were sold as bait for the line fisheries but the bulk was salted and transported to the south as cheap food for the slaves or, after the Civil War, for the poor in the growing cities. Boston, New York and Philadelphia became centres of the herring trade. Salted herring in barrels and later also frozen fish, were transported from ports like Gloucester and Eastport far into the interior. Since the 1880s herrings were also canned, and this production of "Maine sardines" or "Canadian sardines" is still an industry in Maine and Nova Scotia. Young herring for canning are often caught in weirs. Canadian herring fisheries expanded when, in around 1950, production of fish meal and oil began in Newfoundland. In 1967 part of the purse seiner fleet from British Columbia was transferred to the Atlantic coast and the herring catches soon rose to more than 200 000 t. In contrast, the US herring catches remained, until recent times, at a level of only 20 000 to 60 000 t, even during the 1960s when large fleets of European trawlers, particularly ships from the USSR, fished extensively for herring on Georges Bank.

The otter trawl was introduced to New England in 1904 but it was not before the 1920s, with the implementation of filleting and freezing, that trawling was more widely applied. In most states it was strictly regulated to protect fisheries with other gear. The trawlers, of about 150 GRT, were mostly smaller than in Europe as productive fishing grounds in the Gulf of Maine and on Georges Bank were fairly close to the ports. Soon the trawl fishery became the main producer of fresh fish. Cod was the main target in the early years. Later, this fishery concentrated on haddock *(Melanogrammus aeglefinus)* in the Georges Bank area, where a peak was reached at the end of the 1920s (1929: 11 000 t) before catches then declined due to overexploitation. From the mid-1930s grounds farther offshore were also fished, where redfish *(Sebastes* spp.*)* were abundant. Catches of this species reached a peak of more than 100 000 t per year in 1948-1951 but dropped considerably thereafter in spite of an extension of this fishery to the waters of Nova Scotia and the Grand Banks.

Boston, Gloucester, New Bedford and Portland in Maine developed to become centres of the trawl fishery. A major expansion of trawl fisheries in the USA and Canada occurred after World War II. New vessels were made larger and more powerful and an increasing number of stern trawlers participated in the fishery. In Canada, the trawler fleet doubled from 1961 to 1965 to reach more than 250 ships. Groundfish catches, particularly those of cod, haddock, pollock *(Pollachius virens)* and, in Canadian waters, also of redfish increased steeply.

The largest fishery by quantity is for menhaden *(Brevoortia* spp.*)*. It yielded around 1.2 million t annually during recent years. Catches were almost entirely used for the production of fish meal and oil. This fishery began in the coastal waters of southern New England and extended further south with Long Island Sound and Chesapeake Bay as major fishing grounds (Fig. 117). The first colonists already used menhaden, caught with beach seines, as manure. Later, around the mid-19th century, large quantities of menhaden were salted and exported as cheap food for the slaves in the southern states and the West Indies. Many factories along the coast between Maine and North Carolina were built for the processing of menhaden to fish meal and oil. Their number had reached 62 by 1873 and 97 by 1882. The meal was used mainly as chicken feed and the oil in the tanning of leather, the manufacture of paint and for lighting. Annual menhaden landings increased to 200 000 t in 1880, 250 000 t in 1900 and, after some decline during the 1930s, reached 300 000 t in 1940. The catches always showed large fluctuations between years and areas. Almost all catches were made with purse seines, introduced to this fishery in 1826. The seines were operated from dories, carried to the nearshore fishing grounds by sailing vessels and, from the 1870s, also by small steamers.

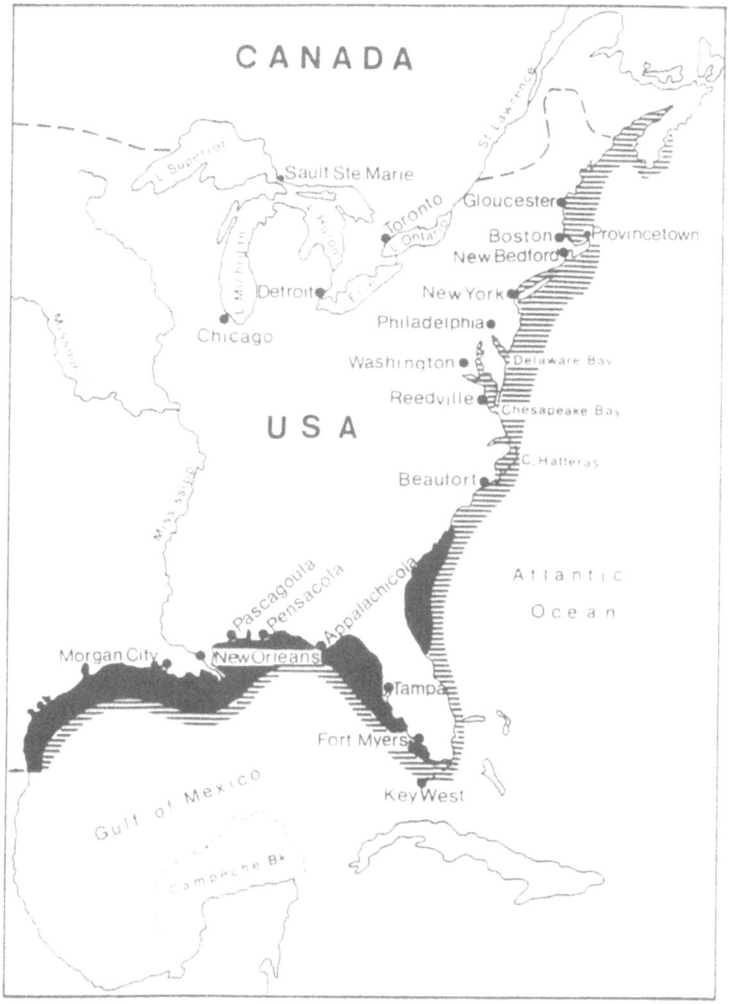

Fig. 117. Atlantic coast of North America with main fishing grounds for shrimps *(dark areas)* and menhaden *(hatched areas)*. Major inland waters, such as the Great Lakes, Mississippi and Missouri, are also shown

When the fish meal and oil industry in California collapsed owing to the lack of sardines, fishing of menhaden was increased in the 1950s. With substantial technical improvements and the application of aircraft to locate fish catches rose steeply. But soon the stocks of Atlantic menhaden *(B. tyrannus)* started to decline and the main centre of the fishery shifted further south to the Gulf of Mexico with its rich resources of Gulf menhaden *(B. patronus)*. Today, about 60 - 70% of the menhaden catches are fished in the Gulf region and are landed in ports like Morgan City, Appalachicola, Pascagoula and Moss Point, where the reduction plants are concentrated. On the Atlantic coast Chesapeake Bay maintained its importance in the menhaden fishery with Reedville as the central landing place. Other important menhaden harbours are Beaufort and Moreland City in North Carolina. A growing part of menhaden landings are used by the pet-food industry. Most of the fish oil is exported to Europe for the production of margarine.

From ports in the Gulf of Mexico, mainly Pensacola, a hook-and-line fishery for red snapper, predominantly *Lutianus blackfordi*, and grouper *(Epinephelus* and related species) developed in the 1930s.

Of greatest value are the crustacean resources. A commercial shrimp fishery with beach seines first developed in Louisiana and expanded along the Gulf coast and also to the South Atlantic states when the introduction of canning opened up a larger market in the 1860s (Fig. 117). In Louisiana, the shrimps were initially mostly sun-dried, like in the Bay of San Francisco, for consumption by the Chinese. The main development of shrimping began when freezing techniques were introduced in the 1930s. The number of boats increased considerably and, in the mid-1960s, about 1850 vessels were employed in this fishery in the Gulf and more than 900 off the Atlantic coast, with a total of about 16 000 fishermen. Annual catches rose to about 100 000 t in the 1950s and even reached 140 000 t in 1986. Landings at such levels made shrimping the most valuable sector of the whole US fishing industry.

Trawling was introduced to this fishery in 1912, and soon beach seining became unimportant. Trawling with a single gear, usually during the night, was modified in 1957 to fishing with a double rig so as to improve the productivity. Initially the catches consisted mainly of white shrimps *(Penaeus setiferus)*, caught in relatively shallow waters. As far too many ships engaged in this fishery and the catches per boat declined drastically, fishing expanded to deeper waters where brown shrimps *(P. aztecus)* and pink shrimps *(P. duorarum)* started to form the bulk of catches. Larger shrimping vessels were built to operate on more distant grounds like the Campeche Bank off Mexico, but the establishment of EEZs put an end to this expansion.

Shrimp fisheries are most important in Louisiana, Texas and on the west coast of Florida. The main ports are Tampa, Morgan City, Key West and Fort Myers. Modern shrimp-processing plants are distributed in all areas

Fig. 118. Fishing grounds for lobster *(dark areas)*, sea scallop *(hatched horizontally)* and clams *(hatched vertically)* along the Atlantic coast of the USA

along the Gulf of Mexico. The shrimp are marketed fresh, frozen, canned, cooked and peeled, and dried. Since the 1950s utilization of fresh and frozen shrimp has increased at the expense of canned shrimp.

Further north, in the region between the Gulf of Maine and Newfoundland, resources of American lobster *(Homarus americanus)* have been exploited since olden times, with catches of 20 000 t in US waters and 30 000 - 40 000 t in Canada during recent years (Fig. 118). Yields in the 1880s reached up to 50 000 t. The lobsters are caught in wooden fish-bai-

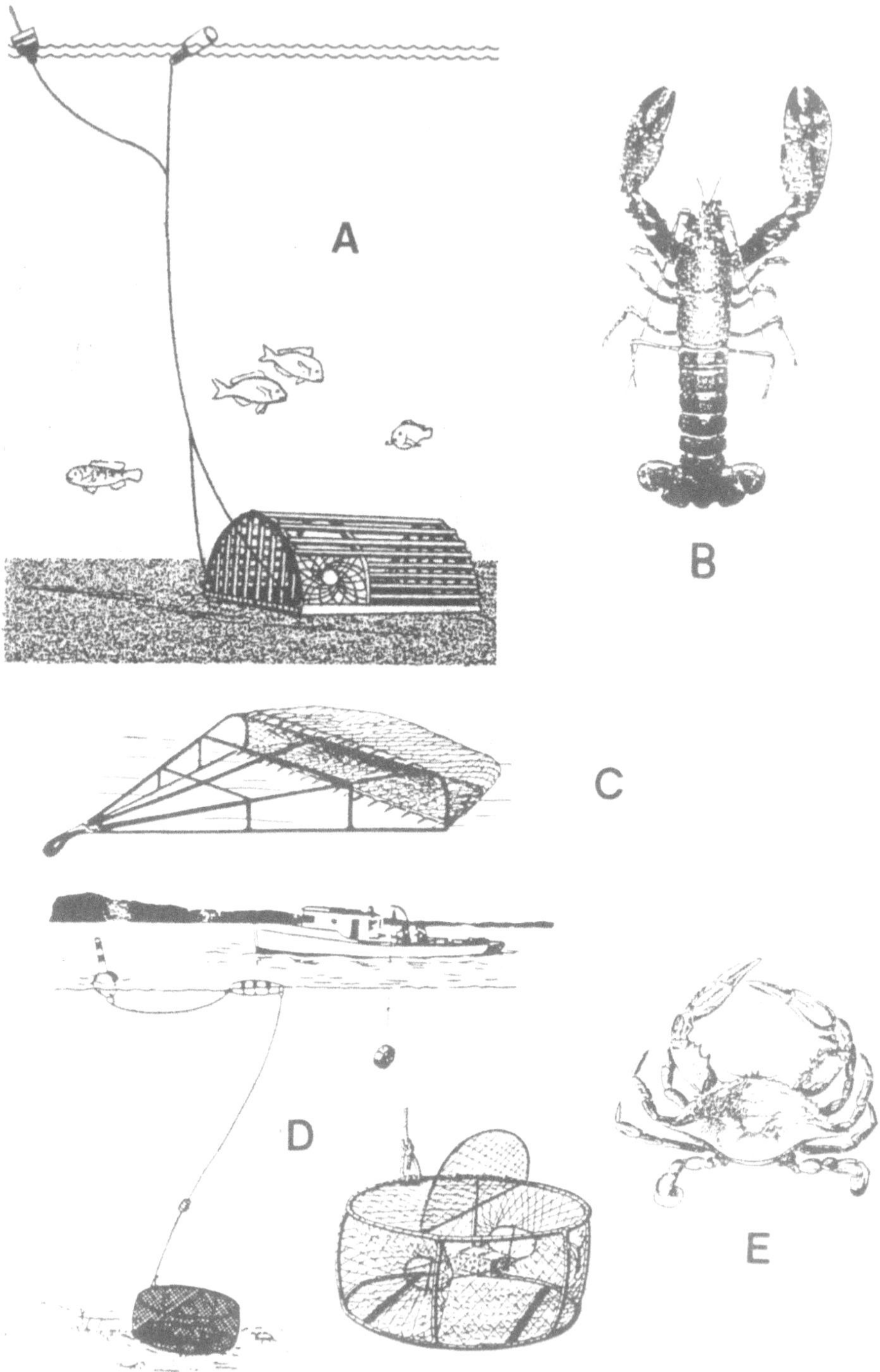

Fig. 119A-E. Lobster and crab fishing in the USA. **A** Wooden lobster trap, baited with fish; **B** American lobster *(Homarus americanus)*; **C** crab dredge; **D** crab fishing with pots; **E** blue crab *(Callinectes sapidus)*. (**A** Walford 1947; **B-E** after figures in Comm Fish Rev)

ted pots under strictly regulated licensing systems and protection measures (Fig. 119). In earlier times most of the catches were canned since transport of live animals was difficult. Today, almost all lobster are marketed alive. Canning is limited only to some remote Canadian areas. A great part of the Canadian lobster production is exported to the USA, the largest crustacean market in the world.

Another valuable crustacean is the blue crab *(Callinectes sapidus)*, abundant from Massachusetts southwards with Chesapeake Bay as the most productive water. This crab was also caught by early colonists but commercial exploitation began in the 1880s when canning factories started to process the catches. The crabs are caught in pots and with trotlines, i.e. baited long lines without hooks. Recent catches of blue crab amounted to 90 000 t per year.

Utilization of mollusc resources is highly developed along the Atlantic coast and is one of the largest in the world. It had already begun in prehistoric times, as indicated by numerous shell mounds. Most important is the American cupped oyster *(Crassostrea virginica)*, distributed from the Gulf of St. Lawrence to Mexico. Main centres of the oyster fishery developed in Chesapeake Bay, Delaware Bay and on the Gulf coast of Louisiana. Heavy exploitation soon led to a serious decline in the oyster resources, particularly in New England. Therefore, a simple but widespread oyster culture was initiated in around 1850. Part of the natural oyster banks were leased by private enterprises, and oyster brood was collected and reared for restocking the private oyster beds. This greatly increased their production as compared to the resources in the public oyster beds. Oyster culture often suffered from pollution, diseases and predators, particularly starfish. Grabs ("tongs") and other simple devices for the collection of molluscs in earlier times were replaced by more efficient gear like dredges, towed since the 1870s by steam vessels and later motor ships, and, in more recent times, large suction apparatus (Fig. 120). In industrial production oysters are opened in shucking plants and frozen for transport to markets like New Orleans and Baltimore. Total oyster production reached a peak towards the end of the last century with about 1.56 million t. In the mid-1950s it still amounted to more than 400 000 t and, since the 1970s, has been 200 000 - 300 000 t per year.

Also of very high value is the fishery for various species of scallops (mainly *Placopecten magellanicus* and *Argopecten gibbus*) which has developed since 1900 in New England and Canada (Fig. 118). This fishery, with New Bedford and Digby (Nova Scotia) as the main ports, pro-

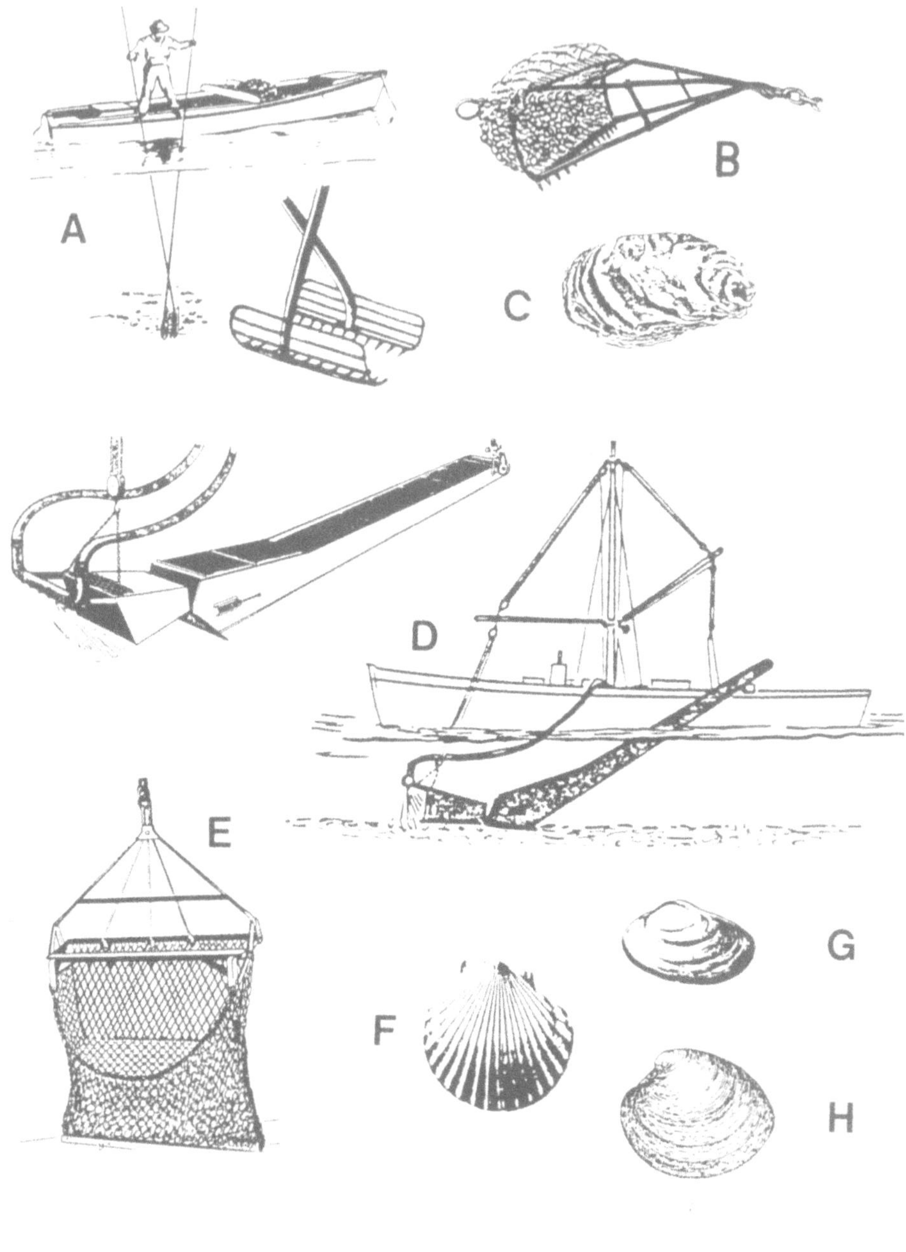

Fig. 120A-H. Harvesting of molluscs in the USA. **A** Oyster collection with hand tongs; **B** oyster dredge; **C** American cupped oyster *(Crassostrea virginica)*; **D** clam fishing with hydraulic clam dredge in Chesapeake Bay; **E** scallops dredge; **F** sea scallop *(Placopecten magellanicus)*; **G** surf clam *(Spisula solidissima)*; **H** ocean quahog *(Arctica islandica)*.
(After figures in Comm Fish Rev 1953, 1962)

duced around 75 000 t in the 1950s and more than 200 000 t during the 1980s. Other mollusc fisheries with substantial production are those of the ocean quahog *(Arctica islandica)* with 150 000 t in recent years, the surf clam *(Spisula solidissima)* with 165 000 t and the hard clam *(Mercenaria mercenaria)* with 30 000 t.

West Coast Fisheries

The development of commercial fisheries along the Pacific coast began later than on the Atlantic side (Fig. 121). Russian fur hunters and traders, who came via the Aleutian Islands and Alaska to northern California from the mid-18th century, relied mainly on the rich salmon resources in rivers and streams for their food. For a regular supply they promoted the traditional fishing of the Indians (see Sect. 1.1.3) and introduced the curing of salmon as a means to preserve the summer catches for the long winter periods. Production of fish was mainly for local consumption and fish trade was rather limited until, in the early 19th century, British-American trading companies, particularly the Hudson Bay Company, contracted Indians fishing in the rivers Fraser and Columbia. In 1823 the first cured salmon were exported to England but most of the production was transported to Hawaii, as a supply for whaling ships, and to Russian Alaska.

Local food demands increased sharply after 1849 when gold was discovered in California and immigration was high. Salmon, entering the Sacramento River in large numbers on their spawning migration, were a most welcome source of food for the people in the mines and in San Francisco, a rapidly growing port where, also, many whaling ships called for re-supply. Large parts of the salmon catches were processed by salting or smoking.

A new development was initiated in 1864 with the establishment of the first salmon canning factory near Sacramento, followed in 1866 by more canneries along the lower courses of the Columbia River and, later, on the river Fraser. When, in 1871, British Columbia became a province of Canada, new salmon canneries were also built along this coast. Generally, there was a gradual shift in the centre of salmon fishing and processing to the north. By the end of the 1880s this industry started to boom in northern British Columbia and in Alaska, which had been acquired by the United States from Russia in 1867. In 1896, 120 salmon canneries existed, of which 30 in Alaska produced 40% of the total output. Bristol Bay and Kodiak Island became centres of the salmon indu-

Fig. 121. Pacific coast of North America with main fishing grounds for sardines *(dark areas)*, herring *(hatched areas)* and tuna *(shaded areas)*

stry in Alaska, mostly controlled by enterprises in Astoria, Vancouver and Seattle.

For the fishing season in summer, fishermen and cannery workers with all their necessary supplies had to be transported from the south each year to Alaska where they lived in camps. While the fishermen were mostly immigrants from southern European countries and Scandinavia, factory workers were Chinese and Indians. In earlier times these people were often recruited by violent means and forced to work under unfavourable contracts.

Since the market along the Pacific coast was fairly small, production originally depended mostly on exports, especially to Great Britain. The

American Civil War created a major incentive to increase production. With the growth of the population in America the home market absorbed, from around 1900, almost all canned salmon, consumed mostly in the east of the US.

During the development of salmon fisheries changes occurred in the composition of the catches. In the early decades of fishing in the rivers Sacramento, Columbia and Fraser efforts concentrated mainly on chinook *(Oncorhynchus tshawytscha)*. Later, with the shift of the fishery to the north, sockeye *(O. nerka)* and pink salmon *(O. gorbusha)* became of greater importance.

The salmon were, and still are, caught mostly with gill nets operated from small boats. Traps and fishing wheels, widely used, for instance, in the Columbia River, have disappeared in most waters since they were banned by regulations. In earlier times beach seines were also applied, often towed with the help of horses. After the introduction of motors before World War I purse seines and troll lines became the major types of fishing gear in coastal waters, particularly in British Columbia and Alaska. Today, trolling is widely used in sport fishing with substantial catches of valuable coho *(O. kisutch)* and chinook.

With progress in refrigeration techniques and improvements in transport facilities it was possible to concentrate canning in fewer factories with larger processing capacities. The number of salmon canneries, with a maximum of 300 during World War I, decreased to 200 in 1936 and to 100 in the 1960s (more than 60 in Alaska).

Salmon catches peaked in 1936 with 445 000 t, from which 80% of the world's canned salmon was produced. However, this high level of exploitation could not be maintained, particularly as human activities, like mining and other industries and the construction of dams for power and irrigation, had detrimental effects on the reproduction of fish. Indications of the decline in the salmon stocks in the river Fraser and other waters gave rise to the establishment of the International Pacific Salmon Commission between Canada and the USA in 1930. This commission was instrumental in the introduction of protective measures, the intensification of artificial fish propagation in numerous hatcheries and the construction of fishways and ladders to facilitate the upstream migration of spawners. With these conservation methods the salmon production of the USA could be stabilized at a level between 200 000 and 300 000 t per year up to the present time. About 10% of this quantity is exported to Europe and Japan as fresh frozen fish. In value salmon production takes first place in all US fisheries. Recent salmon catches in western Canada amounted to almost 100 000 t per year.

The second major fishery, one of the largest in the world during its heyday, exploited the resources of California sardines *(Sardinops caeruleus)* in the waters off California and up to British Columbia (Fig. 121). The fishery developed when, in 1915, tuna canneries began also to process sardines since imports from Europe were lacking during the war. To satisfy urgent need, particularly for animal foodstuffs, e.g. in the raising of poultry, sardines were soon also processed into fish meal and oil. As this turned out to be very profitable, the industry grew rapidly. Landings, showing wide annual fluctuations, reached over 225 000 t by 1933 and the highest catch of 790 000 t was fished during the 1935-1936 season.

At that time 90 reduction plants and 34 canneries were in operation. For the protection of canning state regulations in California prescribed the production of certain quantities of canned sardines as a precondition for the production of fish meal and oil. Catch quota were introduced for sardines in coastal waters. Many enterprises eluded these regulations by employing ships with reduction plants anchored outside the 3-mile-zone; these vessels received the landings from catcher boats at sea. In 1936 about 25% of the total catches were processed on such floating factories.

San Pedro, San Diego, Monterey and places in the Bay of San Francisco were main ports for the sardine industry. The fishery followed the annual north - south migration of the sardines and operated in late summer off Vancouver Island and in winter and spring off southern California. Originally, the sardines were caught with lampara, an Italian type of surface seine, close to the shores, but in around 1930 the more effective purse seine replaced the old gear and the fishery expanded further offshore.

High catches were still obtained during the 1940s (1949: 287 000 t) but, in the early 1950s, a dramatic decline in the yields occurred, first in the north and later also in the south. In the 1960s the annual catches fell to below 1000 t and finally to zero. There has been no revival of this fishery so far, but in the southernmost area of sardine distribution, off Baja California, a Mexican pilchard fishery has developed since 1980; in recent years this has produced between 300 000 and more than 400 000 t annually.

The reasons for the collapse of the Californian sardine fishery are not fully understood but were connected with frequent failures in fish recruitment. There is no doubt that overexploitation contributed greatly to this phenomenon but, evidently, ecological changes also played an important role. Possibly the stocks of the North Pacific anchovy *(Engraulis mordax)*, distributed over the same area, expanded when the sardine resources were depleted, taking their ecological niche. Some catches of anchovy had been used for canning, particularly during World War II, but it was not before 1965-1966 that licences were granted to fish this species for the production of fish meal and oil. Subsequently, catches rose to 150 000 t in 1975. However, since the end of the 1970s they have declined gradually to insignificant quantities for economical reasons. Anchovy contain much less oil than sardines and are, therefore, a less profitable raw material for reduction purposes since oil is the main source of profit in this industry.

Another pelagic fishery further north, in the waters from British Columbia to western Alaska, concentrated on Pacific herring *(Clupea pallasii)*. The first fishing explorations date back to the 1870s. In early times most of the herrings were salted for export. Demands during World War I also created here incentives to increase the production. Since the 1920s curing of herring declined and large parts of the catches were processed into fish meal and oil. Good profits led to rapid expansion of this fishery. The efficiency was improved by the introduction of purse seines. By 1940 the total annual landings had reached 200 000 t, of which 150 000 t were caught off British Columbia. The catches in Alaska declined in the 1960s to about 25 000 t per year, while in British Columbia production remained high with more than 200 000 t until the late 1960s. When the yields began to decline also in Canadian waters and profits from the production of fish meal and oil diminished, part of the purse seiner fleet was transferred, in 1967, to the herring fisheries on the Atlantic coast. Pacific herring landings in the mid-1980s amounted to only 25 000 t in Canada and 50 000 t in Alaska.

Halibut *(Hippoglossus stenolepis)*, as the most valuable component of the flatfish resources, were fished in the Puget Sound - Vancouver area from the 1890s, initially from small dories close to the shore. When the catches declined in coastal waters, activities were extended farther offshore and to the Gulf of Alaska as the main fishing ground. Halibut are caught by bottom long lines of 4 - 5 km length with up to 1000 hooks, operated from vessels of about 30 m length with crews of 8 - 12 men. Seattle and Vancouver are the central ports. The rapid development of this fishery after 1910 resulted in a dramatic decline in catches. Therefore, the fishery was regulated from 1930 by the International Pacific Halibut Commission, established in 1923-1924 by the United States and Canada. Due to effective protection measures an annual catch of about 30 000 t was maintained over a fairly long period. Nevertheless, the yields declined in the late 1960s, probably as a consequence of substantial halibut by-catches made by the Japanese and USSR trawl fisheries. After the banishment of the trawlers from the halibut grounds the catches rose again to around 35 000 t in the US and 5000 t in Canada.

Following the example of the Newfoundland fishery, schooners from San Francisco sailed to the Okhotsk Sea in the 1860s for hook-and-line fishing for Pacific cod *(Gadus macrocephalus)*. When Alaska became US territory this fishery shifted to Alaskan waters, especially around the Shumagin Islands. The cod were salted and transported to San Francisco Bay for further processing. Compared to klipfish from the North Atlantic this product was inferior and thus the Pacific cod fishery remained only at a modest level.

In comparison to other fishing methods trawling was not important until recent times as the shelf along the Pacific coast is mostly rather narrow. However, since the early 1980s new trawlers were commissioned in the US and catches of North Pacific hake *(Merluccius productus)*, Pacific cod and other demersal species increased steeply. Most dramatic was the rise in production of Alaska pollock *(Theragra chalcogramma)* in the Gulf of Alaska and Bering Sea, from negligible quantities in the 1970s to almost 1.4 million t in 1988, i.e. 28% of the total US commercial landings. These catches were made by a new fleet of US stern trawlers, mostly operating from Seattle. Under joint ventures with foreign companies, mainly from Japan and South Korea, the fish were mostly transloaded at sea and shipped to East Asia to supply the production of surimi and other items. Recently, the exports have declined and more Alaska pollock have been processed directly on board the US trawlers. The newest of these trawlers are highly mechanized ships, about 100 m long with 4000 - 6000 hp engines and capable of operating large midwater trawls.

The origin of the tuna fishery goes back to 1903, when immigrants from southern Europe, mainly Portugal and Italy, started to fish for albacore *(Thunnus alalunga)* off California from San Diego and San Pedro. From 1909 tuna were processed in canneries and soon this flourishing industry absorbed almost the total catch. High demands for canned food during World War I gave the impetus to increase production, which amounted to 20 000 t in 1916. Increasingly the fishery also exploited the abundant resources of yellowfin tuna *(T. albacares)* and skipjack *(Katsuwonus pelamis)* while albacore, highly esteemed in America for their white meat, were increasingly imported from Japan as home production was insufficient. With the construction of larger vessels operations expanded southwards into tropical waters during the 1920s and reached the Galapagos archipelago in 1930 and Peru in 1950.

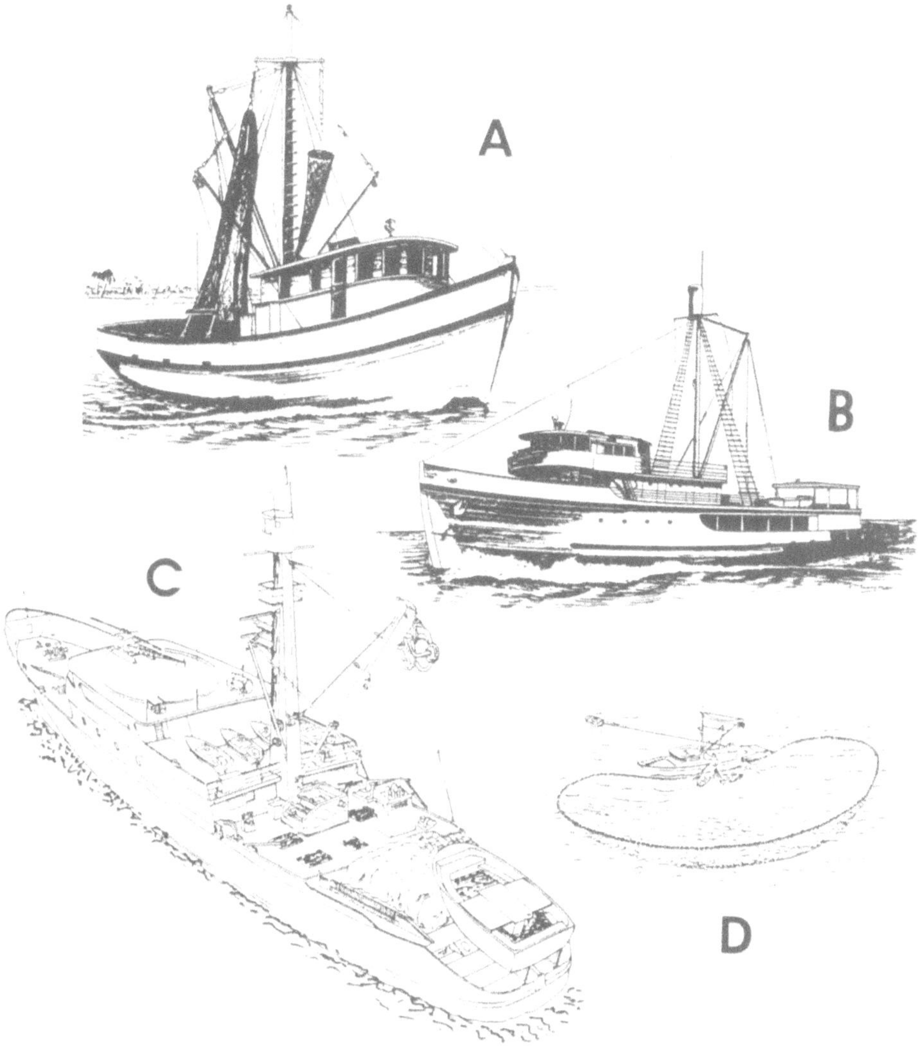

Fig. 122A-D. American fishing vessels. **A** Shrimp trawler from the Gulf of Mexico; **B** pole-and-line vessel of the Californian tuna fishery; **C** modern purse seiner with power-block, skiff and helicopter; **D** tuna purse seining. (**A,B** from Fish Leaflets, US Fish and Wildl Service, **C,D** from FAO 1985)

Most tuna were caught by pole-and-line from tuna clippers equipped with seawater tanks for live bait, usually anchovies and sardines, which were used to attract the tuna (Fig. 122). Trolling developed to become the main technique for catching albacore in surface waters. In the 1950s purse seining was introduced to the tuna fishery, most bait-boats were rapidly converted to purse seiners and the installation of power-blocks facilitated the operation of large nets now made of synthetic material instead of cotton. Modern purse seiners were constructed as sophisticated vessels of up to 65 m length and 1200 GRT, fit for long voyages to distant waters. Aircraft were employed to locate the fish and scientific services were established to increase the efficiency of this fishery.

The tuna landings in the US, which amounted to about 60 000 t per year in the mid-1930s, increased to around 150 000 t during the 1960s and to more than 200 000 t in the 1970s and 1980s. San Pedro and San Diego remained the large centres of this industry. As fish landings in the USA by foreign vessels are prohibited by law, American companies established large tuna-canning factories in American Samoa and Puerto Rico, where most of the deep-frozen tuna are landed by Japanese, South Korean and Taiwanese ships.

The American tuna landings from the Pacific were influenced by the regulations of the Inter-American Tropical Tuna Commission (IATTC), established in 1950, which introduced an international total allowable catch for yellowfin tuna in 1967. A major problem for the US fishery arose with the by-catches of porpoises by purse seiners and when such by-catches were prohibited part of the tuna fleet emigrated to Mexico.

A relatively recent development is the fishery for king crabs *(Paralithodes camtschatica)* in Alaska waters with a centre around Kodiak Island. It began after World War II. Production reached 13 000 t in 1960 and even more than 70 000 t in the mid-1960s, making this the most important king crab fishery in the world. Due to overexploitation the yields declined to half this level in the 1970s and to an average of 15 000 t in the 1980s. The crabs are caught in pots and processed in canneries on land or are deep-frozen. When the resources declined, the fishery increasingly exploited stocks of the tanner crab *(Chionoecetes* spp.*)*, producing around 50 000 t in recent years, and the Dungeness crab *(Cancer magister)*producing about 12 000 t. In Canadian waters a fishery on queen crabs *(Chionoecetes opilio)* developed, reaching around 40 000 t per year in the early 1980s.

Pacific ocean shrimps *(Pandalus* spp.*)* were already being fished from the middle of the last century in the Bay of San Francisco and were mostly dried for export to China until the 1930s. When the shrimp resources declined in these waters, the fishery moved gradually further north and operated along the southern coasts of Alaska. Catches obtained by small beam trawlers rose from a few thousand tons in the 1960s to a maximum of 87 000 t in 1977 but fell thereafter to around 30 000 t. The shrimps are mostly marketed as fresh or frozen-cooked meat, although some are canned.

Utilization of mollusc resources is rather small compared to the situation on the Atlantic coast. The beds of the western flat or Olympia oysters *(Ostrea lurida)* in the Bay of San Francisco were soon exhausted in the 1850s as were, later, the grounds further north. Attempts to transplant the American cupped oyster *(Crassostrea virginica)* from the Atlantic had only limited success but in 1929 the Pacific cupped oyster *(C. gigas)* was introduced from Japan and is grown in hanging cultures, although the brood has to be imported from Japan. Production during recent years was around 25 000 t.

Inland Waters

With the Great Lakes, the Canadian lakes further north and the Mississippi River system the inland waters are very extensive (Fig. 117). Numerous mounds of kitchen middens indicate that they were already being fished by the Indians in prehistoric times. Commercial fishing started at the end of the 18th century on Lake Ontario. In particular, fisheries developed near the transport routes on the isthmuses between the Great Lakes, for instance, in the area of the Sault Ste Marie Falls between Lakes Huron and Superior, where a fishery already existed by around 1840, and later, also at the Strait of Mackinac between Lakes Huron and Michigan. Lake Erie, with a maximum depth of 60 m and the shallowest of the five Great Lakes, turned out to be the most productive, providing about half of the fish catches from these waters. Here, the Detroit River became an important fishing area and in the 1830s Detroit was already a major trading centre for salted fish. Other main fishing grounds developed in Georgian Bay of Lake Huron and in Green Bay of Lake Michigan, from where fish supply was transported to mining settlements. Generally, fish played an important role in the lake region and the Northwest Territory as food for the first colonists. Close to the lakes the fish were consumed fresh but a large part of the catches was salted for transport over greater distances.

Originally, fishing was done chiefly with beach seines but from the early 19th century gill nets came more and more into use. Later, large traps (poundnets) were also introduced. Boats with motors were used from about 1900 and earlier than this fish were transported by steam ships to centres of consumption. Still today, gill nets, reaching lengths of up to

20 km, poundnets and beach seines are the main types of gear in these fisheries. Many activities are devoted to recreational fishing. Lucrative fishing with gill nets under the ice is carried out in winter and often, temporary fishing camps are erected on the ice. The catches are transported by snowmobiles. Most of the Great Lake catches are marketed in Chicago, Detroit, Toronto, New York and many other places.

During the history of these lake fisheries a major change took place in the composition of the catches. In the 19th century sturgeon *(Acipenser fulvescens)* and walleye *(Stizostedion vitreum)* were the most commercially important species. Later, until the 1930s, highly esteemed species like lake trout *(Salvelinus namaycush)*, whitefish *(Coregonus* spp.*)* and cisco *(Leucichthys artedi)* made up most of the catches. After 1930 other, less valuable, "roughfish" species predominated, especially yellow perch *(Perca flavescens)*, alewife *(Alosa pseudoharengus)*, rainbow smelt *(Osmerus mordax)* and carp *(Cyprinus carpio)*, which was introduced from Germany in the 1870s. This change was connected with the gradual immigration of fish species through the Welland Canal, constructed in 1829 to bypass the Niagara Falls. Sea lampreys *(Petromyzon marinus)* entered and damaged the stocks of lake trout, whitefish and other species. Through activities of the International Great Lakes Fisheries Commission the lamprey plague was successfully combatted by electrical fences and selective chemicals but, as the Great Lakes provide major shipping routes to important industrial areas, fish stocks are endangered by pollution.

In contrast to the Great Lakes, fishing in northern Canadian lakes, mainly the Great Slave Lake, still produces sizeable catches of whitefish *(Coregonus* spp.*)* which are mostly exported to the USA.

The total fish catches from inland waters in the USA and Canada remained fairly constant over several decades until the 1970s at around 125 000 t (Canada: 45 000 t). About 50 000 t per year were fished in the five Great Lakes and 40 000 t in the Mississippi River system. In this southern area carp, buffalofish *(Ictiobius* spp.*)* and river catfish *(Ictalurus* spp.*)* form a large part of the catches. During recent years aquaculture of pond-raised catfish has developed rather rapidly and production reached 170 000 t in 1988, raising the total US fish landings from inland waters to more than 300 000 t per year.

5.2.5 Peru and Chile

The seas along the west coast of South America belong to the most productive waters in the world. Very large living resources, available close to the shores, create excellent fishing opportunities for Peru and Chile. Landings in Peru reached a peak of 12.5 million t in 1970 but dropped considerably in later years and during the 1980s fluctuated between 1.6 and 6.6 million t. In Chile fish production rose more steadily to over 5 million t in 1988. In both countries the development of fisheries to large-scale industries began rather late, in the 1950s. In 1939 fish landings in Peru still amounted to only 10 000 t, in Chile to 37 000 t, and for all South America they were estimated at 250 000 t. The fishery in Chile is much more diverse than in Peru, where the industry is almost entirely concentrated on the production of fish meal and oil.

Since the populations in both countries, at present 21 million people in Peru and 12 million in Chile, are fairly small as compared to the large fish production, the markets are limited and 80 - 90% of the production is exported. This provides both countries with an important source of foreign exchange and employment. During the peak of the anchoveta fishery at the end of the 1960s fish exports valued 27% of the total exports from Peru, taking second place after copper. In 1986 the percentage was still 12.6%. A similar percentage was reached in Chile which, in recent years, has exported over 500 000 t of fish meal annually and has replaced Peru as the world's leading country in this sector (Fig. 97).

Basic Conditions for Fisheries Development

The great abundance of living marine resources depends on the Humboldt or Peru Current, a cool continuous water flow along the Pacific coast northwards to 6°S in winter and to 4°S in summer, from where the current runs in a WNW direction south of the Galapagos archipelago (Fig. 123). Prevailing southeasterly trade winds close to the coast cause upwelling of cold deep water, rich in nutrients, to near surface layers. Several centres of upwelling can usually be observed between 25°S and the northern border of Peru. The average water temperature in upper layers is only 17°C. The high content of nutrients leads to the growth of enormous quantities of plankton which form the food basis for large resources of fish, birds and marine mammals. As a result of decomposition of organic material the bottom waters close to shore are extremely poor in dissolved oxygen and certain zones are, therefore, very poor fish-

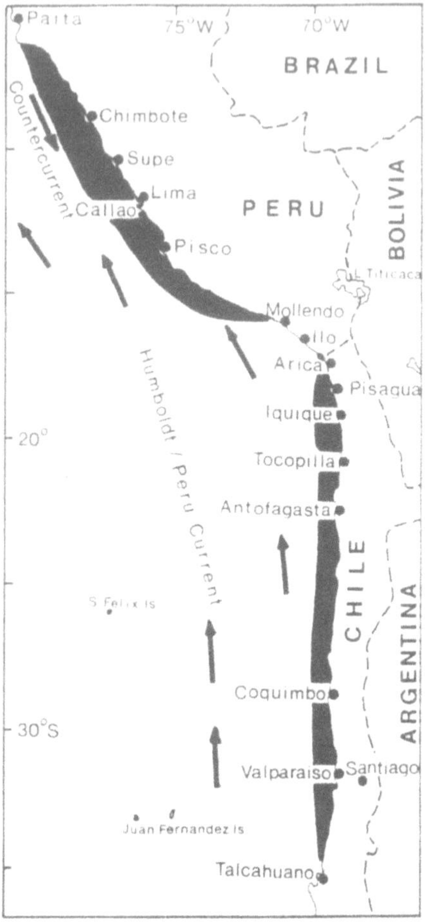

Fig. 123. Main fishing ports in Peru and Chile and major fishing grounds for anchoveta *(black areas)*. System of currents is also shown

ing grounds, in particular in the area of the relatively warm and salty Gunther Under Current, which flows from north to south.

In wide areas along the coast, from about 15°S to the south of Santiago de Chile, the shelf is very narrow and the waters are only suitable for pelagic fishing. Off northern Peru and south of 35°S in Chile the shelves are wider and can be used for demersal trawling. Different environmental conditions prevail in the waters off southern Chile, south of the Island of Chiloe, in an area with many islands, cliffs and inlets and where no upwelling is observed but where high tidal variations create good living conditions for crustaceans, molluscs and seaweeds.

From mid-Chile to the north of Peru oceanographic conditions, the high Andes and prevailing trade winds make the coastal lands extremely dry so that they are mostly very thinly populated deserts. Where warm air meets with cold waters, frequent fogs ("garuas") occur in a narrow zone along the coast.

Although conditions for living resources and fishing are usually most favourable, detrimental natural events occur from time to time causing death to many animals and serious social and economic crises for man. In such events, called "El Niño", which mostly occur in the period from December to May, the cool waters of the Peru Current are held back and warm surface waters (Countercurrent) from the equatorial part of the Pacific invade the coastal regions off Peru, disrupting the upwelling processes (Fig. 123). This results in mass mortality of plankton and, consequently, of many cold-water animals. Fish stocks, especially those of the anchovy *(Engraulis ringens)*, are badly affected even though many fish try to escape to cool waters further south or to deeper waters, where they are, however, out of reach of the fisheries. On the other hand, tropical species migrate into the area and zones usually low in dissolved oxygen improve to provide living conditions and food for bottom-dwelling fish. Thus, "El Niño" has both negative and positive effects. During the 1982-1983 "El Niño", the strongest in this century, the anchovy resources were seriously reduced, less than 2 million of 9 million guano birds survived and many sea lions and fur seals died from starvation. Also, most of the benthos suffered greatly. On the other hand, stocks of shrimps and scallops increased and a large number of sardines *(Sardinops sagax)* migrated south and concentrated close to the Chilean coast, thus giving rise to a dramatic increase in the Chilean landings. "El Niño" events have been observed many times, for instance, in 1891, 1925-1926, 1941, 1957-1958, 1965-1966 and 1972-1973. Scientific investigations indicate that this phenomenon is caused by climatic anomalies, especially low air pressure over Australia and Indonesia and high pressure in the southeastern Pacific (Southern Oscillation) with strong easterly winds moving water from the eastern Pacific to the west.

Artisanal Fisheries, "Pesca Chica"

As described in Section 1.1.3, the living resources along the South American west coast were already being used in prehistoric times. Numerous mounds of kitchen middens ("conchales") and finds from excavations provide evidence of this. The artisanal fisheries, carried out from many small coastal villages (caletas), date back to this tradition. Until recent times the old types of vessels like the reed "caballitos" (Fig. 12), balsa-wood rafts and rafts of inflated sea-lion hides, could still be observed in certain places. Today, almost all have been replaced by plank boats, mostly powered by small motors.

In Peru there are about 4500 craft, operated by approximately 12 000 fishermen who catch around 200 000 t of fish per year for the domestic markets. The landings consist mainly of silverside *(Odontesthes* sp.*)*, mullet *(Mugil* sp.*)* and South Pacific bream *(Seriolella* sp.*)*. Gill nets and long lines are the main gear. In olden times shark meat was dried and sold as "bacalao nacional", a popular Lenten fare.

In Chile's small-scale fishery about 10 000 fishermen operate 6850 boats and provide around 70% of the fish supply for domestic consumption. Beach seines have been widely replaced by long lines and troll lines. In the central area between Coquimbo and Talcahuano the catches consist predominantly of congrio *(Genypterus* sp.*)*, sierra *(Thyrsites atun)* and bonito *(Sarda chiliensis)*. In southern Chile, from Chiloe Island, the Chonos archipelago and further south, the fishery concentrates on the harvesting of "mariscos", a large variety of molluscs and crustaceans, and of seaweeds. Originally, this was a rather simple collecting of animals and algae by hand and with rakes or spades. Since the 1930s harvesting was increasingly done by scuba divers who start each year from Chiloe to work south during the summer season. Transport ships bring the harvests to Calbuco, Puerto Montt, Valdivia and other places where processing factories are concentrated. Besides blue mussels *(Mytilus chilensis)*, Taca clams *(Protothaca thaca)*, false abalones *(Concholepas concholepas)*, sea-urchins *(Loxechinus albus)* and red algae *(Gracilaria* sp.*)* are the main species caught. Landings from the small-scale fisheries have increased considerably since 1976 and amounted to over 286 000 t in 1986. Further south still, in the eastern part of the Magellan Strait, a modern gill net fishery for king crab (centolla, *Lithodes antarcticus*) has developed since the early 1960s when the crab resources around Chiloe were exhausted. This fishery is strictly regulated and has produced around 2500 t annually during recent years. Most catches are canned in factories near Punta Arenas.

Industrial-Scale Fisheries for Human Consumption

A demersal trawl fishery for hake (merluza, *Merluccius gayi*) on the shelf off central Chile developed in 1950 with the arrival of a number of small trawlers with experienced fishermen from Germany. During the first years the hake were processed into fish meal for export since the local market was very limited. Landings reached 100 000 t in the 1960s but fell to an average of 30 000 t in the 1980s. Already, in 1953, the trawlers had also begun to fish for shrimps (camarones, *Heterocarpus reedi*) and squat lobsters (langostinos, *Pleuroncodes monodon*). About 40 000 t per year of these lobsters were caught in the early 1970s but landings decreased to only around 10 000 t due to overexploitation. Most lobsters are exported and the main port for this fishery is Valdivia.

After 1978 the demersal fishery expanded to southern Chilean waters, where 12 modern factory trawlers of up to 3900 GRT, operating from Talcahuano, were employed in fishing for southern hake *(Merluccius polylepis)*, Patagonian grenadier *(Macruronus magellanicus)*, cusk eel *(Genypterus* sp.*)* and other fish.

In Peru the development of larger-scale fishing dates back to the 1940s. At the beginning of World War II, when fish imports were interrupted, a fishery on bonito *(Sarda chiliensis)* began in coastal waters, mainly between Chimbote and Pisco (Fig. 123). A number of canneries processed most of the catches. Annual landings reached almost 90 000 t in the mid-1960s but had declined by the mid-1970s to a few thousand tons in most years. More than half of the catches were exported as canned products. This fishery is carried out mainly with purse seines from October to March. Fishing for yellowfin tuna, skipjack, swordfish *(Xiphias gladius)* and Pacific sierra *(Scomberomorus sierra)* also goes back to the 1940s but landings have never been large.

Demersal trawling for hake *(Merluccius gayi)* on the shelf off northern Peru began in 1965 and in some years produced over 100 000 t (1978 even 300 000 t). The fishery, carried out from Paita, Chimbote and Callao, is now strictly regulated. Licences were also granted to trawlers from Cuba, Poland and Japan. About 25 shrimp trawlers operate from Caleta La Luz.

Guano

A great number of seabirds feed on the rich fish resources, especially on the shoals of anchoveta near the surface. Guanay cormorants *(Phalacrocorax bougainvillii)*, boobies (piquero, *Sula variegata*) and pelicans (alcatraz, *Pelecanus thagus*) breed from October to February on the Islas Pescadores and on other islands near the coast and produce large quantities of guano from their droppings. The use of guano as fertilizer was already known to the old Indian cultures, particularly for the manure of Palta trees. Remains of guano have been discovered in tombs from the fifth century B.C. and guano birds were depicted on many pieces of ancient ceramics (Fig. 16). The exploitation of guano was accompanied by ceremonies for the guano god, Huamancantec, to whom offerings were presented. When, in around 1450, the coastal areas were conquered by the Incas, protection measures were taken and the death penalty was introduced for people who entered the guano islands during the breeding period and who killed the birds.

After the conquest of Peru in 1531 the Spaniards showed little interest in the exploitation of guano for a long time so that it grew to layers 45 m thick. After independence, and with the further development of agriculture, demand for fertilizers increased. From 1840 until 1875 about 20 million t of guano were shipped to North America and Europe. By 1900 the guano deposits had been depleted. A state monopoly was established in 1909 and protective measures were introduced; further exploitation of guano was carefully regulated. From 1909 to 1957 only 6.4 million t were taken, with a maximum of 330 000 t in 1956. The advance of artificial fertilizers reduced the importance of guano. However, the guano industry was still a most lucrative business and, therefore, the management, afraid that fisheries exploitation would take the food of the birds, used all their political and economic power for a long time to prevent the development of a fishery on small pelagic fish, the abundance of which had been known for many years. Conflicts between the interests of the guano and fish meal industries also arose after the dramatic development of fishing.

The number of guano birds was estimated at 45 million in 1956. During the "El Niño" of 1965-1966 their numbers fell from 17 million to 4 million and in 1982-1983 from 9 million to only 2 million. It seems that the decrease in the bird populations is caused by a reduced availability of anchoveta to the birds; this is the result of combined effects of heavy fisheries exploitation and of natural changes during "El Niño" periods.

Pelagic Fisheries for Fish Meal and Oil

Prompted by the lack of fish imports during World War II, a modest canning industry developed and anchoveta *(Engraulis ringens)* and sardines *(Sardinops sagax)* were caught in inshore waters with beach seines to supply the factories. After the war production of canned fish suffered from competition by better quality products from overseas. On the other hand, there was a growing world demand for fish meal as a high protein

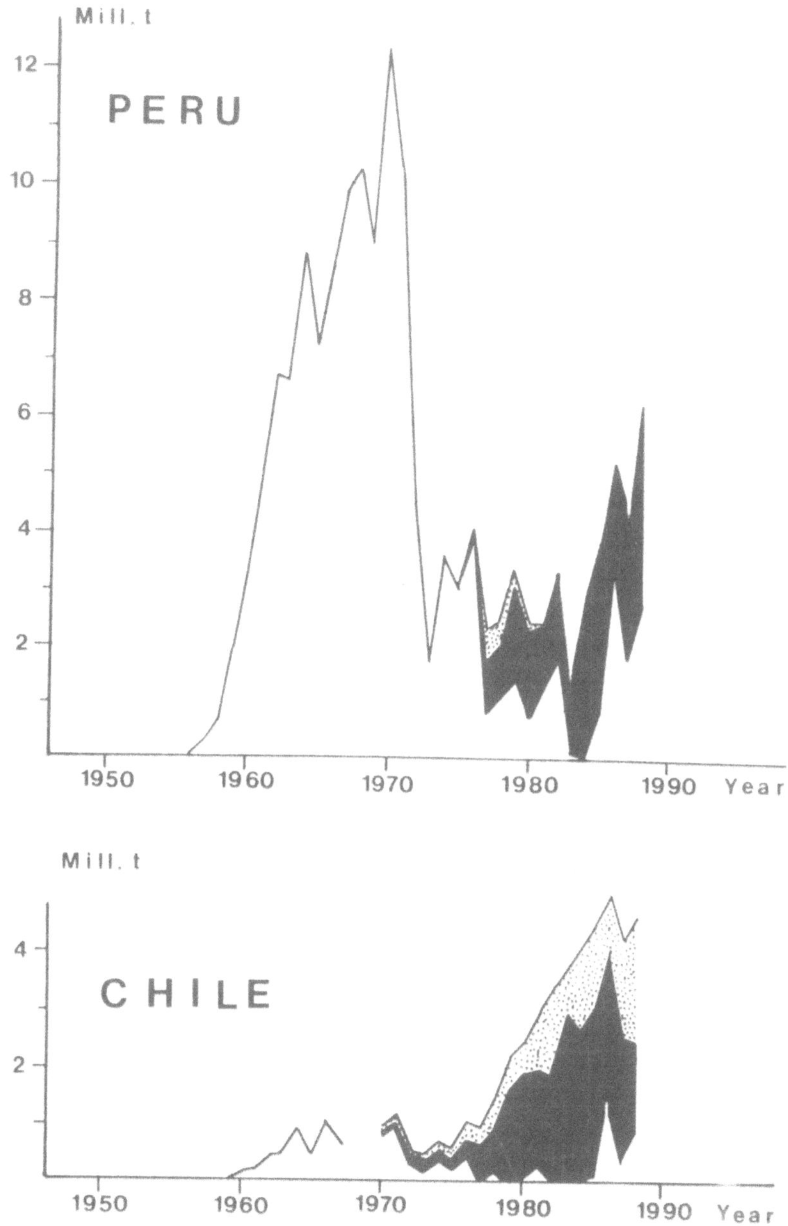

Fig. 124. Landings of pelagic fish for industrial purposes in Peru *(above)* and Chile *(below)*. *White areas*, anchovy *(Engraulis ringens)*; *black areas*, sardine *(Sardinops sagax)*; *dotted areas*, jack mackerel *(Trachurus murphyi)*. (FAO 1947-1989)

additive to animal feedstuff and for oil for industrial purposes, e.g. the production of margarine. Therefore, factories began to produce these commodities in larger quantities from anchoveta catches. Already, in 1940, experiments had been made in Peru to use fish meal as a fertilizer in agriculture. The collapse of the Californian sardine fishery in 1951-1952 was a further major impetus for the Peruvian fish meal industry. Extensive machinery was transferred from California and substantial foreign capital was invested in the Peruvian fishing industry. This led to an unprecedented dramatic increase in anchoveta catches and fish meal production (Figs. 124 and 97).

In 1950 anchoveta catches in Peru amounted to only 438 t, in 1952 to 16 000 t, in 1960 to more than 3 million t, and they peaked in 1970 with 12.3 million t. During the same time fish meal production rose from insignificant quantities to almost 2.25 million t in 1970 (40% of the world production), of which more than 90% was exported. The number of fishmeal factories increased to more than 170, mostly in Chimbote and Callao, and the number of boats to more than 2000. Altogether, the uncontrolled and hectic expansion of this industry resulted in large overcapacities in reduction plants and fishing vessels. Many companies collapsed due to economic problems.

Scientists had warned against overexploitation of the anchoveta resources. The first protection measures were introduced in 1964 with the cessation of anchoveta fishing for 1 month. Later, further regulations were implemented on the basis of scientific investigations and, in particular, included an annual catch quota of 10 million t and two extended closed seasons to protect the recruitment of anchoveta stock. However, all these measures did not prevent a further expansion and, finally, collapse of this fishery.

In 1972 a sudden catastrophic decline in anchoveta catches occurred (Fig. 124) and most of the reduction plants had to close down. Scientific investigations revealed that this was caused by a combination of two events: low recruitment of the heavily fished anchoveta stock, and detrimental effects of a strong "El Niño". Through better management it was possible to maintain an anchoveta catch between 1 million and 3.5 million t in most of the following years but the new "El Niño" event in 1982-1983 again reduced the catches in 1983 and 1984 to low quantities. Since 1977 the fleet has concentrated increasingly on sardine, which was also regulated by catch quota and other management measures. Sardine catches amounted to 3.4 million t in 1988. Also, the stocks of jack mackerel *(Trachurus murphyi)* were exploited for fish-meal production

Fig. 125. Peruvian-built purse seiner, length 25 m, 120 GRT, 335 hp. (Fishing News Int 4/1966, p.51)

and landings of more than 500 000 t were made in 1977. This species is also fished in large quantities offshore by long-distance fleets from the USSR, Cuba, Poland and Bulgaria.

The anchoveta is distributed in cool (<18° C) upwelling coastal waters from northern Peru to mid-Chile, an area of about 2000 km from north to south (Fig.123). Off Peru the main fishing grounds are between Pisco and Chimbote, and off Chile they are between Arica and Coquimbo. In Chile the fish can be caught very close to the shore, often in the breakers zone, while, in Peru, the shoals are concentrated a little farther offshore but usually at a distance of not more than 30 miles. The main fishing season in Peru is from October to February. In Chile there are two main seasons: January - March and September - November.

Fishing is carried out from vessels (bolicheras) that are mostly 18 - 25 m long with crews of 10 - 12, who operate purse seines with the help of small skiffs (Fig.125). New ships with 1200 hp engines reach lengths of 36 m and have a loading capacity of 350 tons. The purse seines (boliches), introduced in 1947, measure about 400 x 50 m and are made of synthetic material. The vessels are fitted with echosounders for fish detection and with power-blocks to facilitate the hauling of large catches. Since 1957 pumps have been used to transport the fish from the net into the fish hold. Around 12 000 of a total of 20 000 Peruvian fishermen are employed in the fisheries for processing purposes.

As the coast is generally unsuitable for the construction of ports, fish landings are concentrated in a few places, usually localities where drinking water, food and other supplies are available in the desert region. Often purse seiners have to be unloaded en route to the reduction plants, which are simple installations with mostly primitive living quarters and from which fish meal and oil are transported to cargo vessels at anchor. Major places for fish-meal production and export in Peru are Chimbote (with annual landings of 3.5 million t in the late 1960s), Supe, Callao, Pisco, Ilo and Mollendo. As a result of the overall decline in fish catches and frequent serious economic problems, including a diminishing demand for fish meal in the 1980s with a drop in price, the number of fish-meal factories has decreased to about 30 in recent years. Large plants process 100 t or more of fish per hour.

In Chile the pelagic fishery for reduction purposes started in 1957, first from Arica and then from Iquique. In earlier times some quantities of anchoveta (in Chile called anchoa) had been salted and dried for human consumption but, now, landings increased rapidly to about 1 million t in 1966 (Fig.124). Besides Arica and Iquique Pisagua, Patillos and Tocopilla also developed to become major places for this industry. As in Peru, the industry suffered from sharp declines in the anchoa landings in the early 1970s. The losses were, however, balanced by increasing catches of sardines and jack mackerel. The "El Niño" in 1982-1983 turned out to be especially advantageous to Chilean fisheries. Landings peaked in 1985-1986 with over 5 million t per year. Sardine catches reached levels of more than 2.5 million t and jack mackerel landings of 2 million t. The latter were caught mostly by larger purse seiners operating from Talcahuano. The Chilean fish-meal production amounted to 8500 t in 1952, reached more than 100 000 t by 1963 and over 1 million t in the 1980s, making Chile the leading country in the world in this sector.

Freshwater Fisheries

Compared to marine fisheries yields from freshwaters are unimportant in both countries. Landings in Peru of a large number of species caught by about 35 000 canoes and boats in the Amazon River system are estimated at 46 000 t. A further 6000 t per year are fished by small-scale and recreational fisheries in Lake Titicaca. Here, the stocks consist of native species of the genus *Orestias*. Trout *(Salmo gairdneri* and *Salvelinus* sp.*)* are produced extensively in aquaculture in floating cages.

In Chile, rivers and lakes are relatively poor in fish and total catches do not exceed a few thousand tons. The most important species are the silversides *(Atherina* sp.*)*. Carp *(Cyprinus carpio)* were imported from Europe early in this century. Other successful introductions were various salmonids like trout *(Salmo fario* and *S. gairdneri)* and coho salmon *(Oncorhynchus kisutch)*, reared on farms.

5.2.6 South and Southwest Africa (Namibia)

Development of fisheries to a large-scale industry in southern Africa is a fairly recent event, starting with the expansion of the pilchard (sardine) fishery in the 1940s. The basis for this development is the abundance of large stocks of pelagic fish such as the southern African pilchard *(Sardi-*

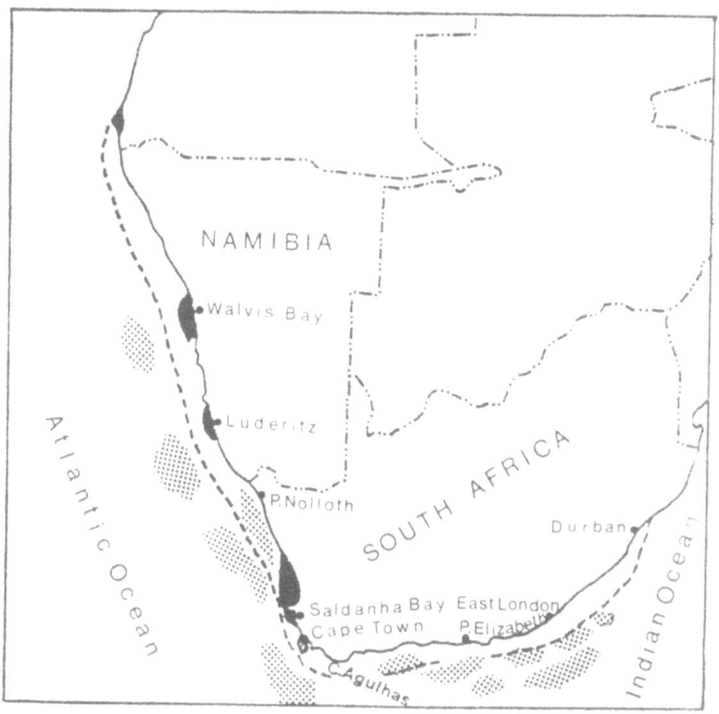

Fig. 126. Fishing grounds and fishing ports of South and Southwest Africa (Namibia). *Black area*, major pilchard fishing grounds; *shaded areas*, trawling grounds

nops ocellatus), the anchovy *(Engraulis capensis)* and the Cape horse mackerel *(Trachurus capensis)*, along the west coast between Cape Agulhas and Angola in the cold waters of the Benguela Current (Fig. 126). Upwelling of cold deep water, rich in nutrients, results in high plankton production and, consequently, favourable feeding conditions for fish, birds and marine mammals. An impressive indication of the huge abundance of fish is the production of guano by fish-eating birds, particularly Cape cormorants, gannets and Jackass penguins, which breed on the islands close to these shores.

In contrast to the west coast, the waters off the east coast with the warm Agulhas Current are much less productive. Fish landings from this area are very small in comparison to the Atlantic side.

Total fish landings in the South African Republic and Southwest Africa/Namibia, which amounted to only 68 000 t in 1938, reached 1 million t per year in the early 1960s and 2 million t in 1968 but decreased thereafter to an average of 1.25 million t during the 1970s and 900 000 t in the 1980s (Fig. 99). During these years the composition of the landings has changed considerably (Fig. 127).

Developments until Around 1900

For a long time fisheries developed at a slow pace as the local market was initially rather small, transport facilities were very limited and the main consumption areas in Europe and North America were far away. Rock paintings prove that the natives, notably the Bushmen, already went fishing in ancient times (see Sect. 1.1.2). The Hottentots, nomadic cattle-breeders who fished in the Orange River and other waters, used fish only as a secondary food and especially at times when they had lost their livestock. In contrast, fish utilization by Bantus was limited by various taboos and still, today, it is difficult to promote fish consumption by this group.

The first fishing activities by Europeans date back to the voyages of discovery undertaken by the Portuguese, who had to supply their crews with food on their way to and from India and further east. In the early 17th century Dutch, British and French ships also appeared in these waters. At this time Dutch merchants organized the hunting of Cape fur seals in Table Bay, Saldanha Bay and around the Seal Islands for the production of blubber oil. This sealing reached a peak in around the 1650s and is still carried out to some extent today (see Sect. 4.2.2).

After the foundation of Cape Town in 1652 on behalf of the Dutch East India Company, an inshore fishery gradually developed in Table Bay, Saldanha Bay, Algoa Bay and False Bay. In the beginning only a few fishing permits were granted in the interests of the Company. Set gill nets, lines and beach seines (introduced in 1665) were used for the capture of mullet *(Mugil cephalus)*, steenbras *(Lithognatus* sp. and *Petrus* sp.*)*, sole *(Austroglossus* spp.*)* and other species. Fish not consumed fresh were dried and salted to serve as staple food for the provision of ships calling at Cape Town. Fish played an important role in the Cape area as food for soldiers and slaves when meat production was still insufficient. Fishing was soon passed from the Dutch settlers almost entirely into the hands of "Cape Malays", Muslim slaves, freedmen and their descendants. As "Easterlings" these fishermen maintained a monopoly until the 1880s. In around 1880 about 2000 of these fishermen in the Cape Province with 350 boats caught 5000 t of fish per year.

Under British dominion from the beginning of the 19th century (1795/1806) all restrictions on commercial fishing were removed but there was, at first, little change in the fishery. In around 1830, when 40 small boats with 200 men were engaged in fishing, some merchants started to export dried fish to Mauritius, a trade that still persists today. After 1890 an increasing demand developed for fish as cheap food for the workers in the gold and diamond mines that had been newly discovered in Transvaal and the Orange Free State. The construction of railroads in the 1890s permitted the transport of fish from Cape Town to the larger centres of consumption in Johannesburg and the Witwatersrand. The immigration of 50 000 Chinese, fish-eating people, in 1905 created further needs. Also, during the following years fisheries development was closely linked to the expansion of fish consumption in the interior and to improvements in railway connections. As demands were larger than the local production, substantial quantities of fishery products were imported from Europe and North America.

In Southwest Africa, a German colony until World War I and later mandatorily governed by South Africa, fishing activities were negligible and limited to some beach seining in Lüderitz Bay, from where fish were transported to Cape Town.

Trawl Fishing

The first trials from Port Elizabeth in around 1878 with a steam tug to trawl in Algoa Bay were not very successful. But good catches from a schooner from the USA, operating with purse seines off the Cape Peninsula in 1889, prompted attempts to establish a trawl fishery. In 1893 the government passed a Fish Protection Act which set the pattern for further fisheries development. A little steam-ship was employed in 1897 to investigate trawling possibilities on Agulhas Bank with promising results. European-type steam trawling developed from 1903 with ships of 210 GRT, mainly imported from England, and in 1906 the fleet already consisted of 15 vessels. The development started from Durban and Cape Town but trawlers also sailed from a number of other ports along the east coast, mainly East London. Landings by sailing ships for line fishing supplemented the catches of the steam trawlers. Since 1922 fishing grounds west and northwest of Cape Town up to Port Nolloth were exploited and the trawl fishery concentrated more and more on this port (Fig. 126). Over the years a number of trawler companies merged to form larger enterprises until one company owned most of the ships. This company developed regular transport of fish by rail in insulated and later refrigerated, wagons to Johannesburg and many other cities in the interior.

In the 1950s the trawler fleet consisted of about 50 vessels, mostly fairly small side trawlers undertaking voyages of up to 5 days and usually operating within 50 nautical miles of the coast. As coal was cheap and easily available, most ships were still steamers. But, in the 1960s, new larger diesel trawlers were put into service and after 1964 stern trawlers were built in Cape Town and Durban. The number of trawlers increased to 70, of which 40 operated from Cape Town where more than 80% of the trawl catches were landed. Today, the trawler fleet consists of 60 deep sea trawlers and 55 vessels for inshore fishing; 5 of these ships are larger than 1000 GRT.

The landings of the trawlers amounted to 20 000 t in 1940 and increased to more than 100 000 t by the mid-1960s. Of these catches 70 - 80% consist of Cape hakes *(Merluccius capensis* and *M. paradoxus)*. Of the many other species only the kingklip *(Genypterus capensis)* should be mentioned as a highly esteemed food fish.

Since 1962 a fishery, mainly for hake, was developed by a number of other countries, particularly Spain and the USSR, in the waters off South and Southwest Africa with production reaching over 500 000 t per year in the late 1960s. Part of the Spanish catches were landed in South Africa. After the introduction of a 200-mile EEZ in 1977 South Africa granted catch quota not only to the national fleet but also to foreign vessels from Spain, Portugal, Israel, Japan and Taiwan against payment of licence fees.

During World War II possibilities to import canned fish products from Europe and the USA were very limited. This led, since 1942-1943, to the utilization of pilchards from home waters. To some extent these sardines were already being caught with beach seines since long ago. Now, the existing rock lobster canneries began to produce pilchard preserves. The new fishery provided excellent earnings for people in hitherto rather poor areas. The government promoted these developments by the formation of the Fisheries Development Corporation of South Africa in 1944. When the Californian sardine fishery collapsed in 1951-1952, at a time when a large demand for fish meal and oil on the world market guaranteed good profits, the pilchard fishery in southern Africa developed very rapidly (Fig. 127). Within a few years the annual landings had increased from 7500 t in 1943 to 500 000 t in 1953 and surpassed 1 million t in 1965. The

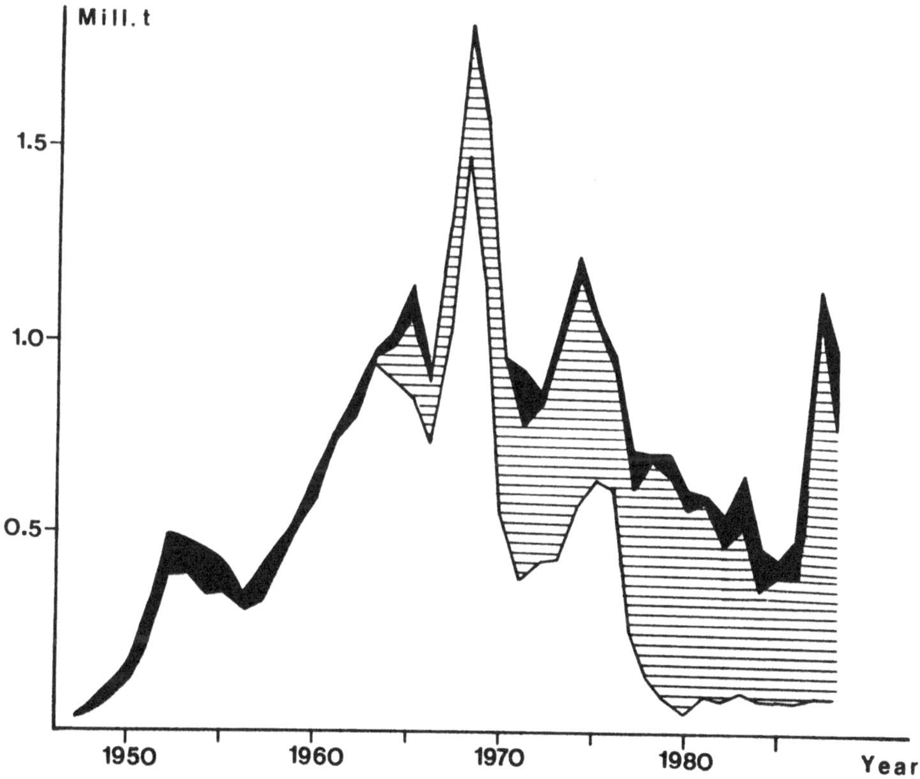

Fig. 127. Landings of pelagic fish for industrial purposes and canning in South and Southwest Africa. *White areas*, pilchard *(Sardinops ocellatus)*; *hatched areas*, anchovy *(Engraulis capensis)*; *black areas*, maasbanker *(Trachurus capensis)*. (FAO 1947-1989)

peak landing of 1.8 million t in 1968 was followed by a decline to between 500 000 and 700 000 t and by another increase in recent years. During the first 20 years the catches consisted mostly of pilchard. Some quantities of maasbanker *(Trachurus capensis)* were caught as well. Since 1963 anchovy were also fished and, after a steep decline in the stocks and catches of pilchard in around 1970, anchovy formed the majority of the landings, particularly after 1977.

As in California, heavy exploitation of the pilchard resources coincided with an increase in the anchovy stocks. As biological investigations indicated that the main increase in anchovy was between 1956 and 1962, i.e. already before the high pilchard catches, it is doubtful whether the rise in anchovy stocks can be attributed to the reduction of pilchard by fishing. Possibly the increase in anchovy had negative effects on the recruitment of pilchard. There is no doubt that seabird populations, especially the jackass penguins which have a rather limited forage range, and guano production were affected by the decline in pilchard resources.

From the early 1950s the coastal fisheries for pelagic fish were regulated by catch quota, closed seasons and limitations on the number of canneries and reduction plants as well as on the number of catcher vessels. However, all these measures did not prevent a serious decline in the pilchard stock. As in California, regulations were partly eluded in the mid-1960s by employing large converted whaling motherships to operate as processing factory ships outside the 12-mile zone.

Pilchard are distributed all along the coast from Natal to Angola but the fishery is concentrated on two small areas off Walvis Bay and St. Helena Bay. Less important fishing grounds are off Natal and Lüderitz (Fig. 126). Most likely, the pilchard concentrations off Namibia and those in the south belong to different stocks. The fishing season runs from April to September in Walvis Bay and from January to the end of July in South Africa. In the south it is preceded in November-December by a fishery on maasbanker and mackerel *(Scomber japonicus)*.

Around 150 ships are engaged in the fishery from the Cape Province and about 80 in Southwest Africa. Boats with lengths of 18 - 25 m are mostly owned by the factories and other vessels fish under contract during the season. Catches are made with purse seines (300 x 50 m), mostly at night and usually not farther than 30 km from the shore. With annual landings of more than 500 000 t Walvis Bay is one of the largest fish-landing places in the world. The fish are unloaded with vacuum pumps over long jetties directly to the plants.

The fish processing industry started towards the end of World War II with the establishment of the first cannery and reduction plant at Lamberts Bay. The first fish-meal plant in Walvis Bay was erected in 1948. The number of factories increased quickly to 43 by 1954 but, later, followed a consolidation with improvements in the efficiency of the installations and, in the 1960s, 15 plants remained in the Cape Province and eight in Southwest Africa. The workers in the factories live in compounds attached to the plants. In the Cape Province they are usually "coloured" people while in Namibia workers are mostly Ovambos contracted for each fishing season and flown to the coast temporarily from their homelands.

The production of fish meal and oil soon became much more important than the canning sector. In the 1960s it reached a peak of over 350 000 t, of which most was exported, mainly to the USA, Great Britain and the German Democratic Republic. In later years fish-meal production gradually declined to less than 150 000 t after 1980 and exports fell even further to insignificant quantities since the meal was needed as animal feed in South Africa.

Rock Lobster Fisheries

The fishery for rock lobster *(Jasus lalandii)* has a long tradition in the Cape Province. Already by around 1890 about 70 t of tail meat were canned and exported every year, mostly to France. In the 1930s, before the mass production of pelagic fish, lobster catches made up 70% of the total value of the fish production. By around 1930 there were 17 factories between Houtbay near Cape Town and Lüderitz; these exported annually 15 million lobsters, mainly as deep-frozen tail meat, to the USA. When the catches reached 15 000 t a regulation was introduced, in 1947, to limit the exports to 2700 t. Later, this quota was reduced to around 1700 t tail meat per year. Other protective measures included closed seasons and minimum lobster sizes. Today, the catches are around 5000 t in South Africa and 1500 t in Namibia. They are supplemented by 1000 t of spiny lobsters *(Palinurus gilchristi)*, caught by trawling in Natal waters and landed mainly in Port Elizabeth. Also from Cape Town a lobster fishery for *Jasus tristani* is operated around Tristan da Cunha and on the Vema Seamount about 500 miles northwest of Cape Town.

The rock lobsters are caught on rocky grounds and in kelp beds close to the coast with baited hoop nets, i.e. net bags with a ring-like iron frame. Small motherships with dinghies, each dinghy with 8 - 10 nets operated by two men, also fish further offshore on deeper grounds. Some of the catches are exported alive by air, mostly to France. Compared to these lobster fisheries some other crustacean exploitations are rather unimportant.

Other Fisheries

Hook-and-line fishing for snoek *(Thyrsites atun)* along the west coast from Mosselbaai to Walvis Bay also has a long tradition. For a long time salted and dried snoek was the only product available for export in larger quantities. During the last 25 years annual catches have fluctuated between 10 000 and 20 000 t. Many snoek are caught by trolling and there is also, now, important sport fishing in this field.

Another traditional fishery from the times of the first Dutch settlers is carried out with beach seines in the warm waters between Cape Town and Natal to catch mullet. In earlier times the catches were used as supply for the ships sailing to and from East India. Today, the mullet fishery is mostly in the hands of fishermen of Indian origin.

Since 1960-1961 a tuna industry has developed with long line catches of albacore (37%), bigeye tuna (19%), southern bluefin tuna (16%) and some other species. Average annual catches since 1977 have been around 42 000 t.

A rather special event is the capture of lantern fish *(Lampanyctodes hectoris)*, of which 42 500 t were caught by purse seine in 1973 for the production of fish meal and oil. Usually these fish live at depths beyond the reach of purse seines. Present catches are only around 4000 t per year.

Gans Bay is a centre for shark fishing. The fins are dried and exported, mostly to Hong Kong, where good prices are fetched for this product.

On rocky grounds along the coast between Cape Agulhas and Lamberts Bay divers collect abalone *(Haliotis midae)*. In earlier times the shells were exported for the production of mother-of-pearl. Today, the meat is canned or frozen for export, mostly to Japan. Present annual catches are around 680 t.

Generally, fisheries in South Africa are more diverse than in Namibia, where the fish industry is limited to mass production of fish meal and to rock lobster fishing.

5.2.7 Europe

Fish landings in Europe increased from about 2 million t in 1900 to 5.7 million t in 1938 (Fig. 128). The two war periods resulted in substantial reductions during the years 1914-1918 and 1939-1945. After World War II production rose steeply, particularly during the 1960s, to around 12 million t. After 1970 the landings fluctuated around a level of nearly 13 million t.

The strong increase during the 1960s was mainly due to a vast expansion of fisheries for pelagic fish caught for the production of fish meal and oil, particularly of the purse seine fishery in Norway and, later, also the young herring trawl fishery for industrial purposes in Denmark. The total landings of pelagic fish rose from a relatively stable level of about 3 million t in the period 1954-1963 to over 6 million t in 1976. During earlier years the catches consisted mostly of herring. When the herring fisheries expanded greatly in the mid-1960s, herring catches dropped from 3.2 million t in 1965 to only 900 000 t in 1969. The decline was due to the collapse of the Norwegian spring-spawning herring fishery. Overexploitation also reduced the herring stocks in the North Sea and adjacent waters and, thus, the total herring catches decreased to less than 400 000 t by 1976. At the same time the industrial fisheries shifted gradually to other pelagic species, mainly mackerel *(Scomber scombrus)*, capelin *(Mallotus villosus)*, sprat *(Sprattus sprattus)*, sandeel *(Ammodytes lancea)*, Norway pout *(Trisopterus esmarki)* and blue whiting *(Micromesistius poutassou)*. Most of the stocks of these species were heavily overexploited but, as the fisheries shifted from one species to the next, the total landings were maintained at high levels. The latest example of the depletion of such a resource is that of the capelin stock in the Barents Sea. The recent decline in total fish production in Norway is a result of this event (Fig. 128). The yields from demersal fisheries have not undergone such extreme variations but all commercial fish stocks in the North Atlantic are heavily exploited and are subject to catch quota, which were introduced in the mid-1970s and which became more and more rigid after serious economic and social problems due to overfishing.

The national fisheries in Europe vary widely in structure, size and economic importance to each country. A brief synopsis may illustrate the main characteristics and differences.

The fisheries of **Great Britain** produced annually about 1 million t with rather stable yields from the beginning of this century (Fig. 128). Until the 1960s roughly two-thirds of the landings were made in England and

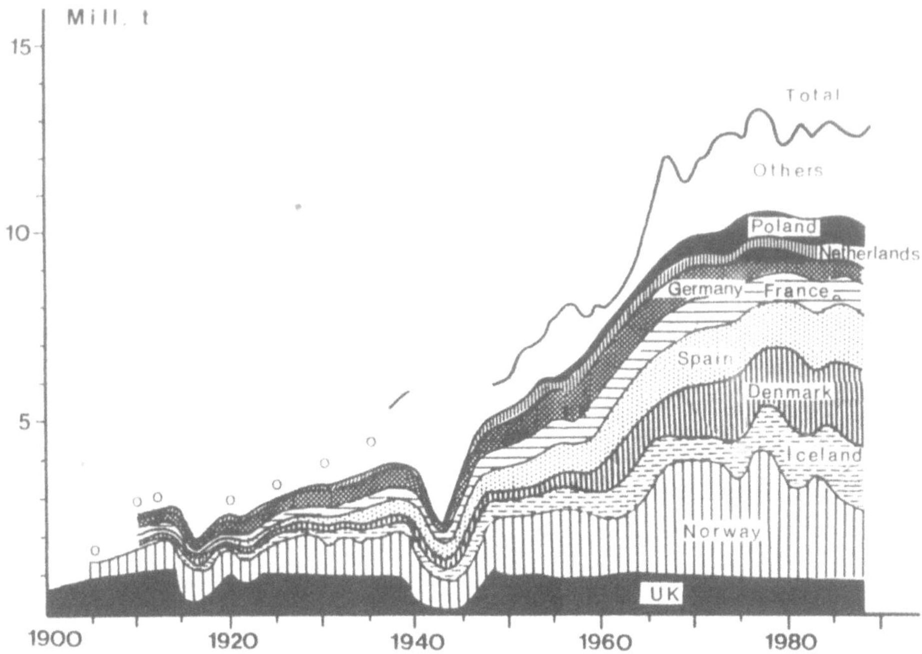

Fig. 128. Fish landings in European countries 1900-1988: circles for total landings 1905-1935 indicate estimates based on ICES data. (ICES 1906-1990; FAO 1947-1989)

Wales and one-third in Scotland. With the loss of fishing grounds around Iceland and in the Barents Sea-Bear Island-Spitsbergen area during the 1970s, as a result of claims by coastal states for exclusive fishing zones, a major change took place, and in the 1980s two-thirds of the catches were landed in Scotland. While, in the 1930s, more than 50% of the catches were made in northern waters, about 40% in the North Sea and the rest in waters west of Britain, now over 50% of the landings come from the North Sea; next in importance are the waters west of Scotland and catches further north are small. Haddock, cod and plaice are the most important demersal species; mackerel and herring dominate the landings of pelagic species. In recent years the annual fish production amounted to about 850 000 t, used almost entirely for direct human consumption. This quantity did not satisfy the large demand. Therefore, 650 000 t fish for direct human consumption and fish meal and oil equivalent to a further 1.4 million t were imported. In contrast, fish exports were in the order of 500 000 t per year. The British fishing fleet decreased from about 15 000 vessels in around 1970 to 7900 ships in 1986. Since new vessels are larger and more powerful, the catching capacity has not declined much. In the present fishing fleet over 6000 vessels are under 12 m in length. Of the remaining larger ships more than half are registered in Scotland. The deep-sea fleet comprises 205 ships. It is rather different from the traditional distant-water fleet which was based in Hull and Grimsby. Most of these modern trawlers operate under Anglo-Spanish management from ports in southwest England and Wales, while only 12 ships are still based on Humberside. Landings from marine fisheries are supplemented by harvests from fish-farming. In 1986 aquaculture produced about 15 000 t of trout. Farming of Atlantic salmon, particularly in Scotland, is expanding rapidly.

The fishery of **Ireland** is mainly an inshore and middle-distance activity, mainly carried out by small trawlers. The fishing fleet comprises some 2800 ships. With assistance by the European Economic Community (EEC) the Irish fishery has developed substantially during recent years. The annual fish landings increased from about 30 000 t in the early 1960s to 250 000 t in the second half of the 1980s. More than 70% of the catches are exported, mostly to Great Britain and Germany. Since the 1970s mackerel has replaced herring as the most important commercial species. Often, mackerel catches are directly transloaded to foreign ships at sea.

Fish landings in the **Netherlands** during the first half of this century remained fairly stable at a level between 100 000 and 170 000 t. After World War II they increased to around 300 000 t and, since the early 1980s, to 400 000 and 500 000 t. Main branches of the fishery are: the inshore fishery with some 580 vessels, catching mainly plaice, sole and cod; beam trawling for flatfish in the North Sea, Skagerrak and Irish Sea; and pelagic and demersal otter trawling for horse mackerel *(Trachurus trachurus)*, mackerel, herring and other species in the North Sea, Celtic Sea and Irish Sea. Traditionally, herring was by far the most important species in Dutch fisheries but, due to the overexploitation of North Sea herring stocks, catches declined steeply after the mid-1960s and, since the late 1970s, other pelagic fish species have been caught instead. Herring landings increased again to up to 100 000 t in the late 1980s after the recovery of the North Sea herring stocks. The Dutch trawler fleet was modernized during recent years and now comprises fairly large vessels of up to 1200 GRT and with very powerful engines of up to 6600 hp, which fish intensively on plaice and sole. The port of Urk on the Zuider Zee developed to become a central place of the fresh fish trade. Other important branches are the shrimp trawl fishery in the Waddensea for *Crangon crangon*, producing around 5000 t per year, and the cultivation of blue mussels *(Mytilus edulis)* with widely fluctuating harvests of about

100 000 t. The Netherlands are very active in fish trading. The exports of fish and fishery products of around 950 000 t per year are much higher than the imports of about 500 000 t.

With only 66 km of coastline **Belgium** has a relatively small fishery. About 200 small and medium-sized motor vessels, operating from the ports of Oostende, Zeebrugge and Nieuwpoort, catch around 40 000 - 50 000 t of fish per year, mainly in the North Sea, the Channel and the Irish Sea. In the 1970s the fishing fleet still comprised more than 300 ships and several fished in waters around Iceland. Beam trawling is the most important fishing method, producing 55% by weight and 70% by value of the catches. The main species caught are plaice, cod, sole, whiting, herring and shrimp. The national production of fish does not cover the demand so that more than 500 000 t are imported, while over 100 000 t are exported or re-exported.

Fisheries production in **France** fluctuated during the years 1907-1939 between 200 000 and over 300 000 t per year. After World War II it rose to about 500 000 t by the early 1950s, to 750 000 t in the mid-1960s, and to almost 900 000 t in the late 1980s. The fishery has a wide range of different branches and catches a large variety of species. The fishing fleet consisted in the mid-1980s of about 13 700 vessels, mostly smaller than 50 GRT. Only some 480 ships were larger than 100 GRT and three were larger than 2000 GRT. The small vessels of the coastal fisheries, usually owner-operated, concentrate on fish of high value and quality. Around 200 larger ships belonging to companies fish in EEC waters, mainly the Gulf of Biscay, the area west of Great Britain and Ireland, the Channel and the North Sea, for saithe, cod, haddock, hake, whiting and other species. The traditional fishery for cod off Newfoundland, West Greenland and Iceland has declined strongly since 1969, and now there are only nine distant-fishing vessels left which trawl for cod and deep-sea prawn *(Pandalus borealis)*. A fleet of 42 ocean-going tuna seiners operate in the waters off West Africa and in the western Indian Ocean. Catches consist mostly of yellowfin tuna and skipjack. In contrast, the traditional trolling fishery for albacore in the Gulf of Biscay has decreased in importance. Most fish landings are made on the Atlantic coast and in the Channel, while catches in the Mediterranean make up only about 5% of the total. Boulogne is the most important fishing harbour, followed by Lorient, Concarneau and La Rochelle. Of high economic value is the culture of oysters, mostly *Crassostrea gigas*, along the coasts of Brittany and further south and producing a harvest of around 100 000 t per year. This culture, with Arcachon as the centre, has developed since the mid-19th century when the natural banks of European flat oysters *(Ostrea edulis)* were exhausted. Other important mollusc fisheries are for blue mussels *(Mytilus edulis)* and scallops *(Pecten maximus)*. Almost all fish production is used for human consumption. A flourishing canning industry concentrates on tropical tuna, sardine and mackerel. Trade with fishery products has greatly expanded during recent years. Imports of over 1 million t per year are considerably larger than exports of roughly 300 000 t.

In **Germany** the annual fish landings reached 180 000 t by 1913. After World War I the quantities increased gradually to over 200 000 t by 1924 and more than 600 000 t in 1937. At that time almost half of all catches were made in the North Sea and 16% each around Iceland and off Norway. With the reconstruction of the fishing fleet after World War II the landings in the Federal Republic of Germany (FRG) rose to between 700 000 and over 800 000 t in the 1950s. These catches included up to 200 000 t of herring fished in the North Sea. After the decline of the herring fishery for economic reasons, the FRG landings fluctuated during the 1960s and early 1970s at around 650 000 t and decreased further to 400 000 t when yields declined as a result of overexploitation of major fish stocks in the Northeast Atlantic as well as in the Northwest Atlantic. The German distant-water trawler fleet was struck very hard when, in 1977, about three-quarters of the traditional fishing grounds were lost due to the introduction of EEZs. The worst was the loss of access to Icelandic waters which were the main fishing area besides the North Sea. Consequently, the number of large stern trawlers, operating from Bremerhaven and Cuxhaven, declined from 140 in 1968 to only 12 in 1988. At the same time landings decreased from 460 000 to 120 000 t. The coastal and offshore fleet of small trawlers in the German Bight, North Sea and Baltic, which comprised more than 900 ships in 1970, did not suffer to the same extent and still consists of 550 ships with catches of about 100 000 t per year. Cod, shrimp and blue mussels are the main targets of the coastal fisheries. The distant-water fleet catches predominantly cod, saithe, redfish, herring and mackerel. Now, as before, the North Sea is the most important fishing area. Catches in the waters around Greenland and west of the British Isles have partly compensated the loss of fishing opportunities around Iceland. Of the German landings 95% are used for human consumption. Landings are, however, much smaller than the demand and therefore more than 900 000 t of fish are imported. Furthermore, the FRG is the largest importer of fish meal in the world. If the 400 000 t of fish meal and 150 000 t of fish oil converted into the weight of fresh fish caught, and if that quantity is added to the imports of fish for human consumption, one arrives at a figure of 2.7 million t. However, about 200 000 t of fish meal are re-exported to other countries in Europe. The fishery of the German Democratic Republic (now united with FRG) reached landings of 380 000 t in 1975, but catches have declined during later years to less than 200 000 t because of the loss of fishing grounds when EEZs were established. In 1972 the distant-water fleet of the GDR consisted of 69 fishing vessels and 21 auxiliary ships, mostly transport and factory ships. They were based in the ports of Rostock and Saßnitz. Like the fleets from other socialist countries, the GDR trawlers operated in many waters of the world's oceans, especially in the Northeast and Northwest Atlantic, off Northwest and Southwest Africa and in the Northeast Pacific. Until the mid-1970s catches consisted mostly of cod, saithe, redfish, Greenland halibut and herring. Later, with less access to these resources, other species from offshore waters, in particular horse mackerel *(Trachurus* spp.*)*, predominated. The latter were mostly landed directly in African ports since they were hardly marketable in the GDR. The coastal fishery of the GDR in the Baltic, mainly for herring, cod and flatfish, was carried out by a fleet of 220 cutters with lengths of up to 27 m. Catches in the Baltic during recent years amounted to around 70 000 t per year.

The fishery in **Denmark** has always been entirely based on small vessels operating in the North Sea, Skagerrak, Kattegat and the Baltic Sea. In 1985 there were 3366 ships larger than 5 GRT, of which 64% were smaller than 20 GRT and only 9% larger than 100 GRT. The main fishing ports are Esbjerg and Hirtshals. The landings up to World War II did not exceed 90 000 t per year. After the war they rose fairly steadily from 200 000 t in 1948 to 2 million t in 1980 and, thereafter, fluctuated around 1.8 million t. Of the catches made by the industrial fishery 70 - 80% were for the production of fish meal and oil and for direct feeding of animals. This fishery began to expand with the start of trawling for young herring in the southeastern North Sea in 1954. Total Danish herring landings reached more than 300 000 t by 1964. When herring catches declined due to overexploitation of the stocks, the industrial fishery shifted increasingly, in the early 1970s, to other species, especially sandeels *(Ammodytes* spp.*)*, Norway pout *(Trisopterus esmarki)* and sprat *(Sprattus sprattus)* in the North Sea. Sandeels formed the bulk of landings for reduction purposes; in recent years catches have reached 800 000 t. Danish industrial landings were supplemented by direct landing in Denmark of substantial catches by vessels from the Faeroe Islands and Sweden. Also, the landings of fish for human consumption, consisting mostly of cod, plaice and other flatfish, and herring, rose considerably and reached 400 000 t in 1980. Other important sectors are the cultivation of blue mussels *(Mytilus edulis)* with 80 000 t and the farming of rainbow trout *(Salmo gairdneri)* with 25 000 t per year. Fish trade plays a large role. Exports comprise over 800 000 t fish for human consumption and more than 1 million t of fish converted into fish meal and oil. Exports are sold mostly to Germany, Great Britain and France. Imports amount to about 650 000 t.

On the **Faeroe Islands**, politically connected with Denmark, fish production rose from 100 000 t in the 1950s to around 350 000 t in recent years. About half of the catches, consisting mainly of cod, haddock, saithe, flatfish and redfish, are made within the EEZ around the islands by a fleet of some 230 vessels (90 larger than 100 GRT). The other half is fished by 50 distant-water trawlers, purse seiners and long line vessels in the Barents Sea, off Greenland and Iceland, west of the British Isles and in the North Sea. These catches are mostly blue whiting, Norway pout, capelin, mackerel and prawns *(Pandalus borealis)*. About 100 000 t per year are landed in Denmark for reduction to fish meal and oil. The economy of the islands is heavily dependent on fisheries. Almost the total fish production is exported and, by value, fish make up 95% of all Faeroese exports.

Favoured by the natural conditions, fisheries in **Norway** developed to become the largest in Europe. Along the coastline of 20 000 km, washed by the relatively warm and nutrient-rich waters of the Gulf Stream, very productive fishing grounds are found on numerous banks and in the fiords, especially during the spawning seasons for cod, herring and other fish in early spring. In earlier times fishing was by far the most important sector of the national economy. This has changed in recent years and now the fishing industry contributes only 1% to the net national income. But fishing and fish processing is still important to many coastal communities where no other means of livelihood exists. Until the mid-1920s Norwegian fish landings fluctuated between 500 000 and 800 000 t per year. From 1928 they exceeded 1 million t in most years. Further increases during the 1950s, and particularly since the mid-1960s, were due to the development of mass production of pelagic fish for the manufacture of fish meal and oil. A decline in the landings during recent years was caused by overexploitation of these resources (Fig. 128). The bulk of catches are made by the coastal fisheries with about 23 000 powered small ships. Before the depletion of the Atlanto-Scandian herring stocks in the late 1960s, herring was by far the most important target, forming up to 70% of the total landings and giving maximum catches of 1.4 million t in 1954. Annual catches have always varied widely and historical records show that since 1500 five periods with good harvests have been observed, each lasting for 50 - 80 years. Fishing for winter herring (storsild) in January-April and for young herring (smasild) in summer and autumn along the west coast are the main components of this industry. Herring catches are made with drift and bottom gill nets, and with purse seines. Up to 200 boats also fished for herring in the open sea between the Faeroes and Jan Mayen and off the north coast of Iceland. After the depletion of the herring stocks annual catches fell to a low of 10 000 t in 1979 but under protection the resources began to recover and in 1986 already more than 300 000 t of herring were caught. Mackerel fishing in the North Sea and Skagerrak also has a long tradition. Another major sector is the cod fisheries, carried out mainly with drift nets, long lines and hand lines; these also exploit haddock, saithe, ling and similar resources (see Sect. 2.4). Cod landings also showed large annual fluctuations but for a long time remained generally at 200 000 - 300 000 t per year. A large fishery for capelin *(Mallotus villosus)* in the Barents Sea developed in the late 1960s to counterbalance the losses in herring. Capelin catches reached a maximum of 2.1 million t but the resources were almost depleted by the late 1980s. Other species, like Norway pout, blue whiting and sandeel, were also caught in large quantities to produce raw material for the production of fish meal. These pelagic fisheries are carried out by a fleet of 200 larger purse seiners run by companies and not owned individually like most of the small coastal fishing vessels. Altogether, more than 1.5 million t of fish per year are reduced to fish meal and oil or are used directly as animal feed. Fish exports, consisting mainly of fresh fish, frozen fillets, stockfish, klipfish, salted herring and fish meal and oil, are equivalent to 2 million t and make Norway, a country with only 4 million inhabitants, one of the largest fish exporters in the world. Spectacular progress has been made during recent years in the farming of Atlantic salmon and trout. Aquaculture production reached almost 90 000 t in 1988.

In **Iceland** fisheries are of paramount importance and this sector employs 15% of the entire labour force on this island with 250 000 inhabitants. Fish landings have gradually increased since the early years of this century to 300 000 t in 1938, roughly 1 million t by 1964 and 1.7 million t in 1988. A fleet of over 700 decked multipurpose coastal vessels are engaged in fishing with long lines, gill nets and trawls, mainly for cod, haddock and saithe. From the late 1950s until the late 1960s herring fishing with purse seines off the north coast in summer and in the southwest in autumn expanded enormously. With the overexploitation of Icelandic herring stocks and of Norwegian spring-spawning herring in the Norwegian Sea catches declined to insignificant quantities in the early 1970s. The fleet of purse seiners turned first to herring fishing in the North Sea and then, after depletion of those stocks and the introduction of an EEZ by the European Community, they changed to a fishery for capelin as the raw material for production of fish meal and oil. Icelandic herring catches reached a maximum of 770 000 t in 1966 and declined to less than 40 000 t in the early 1970s. In recent years they have increased only slightly. On the other hand, capelin catches rose from 50 000 t in 1965 to over 900 000 t in the late 1980s. While the inshore fishery suffered from economic difficulties, the offshore fishery with a fleet of more than 70 stern trawlers prospered, particularly after the exclusion of foreign fishing from the Icelandic EEZ. Catches of demersal fish from deeper waters, especially of redfish, increased substantially. Exports, consisting mostly of frozen fish and fish meal and oil, amounted to around 700 000 t product weight in recent years, or more than 1.5 million t if converted into nominal catches. The main markets are in the USA and EEC countries.

The fishery in **Sweden** is carried out entirely with small and medium-sized vessels operating in the Skagerrak, Kattegat, North Sea and Baltic Sea with trawls, Danish seines and drift nets. Of 4162 ships in 1983 only 67 were larger than 100 GRT. Annual landings were mostly in the order of 200 000 - 250 000 t with a maximum of 387 000 t in 1964. More than half of the production consisted of herring caught mainly by the west coast fishery with its centre at Göteborg. Second in importance are cod and sprat, which are also caught in the Baltic from southern and eastern ports; in addition eels, salmon and Baltic herrings are caught. Up to 70% of the herring catches are exported and landed directly in Denmark to supply reduction plants. Aquaculture is expanding rapidly and produced almost 7000 t of rainbow trout in 1988.

Of some 500 small vessels in **Finland** more than 400 are employed in the salmon fishery with lines and in the capture of Baltic herring with trawls. Herring catches amount to more than half of the total landings, which are in the order of 130 000 t per year. One-third of the Finnish landings are used as feed in mink farming and more than 200 000 t are imported for the same purpose.

With annual landings of 1.2 - 1.6 million t during the last 20 years **Spain** is by far the most important fishing nation in southern Europe. Growth in the fisheries was largest during the 1960s (Fig. 128). The composition of the landings covers a very wide spectrum, especially since Spanish freezer trawlers, tuna seiners and shrimpers operate not only in the Northeast Atlantic but also on many distant fishing grounds off western and southern Africa, off the US Atlantic coast and in the western Indian Ocean under special agreements with the respective coastal states. At roughly 750 000 t, fish form the bulk of the landings. Until the 1970s cod, which was caught by large stern trawlers and smaller pair trawlers mostly off Newfoundland, Greenland and Iceland and all of which was salted and dried to bacalao, was the main target. With the loss of fishing grounds due to the introduction of EEZs hakes, especially the Cape hake from southern Africa, became more important. Fish are now predominantly marketed in fresh or frozen condition. In the inshore fisheries between 100 000 and more than 200 000 t of sardines *(Sardina pilchardus)* are caught by light fishing in lampara seines for the canning industry. Spanish vessels also fish roughly 140 000 t of cephalopods, mainly octopuses and cuttlefish, mostly in West African waters. The most valuable catches are harvested by the tuna fisheries which have expanded during recent years and landed more than 200 000 t in 1988. Catches consist mostly of skipjack and yellowfin tuna, taken in the waters off West Africa and in the western Indian Ocean where Spanish vessels fish from the Seychelles. In home waters 3000 - 5000 t of bluefin tuna are caught, mostly with the traditional large traps (almadrabas). Over 200 000 t of blue mussels are produced each year by cultures in Galicia, Cataluna and other provinces. Vigo, San Sebastian and Cadiz are the main fishing ports. Imports of fishery products at 450 000 t are larger than the exports, which average 280 000 t.

The fishery in **Portugal** was producing 300 000 t by 1950 and the landings increased to 600 000 t in the mid-1960s. Thereafter they declined and in the 1980s varied between 250 000 and 400 000 t. In the inshore and coastal fisheries numerous small boats catch a large variety of demersal fish with long lines, gill nets and other gear. As the shelf is rather narrow, stocks are overexploited and yields have declined over the years. More important is the sardine fishery in these waters where small purse seiners obtain widely fluctuating catches, around 90 000 t in recent years. Further offshore, sardines are also caught with trawls. Half of the production is used for the domestic market and the other half is canned and mostly exported. Irregular landings and strong competition from other producing countries often lead to economic difficulties. Matozinhos and Leixoes are the main ports for the sardine fisheries. Tuna fishing, mainly for skipjack and bigeye tuna, produces high-value catches of about 15 000 t. Harvests of bluefin tuna from the old large traps are now insignificant. As fishing grounds along the Portuguese coast are rather limited, a trawl fishery was developed on the West African shelf off Morocco, Rio de Oro, Mauritania and Senegal. In the 1960s 70 000 t of fish, mostly hake, were caught annually in these waters. With coastal states applying increasing limitations on foreign fishing the Portuguese catches declined to 15 000 t, harvested by 40 trawlers, in the mid-1980s. Some 50 large distant-water trawlers continue the old tradition of cod fishing in the North Atlantic (see Sect. 2.4) but are now forced to concentrate more on horse mackerel, mackerel, redfish and hake. A further 19 trawlers are engaged in fishing off Namibia for hake. Lisbon and Porto are the centres of the trawl fisheries. A major change has taken place in the utilization of the catches: more than 50% of the landings are consumed or are exported in fresh or frozen condition, while the production of dried and salted cod (bacalhau) has decreased to about 15%. Imports of fish and fishery products (average 280 000 t) are larger than exports 93 000 t).

The fisheries in **Italy** have also increased their landings substantially, from 200 000 t in the 1950s to 560 000 t in the late 1980s. Trawling is by far the most common fishing method (68% of the fishing effort), while lines and traps account for 10% and seining for 5%. The greatest part of the catches is taken in the Mediterranean by a large number of small and medium-sized vessels. It was estimated that about 47% of catches are from the Adriatic Sea, 26% from the Ligurian and Tyrrhenian Seas, 18% from waters around Sicily, 6% from the Ionian Sea and 3% from Sardinia. As yields in home waters were quite insufficient to satisfy the demands, a distant-water trawl fishery was developed during the 1960s and 1970s. At times 100 trawlers participated in this fishery off Northwest and Southwest Africa, and in the Northwest Atlantic. The introduction of EEZs restricted these activities. In recent years landings from distant waters amounted to around 47 000 t. The Italian fishing fleet has 30 trawlers larger than 500 GRT. Livorno is the centre of the distant-water fishery and other important fishing harbours are Fiumicino, Anzio, Naples, Pescara and San Benedetto. The landings consist of a wide variety of fish, crustaceans and molluscs. Sardines and anchovies, caught in home waters with purse seines, contribute 60 000 - 90 000 t and hake about 30 000 t. Mariculture produces 85 000 t of mussels *(Mytilus galloprovincialis)* and around 30 000 t of striped venus *(Venus gallina)*. Aquaculture also shows an increase in production with 40 000 t, mostly of trout, eel and bream. Imports of fish and fishery products reached almost 600 000 t in the mid-1980s and exceeded exports of 115 000 t.

Fish landings in **Greece** fluctuate around 100 000 t per year. In the inshore fishery with around 25 000 small vessels about 39 000 t of fish are caught, in particular sardines, anchovies, mackerel, tuna and swordfish. The offshore fishery in Greek waters with 840 ships using Mediterranean trawls or purse seines produces about 56 000 t. Fishing on other Mediterranean grounds, which resulted in catches of 10 000 t in the 1960s, had to be terminated when EEZs were established. Also, the distant-water fishery with large trawlers outside the Mediterranean, mainly off Northwest Africa and off Argentina, suffered from extensions of fisheries jurisdiction. The distant-water fleet shrank from 55 vessels in 1981 to 31 in 1984, and the landings fell to 10 000 t. The bulk of fish production is consumed fresh or frozen; sardines and anchovies are mostly canned.

Imports and exports increased sharply with accession to the EEC; the former amount to about 85 000 t and the latter to 20 000 t per year.

Fisheries in **Yugoslavia** have a production of 70 000 - 80 000 t per year. Around 50 000 t of fish are caught in the coastal waters of the eastern Adriatic Sea by 7000 small boats and 136 decked vessels with sizes of up to 130 GRT (<26 m in length). Sardine, sprat and mackerel constitute 80% of the marine landings, the bulk of which is absorbed by the processing industry. Purse seining in the sardine fishery has, in recent years, been supplemented by midwater pair trawling. Some 100 vessels operate in the trawl fishery between the islands and in deeper waters and land around 2000 t per year. The tuna fishery produces variable catches of 600 - 1500 t. Freshwaters and aquaculture contribute around 25 000 t, of which 13 000 t are common carp from ponds in northern Yugoslavia. Fish imports are around 30 000 t and exports around 22 000 t.

In **Poland** only a very small inshore fishery existed until 1930. From 1931 activities were extended to more distant fishing grounds in the Baltic Sea and a deep-sea fishing company was founded to send drifters to fish for herring in the North Sea. Total landings reached 20 000 t in 1936. After World War II Poland gained a longer coastline on the Baltic Sea and the herring drift net fishery in the North Sea was further developed. By 1955 the landings exceeded 100 000 t; 65 000 t from the Baltic and 40 000 t from the North Sea and other fishing grounds. As in other socialist countries (see Sect. 5.2.2), a fleet of large distant-water trawlers was built during the 1960s for worldwide operations and landings rose steeply to a maximum of 800 000 t in 1975. Of this quantity 320 000 t were caught in the Northeast Atlantic, 125 000 t in the Northwest Atlantic, 130 000 t off Northwest Africa and 115 000 t off Namibia. With the introduction of EEZs the Polish fishery lost many fishing grounds and was forced to operate further offshore. The composition of landings changed considerably; species like cod, saithe, Greenland halibut and mackerel were caught much less. The catches consisted to a great extent of Cape hake from southern Africa, of North Pacific hake, and of southern blue whiting *(Micromesistius australis)*. The landings declined to the present level of about 650 000 t. The distant-water fleet comprises around 100 factory and freezer trawlers, two motherships and 10 carriers. These ships operate from the ports of Gdynia, Swinoujscie and Szczecin. In the Baltic Sea a fleet of cutters (<25 m in length) and other small motorized boats are engaged in fishing for cod, flatfish, herring and sprat. Catches in the Baltic reached over 200 000 t in the mid-1970s but have decreased to less than 140 000 t in recent years. About 30 000 t of fish are produced annually from freshwaters; two-thirds are carp from pond culture.

In **Romania** possibilities for fishing in the Black Sea are very limited and landings from these waters have been only around 10 000 t in recent years, consisting mostly of sprat, anchovy and horse mackerel. Therefore, as in other socialist countries, the government built a fleet of large distant-water freezer trawlers (2600 - 3900 GRT) which operate in the Eastern Central, Southeast and Northwest Atlantic and in other seas worldwide. In 1978 the ocean fishing fleet comprised 37 of these vessels which are based in Tulcea. A large proportion of the catches of this fleet consists of horse mackerel. The total Romanian fisheries production increased in the early 1970s to 100 000 t and in 1982 reached more than 200 000 t. Fresh and brackish waters yield about 50 000 t of fish, particularly in the Danube delta. A large proportion of the harvest comes from pond culture of carp.

A similar situation exists in **Bulgaria**. Catches in the Black Sea were less than 5000 t per year in the 1960s but increased to 15 000 t in the 1980s. Since 1965 a fleet of large stern trawlers, mostly of the *Atlantic* class, was established to fish in the same overseas waters as those fished by the ships from other socialist countries. In 1980 this fleet consisted of 27 vessels, operating from Burgas. As in the Romanian fishery, horse mackerel form a large part of the catches. Total landings amount to 100 000 - 120 000 t. This includes around 12 000 t from freshwaters, almost all common carp.

In spite of great diversity in the fisheries of European countries, some general trends are evident: the number of fishing vessels has greatly decreased over the years, but new ships were made larger, more powerful and more efficient. Therefore, the actual fishing effort has greatly increased and contributed to the heavy overexploitation of the resources in Europe and overseas. The acute problem is how to reduce the existing fishing capacities to levels corresponding to the natural productivity of the resources. A gradual adaptation is necessary despite the economic and social difficulties that this would cause. A solution might be facilitated by another general trend: in all countries the number of fishermen has continuously decreased. General economic development has created many alternatives for making a livelihood in other, often more attractive, sectors.

Twelve countries have joined the European Economic Community: Belgium, Denmark, France, Germany, Greece, Ireland, Italy, Luxemburg, the Netherlands, Portugal, Spain and the United Kingdom. Total landings by the fisheries of these countries amount to about 7 million t per year. In spite of this large production the landings are quite insufficient to satisfy the demands of the population of 322 million people in EEC countries. Therefore, a net import of 1.5 million t (average 1984-1986) supplements the home landings for human consumption. After long negotiations in 1983 the EEC established a Common Fisheries Policy (CFP). A joint EEZ was already declared in 1977 and member states obtained equal rights of access to fishing in EEC waters. Under the CFP a joint management regime was agreed upon. On the basis of scientific research and recommendations total allowable catches (TACs) are fixed annually (in 1988 these were for about 100 fish stocks) with predetermined proportions to each member state. The EEC Commission concludes, on behalf of the members, fisheries agreements with third countries to obtain fishing rights in their EEZs on a reciprocal basis or against financial compensations or access to the EEC market. Such fisheries agreements exist with Norway, the Faeroe Islands, Greenland, Sweden, Finland, Canada and the USA as well as with 15 coastal countries in Africa and on islands in the Indian Ocean. The latter agreements are of benefit to the French, Portuguese and Spanish fisheries. Furthermore, the CFP comprises regulations for the surveillance of fishing operations, for the marketing of fish and fisheries products, and for the restructuring and modernization of the fishing fleets. Much remains to be done, but certainly the European Community has developed great economic strength and an attractive reason for other European countries to join. Further accessions should be welcomed and hopefully also Norway, Sweden and Iceland will enter the EEC, if the problem of access to their fisheries resources could be overcome.

5.3 International Law of the Sea

Legal provisions have always been of great importance to fisheries development. The original practice whereby everybody was free to utilize the food resources in all public waters was already abrogated in many places a long time ago. Instead, fishing rights and other related prerogatives were granted by kings or other sovereigns to their feudal tenants, to communities or to institutions of the church, mostly monasteries. With the evolution of cities in the Middle Ages independent urban communities also gained the right to invest fishermen, merchants and others with privileges in fishing and fish trading. Often quarrels arose and even wars were fought to secure fishing rights and economic advantages. However, up to the present day legal regulations have formed a major basis for the orderly development and conduct of fishing, both at national and international levels.

Many existing privileges and regulations for fresh and inshore waters date back many centuries. However, major changes have taken place in the high-seas regime. Still in the 15th and 16th centuries the opinion prevailed that the sea was just as much the subject of property rights as was the land. In the Treaty of Tortesillas (1494), based on a Papal Bull issued by Pope Alexander VI, Spain and Portugal took possession of the newly discovered and still unexplored seas and divided the areas of their respective powers. The English did not accept these claims and insisted that large parts of the North Sea and of the Atlantic Ocean were under their rule. Denmark claimed most of the North Sea not held by England, Sweden claimed the Baltic, France large parts of the Mediterranean, Venice the Adriatic, and so on. But with the rapid development of trade on the world's oceans cognition succeeded that it was beneficial to all to agree on free movement of ships.

The principle of the Freedom of the Sea was formulated by the Dutch lawyer H. Grotius (1583 - 1645) in his book *Mare liberum* (1609) and was accepted until well into the 1900s. From the 18th century coastal states claimed only a territorial zone of 3 nautical miles along their shores. The width of this zone, which was also considered an area of exclusive fishing rights, was defined by the distance of a cannon shot reached at that time. Still, in the 1880s, territorial waters of 3 miles were included in international treaties but several countries had already claimed wider zones, mostly in the interest of specialized fisheries. In 1909 Russia declared a 12-mile territorial sea.

Between the World Wars the need was felt for more uniform regulations and the League of Nations convened in 1930 in The Hague for a conference on Codification of International Law. However, participants from 47 countries could not reach agreement on the width of territorial waters.

New demands by coastal states were stimulated when, in 1945, President Truman proclaimed that the living and mineral resources of the sea-bed and subsoil of the continental shelf adjacent to the USA to a depth of 200 m were subject to US control and jurisdiction. In the following years a number of states, especially Chile, Peru and Ecuador, claimed 200-mile zones off their coasts. At the same time extension of territorial waters to 4 miles by Norway and Iceland led to economic warfare between Great Britain and these countries.

Based on recommendations of the International Law Commission established by the United Nations in 1949, a first UN Conference on the Law of the Sea was held at Geneva in 1958. During this conference four conventions were completed. One of them was the Convention on Fishing and Conservation of the Living Resources of the High Seas. In this convention it was agreed to base the protection and management of fisheries on the principle of the maximum sustainable yield, i.e. the largest annual catch that can be taken continuously from a stock without endangering its recruitment. This convention greatly influenced subsequent international agreements and management schemes but the representatives of 86 countries could not agree on the width of the area beyond the territorial sea where the coastal state had preferential rights to fishery resources. Furthermore, they could not reach consensus on the width of the territorial sea.

After the failure in 1958 a second UN Law of the Sea Conference was organized in Geneva in 1960. Unfortunately this conference could not resolve the problems either.

In the meantime fast technological progress was made on exploring and exploiting mineral resources such as oil and gas from the sea-bed and subsoil in offshore regions. There was also growing concern about the pollution of the sea. With the worldwide intensification of fishing and overexploitation of many living resources it became apparent that the ocean resources were by no means inexhaustible. Furthermore, several international fishery commissions had demonstrated their inability to protect and manage fishery resources efficiently. All these concerns increased efforts by states to secure through claims their future access to living and non-living resources and also for them to press for international standards to secure claims and to minimize conflicts between nations over resources and shipping.

This situation led the UN General Assembly in 1970 to establish a Seabed Committee for the preparation of a third UN Conference on the Law of the Sea. This conference began in 1974 with a session in Caracas and ended in 1982 after a further 10 sessions held alternately in New York and Geneva. The new convention resulting from these long and complex negotiations was adopted by 130 states with four votes against and 17 abstentions. The Treaty was signed by 117 nations in November 1982 in Montego Bay, Jamaica and after ratification by 60 states the convention will come into force. By early 1991 45 states had ratified the convention.

Under the convention it was agreed that the territorial sea will extend to 12 nautical miles from the coastline and that in a contiguous zone of another 12 miles coastal states would have further special rights. Within a 200-mile Exclusive Economic Zone (EEZ) coastal states have full control and jurisdiction over all living and non-living resources, pollution abatement and scientific research. Coastal states are responsible for protecting and managing the living resources in their EEZ to maintain levels of maximum sustainable yields. They are also committed to granting access to fishermen from other nations when their own fishing industries are unable to harvest the surplus. If fish stocks occur within the EEZs of two or more countries, they are to work together or through regional organizations to coordinate and ensure the conservation and development of the resources.

In anticipation of the convention most coastal states had already extended their jurisdiction to EEZs during the late 1970s, mainly in 1977. High expectations concerning the results of such extensions were not always realized. In many countries, particularly developing countries, there was, and still is, a serious lack of trained personnel, adequate institutional structures and funding to ensure proper conservation and management of the living resources and fisheries. In many instances the implementation of the new regime led to conflicts between states and fishermen from different nations. Often, the problems which existed previously with the presence of foreign fleets were replaced by new domestic conflicts between competing fishermen or other users from the coastal state when the national fishery was further expanded in the EEZ.

However, gradually a system of coexistence between national fisheries and certain foreign fishing activities, limited by special licences, fees and catch quota, developed. Access to fish resources in EEZs was often granted against concessions of preferential duties or other trade facilitations. It is clear that coastal states will, in the long run, tend to use all the living resources for their own nationals so that there will be no surplus for others. Of course, coastal states with adequate competence for proper management and good facilities for the surveillance of their EEZ gained much under the new convention. Examples are the USA, Canada, Norway and Iceland.

During the first years (1977-1979) after the introduction of 200-mile zones increases in world fish production declined due to loss of many fishing grounds by industrial countries with long-distance fleets (Figs. 94 and 99). But after the adaptation of fishing strategies to the new conditions the level of fish production rose again steeply and world trade of fish and fishery products, particularly crustaceans and molluscs, boomed.

With the establishment of the new legal regime about 40% of the world's ocean and more than 90% of the marine living resources were put under

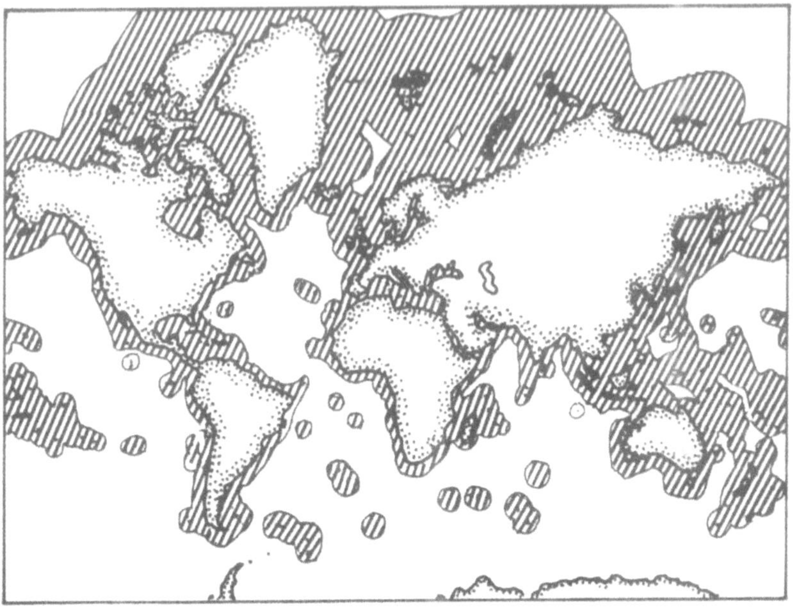

Fig. 129. Regions of the world's oceans under national jurisdiction after the introduction of 200-mile Exclusive Economic Zones (EEZs)

the jurisdiction of coastal states (Fig. 129). This had far-reaching consequences for fisheries development and for the overall economy of many countries. States with long coastlines gained substantially while geographically disadvantaged countries suffered serious losses. The degree of gains or losses varied widely from one country to the other. It is fair to say, however, that the formation of the new regime was the largest redistribution in history of the world's territories and resources.

6 Fisheries Science

The development of fisheries was accompanied by progress in related scientific research. The formation of specialized fisheries investigations is, however, only about 100 years old. Fisheries science developed greatly during this period. Today, all countries with sizeable fisheries possess research laboratories, usually governmental, with specialized staff and many maintain research vessels for the exploration of living resources and investigation of fish stocks and their environment. Fisheries research is concerned not only with biological work but also with experiments to develop and improve fishing gear and operations with the aim of increasing the profitability of fishing. Another sector is concerned with food technology to ameliorate the quality of fish landings and products and to develop new processing techniques and new types of fisheries products.

The oldest written record of fish investigations which has come down to us is by Aristotle (384 - 322 B.C.). He collected all the available biological knowledge of his time, supplemented it with his own versatile observations and prepared a first systematic account. Of the more than 400 animal species described in his *Historia animalium* about 20% are fish. Aristotle wrote essays on the anatomy, reproduction biology and physiology of fish, and his work was the main scientific basis for many centuries.

A phase of ichthyological research began in the 16th century. In Paris P. Bélon described in his *Histoire naturelle des étranges poissons* (1551) 110 fish species, including 17 from freshwaters, as well as whales, seals, birds, etc. Around the same time, 1554-1557, H. Salviani published in Rome his *Aquatilium animalium historia* with descriptions of fish presented in 76 excellent copperplate engravings. Salviani also commented on the use and yields of fishing. Another ichthyological review by G. Rondelet was published in Lyon in 1555. It covered 250 fish species (47 from freshwaters) and many other aquatic animals and contained information on the anatomy and life history of the animals. Also famous is the publication *Historiae animalium liber qui est de piscibus et aquatilibus* by C. Gesner (Zürich 1558). Books by G. Mangolt on fish in Lake Constance, by

S. Schonevelde on fish in Schleswig-Holstein (1624) and by L. Baldner on fish in the Rhine river (1666) widened the knowledge of the ichthyofauna in certain areas. The German physician G. Margrav described 100 so far unknown fish species from Brazil (1648). In England J. Ray and F. Willughby prepared an illustrated catalogue of 420 fish species which were already classified in accordance with anatomical characteristics (Oxford 1686). It was, however, not until the work of P. Artedi and his friend C. Linnaeus in Sweden that the modern system of scientific nomenclature was introduced. Artedi's *Ichthyologia* was published in Leyden in 1738. Another 18th century ichthyologist was M.E. Bloch. His famous publication *Oeconomische Naturgeschichte der Fische Deutschlands* (Berlin 1782-1784) contains many beautiful illustrations. In France extensive reviews of fish species were written by B. Lacépède in his *Histoire des poissons* (Paris 1798-1809) and by G. Cuvier and A. Valenciennes in their 20 volumes *Histoire naturelle des poissons* (Paris 1828-1849). Cuvier stimulated a considerable number of ichthyological expeditions overseas. The knowledge of fish was greatly enhanced, and the *Catalogue of Fishes* by A.C.L.G. Günther, published in 1859-1870, comprised over 8500 species. Since then the number of species known has increased by more than threefold and the ichthyological literature has expanded enormously. A modern worldwide catalogue of fish of commercial interest has been created since the 1970s with the series *FAO Species Identification Sheets for Fishery Purposes*.

Around the mid-19th century interest turned to the wider field of marine biology. In particular, P.H. Gosse in England and L. Agassiz in the USA raised, with their books and lectures, much public interest in such studies and in the construction of aquaria. A number of marine biological stations were founded, first in Concarneau, Arcachon and Roscoff in France (1859-1871) and during the 1880s and 1890s also in Great Britain, the USA, Norway, Germany, Monaco and other countries. A. Dohrn, founder of the Zoological Station at Naples, was a successful promoter of such research laboratories.

The hypothesis by the Scottish naturalist E. Forbes that there could be no living organisms deeper than 500 m in the oceans created an incentive for several expeditions to the high seas. The most important was the voyage around the world of the *Challenger* (1872-1876) under the leadership of C. Wyville Thomson; this provided, by large collections of marine fauna and flora, a wealth of new knowledge and proved by deep-sea dredges, not only that life could be found in the depths of the oceans but also that the deep-sea fauna was not a relict from earlier geological periods, as had been thought so far. Already, the earlier *Beagle* expedition (1831-1836) by

C. Darwin had obtained much new information on marine life. After the *Challenger* voyage a large number of further scientific expeditions were mounted in various countries, including Belgium, Denmark, France, Germany, Great Britain and the USA. In connection with these studies fast progress was made in the development of methodology and instrumentation for oceanographic research and for the sampling of plankton, benthos and fish.

Fisheries science literally developed during the 1880s and 1890s when the fisheries in the North Sea were strongly intensified and the yields started to decline (see Sect. 3.1). Scientific advice was needed to help with the exploration of new fishing grounds and to investigate the reasons for the decrease and change in catches. It was soon realized that fisheries investigations on the high seas required international collaboration. After an international conference in Stockholm in 1899 the International Council for the Exploration of the Sea (ICES) was founded in 1902 as a body to initiate and coordinate studies on fish stocks.

Methods were developed to investigate fish populations, and included, especially, the use of length measurements to establish year-classes (Petersen method), the determination of fish age from annual growth rings in scales, otoliths and bones, and the tagging of fish to study their migration, growth and mortality. Research was initiated on the fecundity of fish and on the effects of environmental factors and food availability on the production and distribution of fish brood to explain the large annual fluctuations in the abundance of fish and, subsequently, in the catches of commercial fisheries. Systems were developed for the collection and compilation of fishery statistics, needed on a regular basis to quantify the landings made by the various branches of the fisheries and to assess the effects of fishing on the stocks.

During the years 1905-1914 countries cooperating in ICES, mainly Denmark, Germany, England, Scotland, Norway, Sweden and Russia, concentrated their research activities primarily on the plaice stocks and on the question of how many small fish were destroyed by the trawl fisheries on the young fish grounds. Proposals already made in 1895 to protect the resources by the introduction of minimum sizes of landed fish and to design trawls which would allow small fish to escape through the meshes were reiterated but the outbreak of World War I prevented further work.

Based on the experience of quick recovery of North Sea fish stocks during the war years, simple fish population models, with assumptions on the effects of fisheries exploitation on the stocks, were developed during the

1930s. Investigations on the detrimental effects of capture of young fish were resumed and further proposals elaborated for the conservation of stocks through the introduction of minimum mesh sizes in trawls and size limits of fish in the landings. In 1939 ICES held a conference on overfishing problems but World War II prevented the timely introduction of international conservation measures.

The Convention for the Regulation of the Meshes of Fishing Nets and the Size Limits of Fish, concluded in London in 1946, was a major step forward and measures were taken after a Permanent Commission had been established in 1954. Scientific advice was provided regularly by ICES. The requirements for this advice stimulated much further progress in fish population analysis and stock assessment. However, the convention was not successful since it was an obstacle to the development of fisheries on small species, and with the strong increase in fishing intensity during the 1960s the regulation of mesh sizes alone proved to be quite insufficient to prevent overexploitation of the resources. Unfortunately, the new Northeast Atlantic Fisheries Convention of 1959, enforced in 1963, did not empower the commission (NEAFC) to introduce other more rigid conservation measures, like catch quota and effort restrictions. Thus, in spite of frequent warnings by the scientists, many fish stocks were depleted and it was not until 1975, after a revision of the convention, that the first total allowable catches (TACs) were introduced in the Northeast Atlantic region. For Northwest Atlantic waters the International Commission for Northwest Atlantic Fisheries (ICNAF) had agreed on catch quota for the main commercial fish stocks since 1973. In contrast to NEAFC, scientific advice in ICNAF (now Northwest Atlantic Fisheries Organization, NAFO) is provided directly by scientists from member states collaborating in the scientific council.

Similar to Europe fisheries science in North America also developed as a consequence of the decline in catches by commercial fisheries. Such problems were observed first in the Pacific halibut fishery in Puget Sound and in the salmon fisheries in the river Fraser and neighbouring waters. The Convention for the Preservation of the Halibut Fishery of the Northern Pacific, Including Bering Sea (later revised and the commission renamed as the International Pacific Halibut Commission, IPHC) between Canada and the USA was already agreed in 1923. With scientific guidance by the staff of the commission this convention was quite successful in rebuilding the halibut stocks. Likewise, an International Pacific Salmon Fisheries Commission (IPSFC) was established in 1930, and this body with its scientific staff also worked successfully for the recovery of salmon resources. In contrast to conditions in the North Atlantic region, these commissions, and also the International North Pacific Fisheries Commission (INPFC), enforced in 1953 (see Sect. 5.2.1), were entrusted with responsibilities for the allocation of fish catches at an early stage.

With the worldwide expansion of fisheries a number of fisheries councils and commissions were also formed in other regions, mostly under the auspices of the FAO of the United Nations; these included the General Fisheries Council for the Mediterranean (GFCM), the Indian Ocean Fishery Commission (IOFC), the Indo-Pacific Fisheries Council (IPFC) and the Fishery Committee for the Eastern Central Atlantic (CECAF). Independent organizations, like the Inter-American Tropical Tuna Commission (IATTC), the International Commission for the Conservation of Atlantic Tunas (ICCAT) and the International Commission for the Southeast Atlantic Fisheries (ICSEAF) were also formed. All these organizations provide fora for the coordination of scientific activities and the discussion of results in the various regions.

The introduction of 200-mile EEZs and the fact that coastal states have taken over sole responsibility for the management of resources in their EEZs did not cause much change in the scientific work required to provide the necessary advice on a continuing basis. International cooperation between scientists is very extensive and includes many joint research programmes and closely coordinated voyages by research vessels with mutual evaluation of the results. Ichthyological research and studies of the life history of species still play an important role. The contribution of research to the development of aquaculture is invaluable. Environmental investigations for the abatement of pollution have been initiated during recent years. As most resources are now fully utilized, explorations of new fishing grounds and for little-known or unknown stocks become less productive. The largest demands for proper scientific advice come, however, from the various bodies concerned with fisheries management. Methods for stock assessment, catch predictions and elaboration of management options have been further refined and now form an agreed standard tool in fisheries science worldwide. More and more studies have been concerned with the interrelationships within ecosystems and with the development of multispecies models since it was realized that good resources management must protect the ecosystems and not only single economically important species. This aspect is incorporated in newly drafted resource management conventions such as the Convention for the Conservation of Antarctic Living Marine Resources (CCAMLR), agreed in 1980.

Fortunately, scientific advice is now increasingly accepted by fisheries administrators and managers, who have had bad experiences in disregarding scientific recommendations in the past. This puts great responsibility on the shoulders of the stock assessment experts since the problem is not only to ensure the necessary protection of stocks but also to enable, in the long run, the fisheries to utilize fully the available resources at permissable levels to produce food.

7 Perspectives for the Future

After almost continuous increase world fisheries production had reached nearly 100 million t per year by 1988 (Fig. 94). This result was achieved by manifold technical improvements and by an enormous expansion of fishing activities. In particular the construction of large fleets of powerful vessels and the introduction of efficient gear, like purse seines and trawls, contributed to this development. Location of fish by hydroacoustic methods and by other means helped to increase the production, and it was possible to trawl deeper and deeper on the fishing grounds in all parts of the world.

The remarkable progress in production of valuable food also met, however, with considerable problems. With the increase in technical requirements more and more capital was needed, and the relation between the necessary investments and running expenses to the economic returns became less favourable. The most serious problems arose, though, when the fishing effort exceeded the natural productivity of the resources. There are many examples in various regions of the world to show that, as a result of excessive fishing, one fish stock after another was fished down, usually with rather bad economic and social consequences. This situation was experienced not only by the fleets of large high-seas vessels but also by intensive coastal fisheries, such as that off Peru.

The reduction in resources is not always the result of fishing. Natural fluctuations, influenced by changes in environmental conditions, also play an important role, both for the decline and the increase in stocks. Examples of such events are the "El Niño" effects on the anchoveta off Peru and the large fluctuations in the stocks of Japanese sardine. In 1936 this species provided catches of 1.6 million t and its resources then declined rapidly so that no catches could be made in 1943. For reasons not yet fully understood, the stocks have increased again since the mid-1960s and, today, more than 5 million t of these sardines are fished every year.

Concern about overfishing has been expressed since ancient times, e.g. in China in around 300 B.C. in connection with fish ponds, or in England in the Middle Ages when trawl fishing was introduced. The first real problems arose at the end of the last century with the expansion of North Sea trawl fisheries, and this was one of the major reasons for the formation of the International Council for the Exploration of the Sea (ICES) in 1902 to promote joint research on this matter. As problems of overexploitation increased in many parts of the world, a number of regional and international fishery commissions were established to engage in the protection and management of the resources. At the same time substantial scientific investigations were implemented and bodies were formed to secure the necessary scientific advice. Many protective measures, like minimum mesh sizes, size limits for fish landed, closed areas and seasons to facilitate reproduction, were enforced and have been tightened up over the years. But, in most cases, these measures were insufficient to halt the depletion of resources. Progress was achieved with the introduction of catch limitations in the form of quota (total allowable catch; TAC) and effort restrictions, e.g. laying up of vessels. However, under the pressure of economic interests catch quota were often set at levels considerably higher than those recommended by the biologists, with the result that the stocks were further reduced and fisheries became uneconomic. Fishing of certain stocks, for instance, herring in the North Sea from 1977 until 1980, had to be totally banned. Fortunately, scientific monitoring proved that the protective measures were effective. The herring stocks recovered and the fishery could be re-opened.

After such experiences systems of resources and fisheries management are gradually improving. Now all commercial fish stocks on both sides of the North Atlantic, and also resources in many other waters, are under management control. Standard methods of stock assessment were developed by the biologists and have been further improved. Since the introduction of EEZs the management of resources within the waters up to 200 miles offshore are under the sole responsibility of coastal states and it has proved easier to take sometimes harsh decisions at a national level rather than a multilateral level. Proper management must include attempts to convince fishermen and fishing industrialists that protective measures need to be taken in their own interest. In any case, adequate resources management cannot be secured without effective surveillance of fishing activities and monitoring of commercial stocks. It would be beyond the scope of this book to deal with this important subject in more detail. The reader is, therefore, referred to the large number of existing publications, some of which are listed in the bibliography.

Although, in most areas, introduction of proper management still has a long way to go, recent progresses give hope that the situation will gradually ameliorate in the future. This is most important since almost all major stocks which can be utilized by conventional means are now being fished, and there is little room for the development of hitherto untouched or underfished wild resources. Substantial additional landings can be achieved mainly by better management of the resources and their environment.

An acute problem arose with the recent development of high-seas drift netting. Large fleets of vessels from Taiwan, Japan and South Korea, estimated at 1500 ships, are fishing in the North Pacific surface waters with very long drift nets. The main targets of this fishery are squid and tuna, but salmon are also caught. It was estimated that the total length of the drift net walls used by the fleets might be around 20 000 nautical miles. The nets, which usually extend to depths of 8 m from the water surface, are made of thin synthetic filaments, hardly visible in the water, and do not decompose for a long time. The drift nets are usually uncontrolled during the night, and pieces of netting are often lost during storms or as a result of other damage and remain floating as "ghost" nets. Unfortunately, large numbers of seals, small whales, particularly dolphins, sea birds and marine turtles get entangled in the drift nets and drown. The oceanic surface drift net fishery is also expanding in the South Pacific and starting in the North Atlantic. These developments raised worldwide concern and international negotiations under the auspices of the UN are expected to lead to a ban of such fisheries. In some EEZs the application of long drift nets has already been prohibited.

Figure 130 shows, for the various regions of the world oceans (in accordance with the subdivisions given in the FAO *Yearbook of Fishery Statistics*), the recent annual landings and also the respective estimates by FAO for the potential of further increases by conventional fishing. Of course, such estimates are subject to considerable uncertainties, especially for areas with resources not yet fully investigated.

The most productive waters are found in the Northwest Pacific (recent landings 26 million t), the Northeast Atlantic (11 million t), off the west coast of South America (10 million t) and in Southeast Asia (6 million t). The resources in the Mediterranean and the Black Sea are, with annual landings of 2 million t, already overexploited. Development potential in waters off Northwest and Southwest Africa is also limited almost exclusively to better management of the fully exploited resources. In the North-

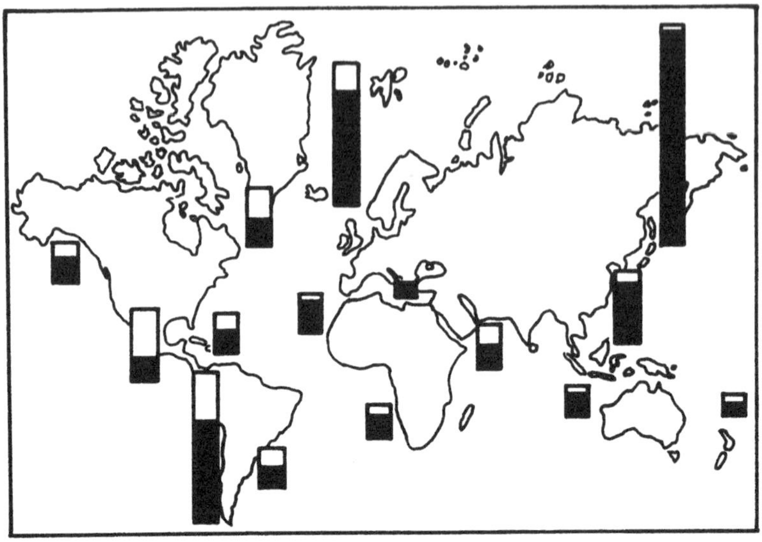

Fig. 130. Annual landings of fish and other aquatic organisms from the various regions of the world's oceans *(black)* and estimates of potential for further increases by conventional fisheries *(white)*. (based on data published by FAO)

west Atlantic, where rigorous catch limitations by Canada and the USA have caused landings to decline from 4.2 million t to less than 3 million t, new development possibilities are apparent as a result of successful fisheries management. The largest potential for further increases may exist in the seas around Latin America. The Northeast Pacific, Caribbean and Northwest Indian Ocean also still offer some possibilities.

Adding up these estimates one arrives at a figure of about 120 million t per year as the potential yield from conventional sea fisheries. A large part of the potential for further increases consists of small fish similar to herring and horse mackerel which are suitable for industrial purposes (fish meal and oil, fodder, etc.); however, markets and trade links would first have to be developed if such fish were used for human consumption on a larger scale. Fundamentally, it is most desirable to utilize the maximum of fish landings for direct human consumption and to reduce to fish meal only such fish as are unsuitable for human food or which could otherwise not be marketed. If fish is utilized indirectly as animal feed to produce meat or poultry for human consumption, about 60-70% of its calorific value is lost.

It was estimated that about 4 - 6 million t of fish are lost every year as they are discarded after having been caught because they are considered unmarketable. Only a small fraction of such discarded fish survive. This problem is largest in the shrimp fisheries where, often, more than 80% of the catches consist of fish and fishermen are mostly interested only in the very valuable crustaceans. It would be desirable to find ways and means to use also the by-catches.

At least 5 - 6 million t of fish are lost after harvest due to spoilage during handling and transport or because storage facilities are insufficient or fish are infested with parasites and insects. Much can be done to improve the situation, particularly in developing countries with tropical climates.

Much larger is the potential to increase production from inland waters and especially from aquaculture. The catch from lakes, reservoirs and rivers could probably be raised to several million tons per year, particularly if flood-plain fisheries were expanded in Latin America and Africa. But a large increase in this sector is inhibited by other uses of water, e.g. for irrigation and power generation, by flood control and by low water quality due to pollution. In contrast, experience with aquaculture has proved that very substantial further increases in the production of fish, crustaceans, molluscs and seaweeds can be expected (Fig. 98). This holds true for freshwater aquaculture as well as for brackish water aquaculture and mariculture. It is estimated that aquaculture production will reach 22 million t by the year 2000 and could be increased in the longer run to 40 million t per year or even more.

Such an enormous growth is, of course, subject to further expansion of the necessary installations and to improvements in methods and techniques. With the often limited space on land it will be necessary to extend the aquaculture of fish more into lakes, reservoirs and the open sea. This may imply higher investments and running costs in the future but could lead to a greater diversification in the industry and markets so as to reduce the risks of overproduction of certain species. Development of warm-water shrimp culture has been spectacular during recent years and a number of developing countries have high economic benefits from the export of such products. However, there may soon be overproduction with the risk of a drop in price. Therefore, in future countries might rely more on the needs of national markets than on exports.

The further expansion of aquaculture is also predominantly dependent on adequate water quality. Preventing and combating pollution is of the utmost importance, not only for the culture of aquatic organisms but also for fisheries on wild stocks in freshwaters and coastal areas.

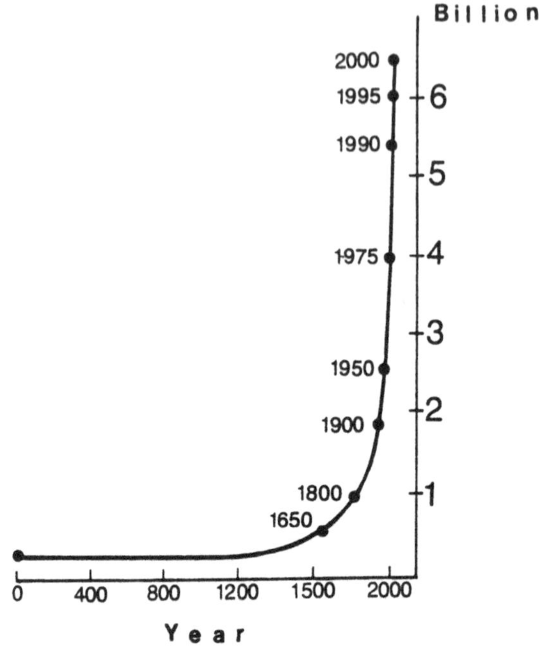

Fig. 131. Estimated increases in the human population

As the human population is growing very rapidly and will reach 6.4 billion by the year 2000 (Fig. 131), the question arises as to whether and for how long fisheries production will be sufficient to meet the needs with the means presently available. Of the present landings roughly 70% or 69 million t are used directly for human consumption while 30% or 28 million t are reduced for industrial purposes (Fig. 94). This means that, on average, 13 kg per year of food fish are available to each of the 5.3 billion people living today. In practice there are, of course, large differences in the per caput fish consumption, e.g. between Japan with 86 kg per year and Nepal with 0.3 kg per year. Assuming that the percentage of fish production for industrial purposes remains fairly constant, and further assuming that the per caput consumption may increase somewhat as a result of higher purchasing power in developing countries, the likely demand for food fish in the year 2000, with a population of 6.4 billion, can be estimated at around 85 million t with a total fisheries production of 120 million t. Recalling that the potential for conventional sea fisheries landings is of the same order and that further steep increases in aquaculture production can be expected, it is safe to say that the needs can easily be satisfied. Considering that the overall potential for fisheries production by conventional means might be around 170 million t per year (Fig. 132), there is, furthermore, hope that at least 120 million t of food fish

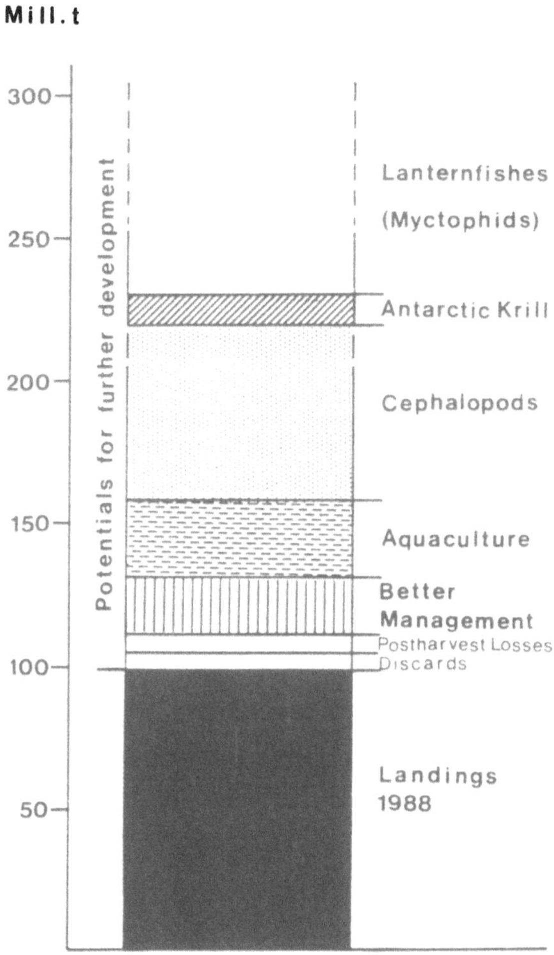

Fig. 132. Possibilities for future utilization of fisheries resources, including unconventional fishing. (based on estimates published by FAO)

could be supplied in later years to satisfy the needs of a population of perhaps 8 - 9 billion people.

Such rough calculations seem to present a fairly optimistic picture. But, looking at the fast acceleration in the growth of the human population, it appears urgent to develop additional food sources. Three unconventional resources seem to offer possibilities: (1) squid and other cephalopods, (2) Antarctic krill; and (3) lantern fishes and other small mesopelagic and bathypelagic species (Fig. 133). The following considerations, based on FAO publications, are somewhat speculative but crude estimations are, nevertheless, presented in Fig.132.

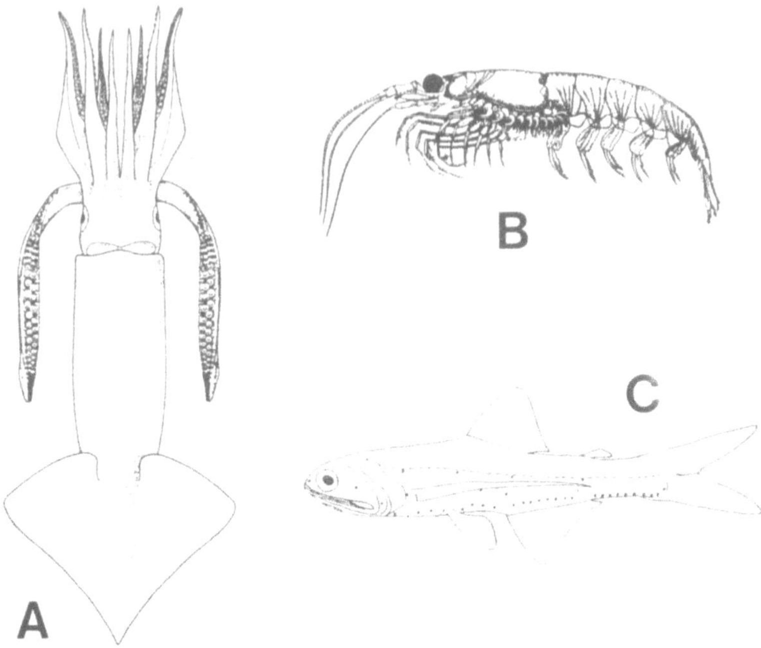

Fig. 133. Unconventional resources offering possibilities for increased production. **A** Oceanic squid; **B** Antarctic krill *(Euphausia superba)*; **C** lantern fish (myctophids)

The total world landings of cephalopods presently amount to about 2 million t per year. Almost half of this quantity is caught by Japanese vessels. Only a few other countries, like Korea, China, Thailand and Spain, participate to a larger extent in this sector of fishing. It was estimated that about 8 - 12 million t per year could be caught over the continental slopes of the world's oceans. Far more difficult is an assessment of squid resources in the open oceans, which are virtually unfished. The abundance of such squid is especially apparent from remains found in the stomachs of sperm whales. Estimates of the fishing potential for oceanic squid vary widely and some go beyond a figure of 100 million t per year. It is, however, doubtful whether a commercial fishery on these resources could be developed to any larger extent. Experience has shown that the world market for cephalopods has only a limited capacity for certain countries and serious drops in price are observed frequently, even with the present catches. Another reason is that squid are very active swimmers and are difficult to catch with conventional gear.

The possibilities to exploit Antarctic krill *(Euphausia superba)* in the Southern Ocean have been overestimated. International investigations have shown that the biomass of krill around the Antarctic may be as large as 200 million t, or even 500 million t, but research has also indicated that the annual natural production of krill stocks is relatively small in relation to their total biomass. Therefore, in the interest of protection of the Antarctic marine ecosystem as a whole, only a limited utilization of krill by man should be permitted. With proper management of the fisheries an annual catch of up to 10 million t might be possible. Landings by the commercial krill fishery, which started in the austral summer of 1972-1973, reached 530 000 t in 1982-1983. During the period 1983-1985 a great reduction in the catches occurred as the USSR fishery, taking 80% of the catches, met with difficulties in the processing and marketing of krill. These problems were resolved and since 1985-1986 krill catches have stabilized at around 400 000 t per year, a small quantity compared to the potential. In view of the extreme distances between krill fishing grounds and centres of krill consumption and the high costs of operating fleets of factory trawlers in such remote waters, it is difficult to guess to what extent the krill fishery will further develop. The international Commission for the Conservation of Antarctic Marine Living Resources (CCAMLR) has initiated negotiations which may lead to the introduction of a total allowable catch for krill, as a protective measure against a sudden dramatic increase in this fishery. As a first step, the commission has set, in 1991, a precautionary catch limit of 1.5 million t of krill per year in the South Atlantic.

The third source to be considered here are the lantern fish (myctophids) and light fish (gonostomatids); these are small and abundant but usually widely scattered fish of the open oceans. They constitute a major part of the biomass in oceanic areas and are found in all oceans from the Arctic to the Antarctic, with the highest production in tropical and subtropical waters. By speculation, rather than by estimation, the potential annual catch of these resources was assumed to be in the order of 100 million t. So far, only vessels from South Africa and the USSR have fished for lantern fish. South African catches of *Lampanyctodes hectoris* in the waters from Walvis Bay to Angola reached 42 000 t in 1973 and 31 000 t in 1985 but ranged between a few tons and a few thousand tons in other years. The USSR fishery processed the Indo-Pacific lantern fish *Diaphus coeruleus* and possibly also *Gymnoscopelus nicholski*, a subantarctic species, to a product like canned sardines for human consumption. Annual catches so far do not exceed a few thousand tons. The myctophid family comprises some 220 species, of which only a few would be exploitable. The largest fish are about 30 cm long. Lantern fish usually live in light shoals between 200 and about 1000 m water depth during the day and come up to the surface at night during diel migrations. Only occasionally do they form denser concentrations near the shelf slopes or at fronts with special hydrographic conditions. A great effort would be required to catch such fish in big quantities by large midwater trawls and high fuel costs would certainly reduce the economic feasibility. It would be difficult to introduce lantern fish products on the world market in the required large quantities. Most likely, the fish would predominantly be processed to fish meal and oil as shown by the South African experience. It is not very likely that a large-scale utilization of these resources will begin in the near future.

In summary, there are some possibilities, beyond the range of conventional fisheries, to obtain also harvests from unconventional resources. One should, however, not be over-optimistic in this respect. The solution to the problem of an increasing demand for food in the world can, in the long run, only be found in better human population control, for which promising onsets are being made in China and some other countries.

Fishing has come a very long way from the collection of food in prehistoric times to the present mass production of a large variety of commodities. Most of the basic methods have remained the same over thousands of years but human ingenuity and diligence have allowed the development of technical means to provide not only sufficient nourishing food for the ever increasing human population but also labour and income for millions of fishermen and other people in fishing industries and trade with their families. Let us hope that man will learn to manage the natural resources and the environment in a better way for the benefit of future generations.

Bibliography

General

Anon (1964) Modern fishing gear of the world, vol 2. Fishing News (Books), London

Bartz F (1964,1965,1974) Die großen Fischereiräume der Welt, vols 1-3. Steiner, Wiesbaden

Burgess J (1966) Fishing boats and equipment. Fishing News (Books), London

Cushing DH (1988) The provident sea. Cambridge Univ Press, Cambridge

Cutting CL (1955) Fish saving. A history of fish processing from ancient to modern times. Hill, London

Duhamel du Monceau (1769) Traité général des pêches et histoire des poissons, vols 1-3. Saillant & Nyon, Desaint Paris

FAO (1972) Atlas of the living resources of the seas. FAO, Rome

FAO (1975-1991) Fishery country profiles. FAO, Rome

FAO (1985) Definition and classification of fishery vessel types. FAO Fish Tech Pap 267

Fischer W and various co-authors (eds)(1973-1987) FAO species identification sheets for fishery purposes. FAO, Rome

Gulland JA (ed) (1971) The fish resources of the ocean. Fishing News (Books), Farnham

Koers AW (1973) International regulation of marine fisheries. Fishing News (Books), Farnham

Kreuzer R (1974) Fish and its place in culture. In: Kreuzer R (ed) Fishery products. FAO, Rome; Fishing News (Books), Farnham

Kristjonsson H (ed) (1959, 1971) Modern fishing gear of the world, vols 1, 3. Fishing News (Books), London

Lundbeck J (1972-1979) Die Fischerei von den Naturvölkern bis zur modernen Technik und Wirtschaft. Mitt Inst Seefisch Hamburg 13,14,15,18,22,24,28

Nédélec C (1982) Definition and classification of fishing gear categories. FAO Fish Tech Pap 222

Traung JO (1955,1960) Fishing boats of the world, vols 1, 2. Fishing News (Books), London

Von Brandt A (1975) Das große Buch vom Fischfang - international. Pinguin, Innsbruck

Von Brandt A (1984) Fish catching methods of the world, 3rd edn. Fishing News (Books), Farnham

1.1 Stone, Bronze and Iron Ages

Amano Y (no date) Redes teñidas. In: Pesca Peru La pesca en el Peru prehispanico. Ed Unidas, Lima, pp 79-81

Anell B (1955) Contribution to the history of fishing in the southern seas. Stud Ethnogr Upsaliensis IX

Breuil H (1945) Sea animals amongst the prehistoric rock paintings of Ladybrand. S Afr J Sci XLI: 353-360

Breuil H (1950) Les peintures et gravures de la caverne de Niaux (Ariège). Bull Soc Prehist Ariege

Breuil H, Berger-Kirchner L (1960) Franko-kantabrische Felskunst. In: Die Steinzeit, Kunst der Welt, Holle, Baden-Baden, pp 9-68

Breuil H, de Saint-Perier R (1927) Poissons, bactraciens et reptiles dans l'art quaternaire. Arch Inst Paleontol Hum Mem 2: 169 pp

Brjussow AJ (1957) Geschichte der neolithischen Stämme im europäischen Teil der UdSSR. Berlin

Brögger AW (1931) Die arktischen Felsenzeichnungen und Malereien in Norwegen. Jahrb Prähist Ethnogr Kunst. De Gruyter, Berlin

Buse de la Guerra H (no date) Pesca y recoleccion de mariscos. In: Pesca Peru La pesca en el Peru prehispanico. Ed Unidas, Lima, pp 17-37

Clark JGD (1948) The development of fishing in prehistoric Europe. Antiquaries J 27: 45-85

Cleyet-Merle JJ (1990) La préhistoire de la pêche. Collection des hesperides. Errance, Paris

Dabry de Thiersant PD (1872) La pisciculture et la pêche en Chine. Paris

Del Busto JA (no date) Vision del Peru prehispanico. In: Pesca Peru La pesca en el Peru prehispanico. Ed Unidas, Lima, pp 83-101

Douglas FH, d'Harnoncourt R (1941) Indian art of the United States. Mus Modern Art, New York

Ellmers D (1976) Kogge, Kahn und Kunststoffboot. Führer Dtsch Schiffahrtsmus 7, Bremerhaven

Friedel E (1881) Geschichte der Fischerei. Amtl Ber Int Fisch Ausst Berlin 1880

Gjessing G (1936) Nordenfjelske ristninger og malinger. Oslo

Gruvel A (1928) La pêche de la préhistoire, dans l'antiqitée et chez les peuples primitives. Soc Etud Geogr Mar Colon, Paris

Haberland W (1965) Nordamerika. Kunst der Welt. Holle, Baden-Baden

Hallström G (1938) Monumental art. Stockholm

Hornell J (1929) A biologist as ethnologist. Nature (London) 123

Hornell J (1950) Fishing in many waters. Cambridge Univ Press, Cambridge

Hurum HG (1976) A history of fish-hook and the story of Mustad the hook maker. Black, London

Indreko A (1964) Mesolithische und frühneolithische Kulturen in Osteuropa und Westsibirien. Viterbets Hist Antikvitets Akad Handl, Antikv Ser 13, Stockholm

Krause E (1904) Vorgeschichtliche Fischereigeräte und neuere Vergleichsstücke. Z Fischerei 11: 133-300

Kühn H (1952) Die Felsbilder Europas. Kohlhammer, Stuttgart

Landberg LCW (1965) The Chumash Indians of southern California. SW Mus Pap Los Angeles 19

Lhote H (1959) The story of prehistoric rock paintings of the Sahara. Hutchinson, London

Lindig W, Münzel M (1978) Die Indianer. DTV, München

Nordenskiöld AE (1882) Die Umsegelung Asiens und Europas auf der Vega. Mit einem historischen Rückblick auf frühere Reisen längs der Nordküste der Alten Welt, vol 1. Leipzig

Pager H (1975) Stone age myth and magic. Cape Town

Pesca Peru (no date) La pesca en el Peru prehispanico. Ed Unidas, Lima

Piette E (1907) L'art pendant l'age du renne. Paris

Rau C (1884) Prehistoric fishing in Europe and North America. Smith Contrib Knowledge XXV, 1,2

Raudonikas WJ (1936-1938) Les gravures rupestres des bords du lac Onéga et de la Mer Blanche, vols 1, 2. Moscow

Rudner J, Rudner I (1970) The hunter and his art. Struik, Cape Town

Schoonraad M (1962) Rock painting depicting prehistoric fishing found near Maclear. S Afr J Sci 58: 141-143

Schultze M (1914) Frühneolithische Jagd- und Fischereigeräte der Provinz Posen. Arch Fischereigeschichte 2: 109-132

Thomazi A (1947) Histoire de la pêche des ages de la pierre à nos jours. Payot, Paris

Timmermann G (1956) Vom Einbaum zum Wikingerschiff. Vergleichende Betrachtung der Bodenfunde von Schiffen als Grundlage der Schiffsgeschichte. Schiff und Hafen 8 (2): 130-138, 218-226, 336-342, 403-412, 545-549, 602-612

Torke W (1978) Fischreste als Quellen der Ökologie und Ökonomie in der Steinzeit Südwest-Deutschlands. Thesis Univ Hamburg

Vanstone JW (1974) Athapascan adaptations. Hunters and fishermen of the subarctic forests. Chicago

Von Brandt A (1970) Vor- und frühgeschichtliches Netzwerk. Protokolle Fangtechnik XII,55: 107-128, Hamburg

Von Erdberg-Consten E (1958) Das alte China. Große Kulturen der Frühzeit. Kilpper, Stuttgart

Von Tröltsch E (1902) Die Pfahlbauten des Bodenseegebietes. Stuttgart

Woodhouse B (1984) When animals were people. Van Rensburg, Cape Town

Yamaguchi, K (1947) The history of fishing in Japan. Tokyo (in Japanese)

1.2 Mesopotamia and Egypt

Chimits P (1957) Tilapia in ancient Egypt. FAO Fish Bull X(4): 211-215

Gamer-Wallert I (1970) Fische und Fischkulte im alten Ägypten. Ägyptologische Abhandlungen. Harassowitz, Wiesbaden

Köster A (1925) Seefahrten der alten Ägypter. Meereskunde XIV,9 (165): 1-31

Lucas A (1948) Ancient Egyptian materials and industries. London

Radcliffe W (1921) Fishing from the earliest times. Murray, London

Salonen A (1970) Die Fischerei im alten Mesopotamien. Ann Acad Sci Fenn, Ser B 166: 1-314

Sanati-Müller S (1989) Texte aus dem Sinkasid-Palast. II Fischtexte und Bürgschaftsurkunden. Baghdader Mitt 20: 225-313

Schiemenz F (1933) Fischerei vor 4500 Jahren im alten Reiche der Ägypter. Dtsch Angler Z 8: 1-7

Strommenger E (1980) Habuba Kabira. Eine Stadt vor 5000 Jahren. Von Zabern, Mainz

Wilkinson JG (1878) The manners and customs of the ancient Egyptians. Revised by S Birch, London

1.3 Mediterranean Cultures

Bayer E (1983) Fischerbilder in der hellenistischen Plastik. Habelts Dissertationsdrucke, R Klass Archäol, Bonn

Bohlen D (1937) Die Bedeutung der Fischerei für die antike Wirtschaft.(ein Beitrag zur Geschichte der antiken Fischerei). Thesis, Univ Hamburg

Brennfleck S (1926) Die Fischerei in der griechischen und römischen Literatur. Arch Fischereigesch 10: 102-115

Callender MH (1965) Roman amphorae. London

Casson L (1971) Ships and seamanship in the ancient world. Princeton

Edey MA (1973) Die verlorene Welt der Ägäis. Time Life Int Neth

Faure P (1973) La vie quotidienne en Crète au temps de Minos (1500 avant Jesus-Christ). Hachette, Paris

Flacelière R (1959) La vie quotidienne en Grèce au siècle de Périclès. Hachette, Paris

Gamer G (1987) Antike Anlagen der Fischverarbeitung in Hispanien und Mauretanien. Antike Welt 18 (2): 19-28

Gerlach R (1950) Die Fische. Claasen, Hamburg

Gianfrotta PA, Pomey P (1980) Archeologia subacquea, storia, tecniche, scoperte e relitti. Mondatori, Milan

Hosius C (ed) (1967) Die Moselgedichte des Decimus Ausonius und des Venantius Fortunatus. Olms, Hildesheim

Kahrstedt U (1944) Kulturgeschichte der römischen Kaiserzeit. Bruckmann, München

Keller W (1955) Und die Bibel hat doch recht. Econ, Düsseldorf Wien

Kirk R (1987) A history of marine fish culture in Europe and North America. Fishing News (Books), Farnham

Klauser T (ed) (1969) Reallexikon für Antike und Christentum vol 7. Hiersemann, Stuttgart

Köster A (1934) Studien zur Geschichte des antiken Seewesens. Klio Beih 32. Leipzig

Lesky A (1947) Thalatta - der Weg der Griechen zum Meer. Wien

Matz F (1956) Kreta, Mykene, Troja, die minoische und die homerische Welt. Phaidon, Stuttgart

Mocchegiani Carpano C (1986) Archeologia subacquea. Note di viaggio nell'Italia sommersa. Palombi, Rome

Nun M (1964) Ancient Jewish fishery. Hakibbutz Hameuchad (in Hebrew)

Nun M (1989) The Sea of Galilee and its fishermen in the New Testament. Kinnereth, Kibbutz Ein Gev

Oliver Narbona M (1982) Almadrabas de la costa alicantina. Univ Alicante, C Ah Prov Al, Alicante

Pauly (1910) Real-Encyclopädie der classischen Altertumswissenschaft 13. Stuttgart

Radcliffe W (1921) Fishing from the earliest times. Murray, London

Rostovcev M (1929) The social and economic history of the Roman Empire, vols 1, 2. Clarendon, Oxford

Smidth JK (1876) Historical observations on the conditions of the fisheries among the ancient Greeks and Romans and on their mode of salting and pickling fish. Rept US Comm Fish Fisheries 1873-74 and 1874-75, Washington

Thompson d'Arcy W (1947) A glossary of Greek fishes. Oxford Univ Press, London

Throckmorton P (ed) (1987) History from the sea. Mitchell Beazley, London

Toutain J (1930) Economic life of the ancient world. London

Vegas M (1975) Die augustische Gebrauchskeramik von Neuss. Limesforschungen 14. Novaesium VI

Wehrhahn-Stauch L (1972) Christliche Fischsymbolik von den Anfängen bis zum hohen Mittelalter. Z Kunstgesch 35: 1-68

Willetts RF (1969) Everyday life in ancient Crete. London New York

2.1 Mass Migration and the Vikings

Capelle T (1987) Die Eroberung des Nordatlantik. Archäologie am Rande des Meeres. Wachholtz, Neumünster

Graham-Campbell J (1980) The Viking world. Lincoln Frances, London

Jankuhn H (1986) Haithabu. Ein Handelsplatz der Wikingerzeit, 8th edn. Wachholtz, Neumünster

Magnusson M (1976) Hammer of the North. Orbis, London

Oxenstierna E (1959) Die Wikinger. Kohlhammer, Stuttgart

Oxenstierna E (1965) Die Nordgermanen. Große Kulturen der Frühzeit. Kilpper, Stuttgart

2.2 Freshwaters and Fish Ponds

Anon (1987) Fisch, Fischfang, Fischhandel. Lexikon des Mittelalters IV. Artemis, München Zürich, pp 493-502

Cnopf W (1927) Die Entwicklung und wirtschaftliche Bedeutung der Teichwirtschaft in Mittelfranken. Arch Fischereigesch 11: 1-155

Hünemörder C (1975,1981,1986) Fischfang und Fischkunde im Mittelalter. Dtsch Schiffahrtsarch Bremerhaven 1: 185-200; 4: 183-190; 9: 189-198

Koch W (1925) Die Geschichte der Binnenfischerei von Mitteleuropa. Handbuch der Binnenfischerei Mitteleuropas IV, 1, Schweizerbart, Stuttgart

Louda J (1966) European civic coats of arms. Hamlyn, London

Von Severus E (1980) Klösterliche Sachkultur im Rituale Hyparchiae des Abtes Johann Augustin Machhausen von Laach (1553-1568). In: Klösterliche Sachkultur des Spätmittelalters. Österr Akad Wiss, Phil-Hist Kl 367, Wien, pp 247-265

2.3 Herring Drift Net Fisheries

Andersson KA (1956) The great periods of herring fishing on the west coast of Sweden and the herring which gave rise to them. Inst Mar Res Lysekil Lund Ser Biol Rep 5

Andersson KA (1960) On the causes of the great fluctuations in the herring fishery on the west coast of Sweden. Inst Mar Res Lysekil Lund Ser Biol Rep 12

Beaujon A (1884) The history of the Dutch sea fisheries; their progress and revival especially in connection with the legislation on fisheries in earlier and later times. The Fisheries Exhib Lit, vol 9: Prize Essays, pt 2. London

Coull JR (1983) The herring fishery in Shetland in the first half of the nineteenth century. N Scot 5: 123-140

Cushing DH (1968) The East Anglian herring fishery in the eighteenth century. J Cons perm int Explor Mer 31 (3): 323-329

De Caux JW (1881) The herring and the herring fishery. Hamilton Adams, London

Devold F (1963) The life of the Atlanto-Scandian herring. Rapp P-v Réun Cons int Explor Mer 154: 98-108

Fulton TW (1911) The sovereignty of the sea. Blackwood, Edinburgh

Hodgson WC (1957) The herring and its fishery. Routledge & Kegan Paul, London

Höglund H (1972) On the Bohuslän herring during the Great Herring Fishery period in the eighteenth century. Inst Mar Res Lysekil Lund Ser Biol Rep 20

Jagow K (1915a) Die Heringsfischerei an den deutschen Ostseeküsten im Mittelalter. Arch Fischereigesch 5: 1-45

Jagow K (1915b) Der Hering im Volksglauben und in der älteren Forschung. Arch Fischereigesch 6: 213-247

Jenkins JT (1927) The herring and the herring fisheries. King, London

Jensen AJC (1963) On the fluctuations of the herring fisheries in the Skagerak. Rapp P-v Réun Cons int Explor Mer 154: 213-214

Johansen AC (1924) On the summer- and autumn-spawning herrings of the North Sea. Medd Komm Havunders Ser Fisk 7, 5

Kyle HM (1929) Die geschichtliche Entwicklung der britischen Fischerei. Handbuch der Seefischerei Nordeuropas VI, 1. Schweizerbart, Stuttgart

Le Danois E (ed) (1937) Die französische Seefischerei. Handbuch der Seefischerei Nordeuropas VII, 1. Schweizerbart, Stuttgart

March EJ (1952) Sailing drifters. Perceival Marshall, London

Meyer P-F (1947) Die geschichtliche Entwicklung der deutschen Ostseefischerei. Fischwirtschaftskunde III (3,4). Keune, Hamburg

Mitchell JM (1864) The herring, its natural history and national importance. Edmonston & Douglas, Edinburgh

Peesch R (1961) Die Fischerkommünen auf Rügen und Hiddensee. Berlin

Rosén N (1953) Die schwedische Westküstenfischerei. Handbuch der Seefischerei Nordeuropas VIII, 3a. Schweizerbart, Stuttgart

Sahrhage D (1957) Unsere Loggerfischerei. Heringsfischerei mit Treibnetzen. Kosmos 6/1957: 290-295

Samuel AM (Lord Mancroft) (1918) The herring. Its effect on the history of Britain. Murray, London

Saul A (1981) The herring industry at Great Yarmouth, c 1280 - 1400. Norfolk Archaeol 38: 33-43

Schäfer D (1887) Das Buch des Lübeckischen Vogts auf Schonen. Hansische Geschichtsquellen 4. Halle/Saale

Schwinden H (1772) The history and antiquities of the ancient burgh of Great Yarmouth in the county of Norfolk. Norwich

Smith P (1985) The Lammas Drave and the winter herrin'- A history of the herring fishing from East Fife. Donalds, Edinburgh

Spiess K-H (1988) Beziehungen nord- und nordostdeutscher Zisterzienserklöster zum Ostseeheringsfang und dem dortigen Heringshandel. Walkenrieder H Beih 1988/1

Tesch J, de Veen J (1933) Die niederländische Seefischerei. Handbuch der Seefischerei Nordeuropas VII, 2. Schweizerbart, Stuttgart

Thomson J (1849) The value and importance of the Scottish fisheries. Smith Elder, Cornhill London

Tomfohrde T (1914) Die Heringsfischereiperiode an der Bohus-Len-Küste von 1556 - 1589. Ein Beitrag zur Klärung der Frage nach dem Untergang des deutschen Heringshandels in der Ostsee und dem Übergewicht der Holländer in der Nordseeheringsfischerei um die Wende des 16. Jahrhunderts. Arch Fischereigesch 3: 1-192

Verbrugghe L (1932) Die belgische Seefischerei. Handbuch der Seefischerei Nordeuropas VII, 3. Schweizerbart, Stuttgart

Wilson G (1965) Scottish fishing craft. Fishing News (Books), London

Wolf T (1986) Tragfähigkeiten, Ladungen und Maße im Schiffsverkehr der Hanse (vornehmlich im Spiegel Revaler Quellen). Quellen und Darstellungen zur hansischen Geschichte NF 31. Böhlan, Köln Wien

Zimmerling D (1976) Die Hanse. Handelsmacht im Zeichen der Kogge. Econ, Düsseldorf Wien

2.4 Old-Time Cod Fisheries

Alward GL (1911) The development of the British fisheries during the nineteenth century with special reference to the North Sea. Grimsby News, Grimsby

Alward GL (1932) The sea fisheries of Great Britain and Ireland. Gait, Grimsby

Beaujon A (1883) The history of the Dutch sea fisheries; their progress and revival especially in connection with the legislation on fisheries in earlier and later times. The Fisheries Exhib Lit, vol 9: Prize Essays, pt 2. London

Bojer J (1955) Die Lofotfischer. Fischer Bücherei 75. Fischer, Frankfurt Hamburg

Bruns F (1900) Die Lübecker Bergenfahrer und ihre Chronistik. Hanseat Geschichtsquellen NF 2, Berlin

Cochin (1935) Introduction à l'histoire des pêches maritimes en France. Mem Off pêches Marit Paris Ser Spec 9

Cushing D (1988) The provident sea. Cambridge Univ Press, Cambridge

De la Morandière C (1962-1966) Histoire de la pêche francaise de la morue dans l'Amérique septentrionale, vols 1-3. Maisonneuve & Larose, Paris

De Loture R (1949) Histoire de la grande pêche de Terre -Neuve. Gallimard, Paris. History of the great fishery of Newfoundland (transl by C.C. Taylor). Spec Sci Rep Fish 213, US Fish Wildl Serv (1957)

Duhamel du Monceau (1769) Traité général des pêches et histoire des poissons, vols 1-3. Saillant & Nyon, Desaint Paris

Gilcich (1883) Der Fischfang der Gascogner und die Entdeckung von Neufundland. Z Ges Erdk 18

Graham M (1943) The fish gate. Faber & Faber, London

Innis HA (1954) The cod fisheries. The history of an international economy. Univ Press, Toronto

Klein RT (1950) The codfish industry in northern Portugal. US Fish Wildl Serv, Fish Leaflet 367

Le Danois E (ed) (1937) Die französische Seefischerei. Handbuch der Seefischerei Nordeuropas VII, 1. Schweizerbart, Stuttgart

Lounsbury RG (1934) The British fishery at Newfoundland, 1634 - 1763. Yale Univ Press, New Haven

Sund O. (1938) Die norwegische Seefischerei. Handbuch der Seefischerei Nordeuropas VIII, 1a. Schweizerbart, Stuttgart

Tesch JJ, de Veen J (1933) Die niederländische Seefischerei. Handbuch der Seefischerei Nordeuropas VII, 2. Schweizerbart, Stuttgart

Villiers A (1951) The quest of the schooner Argus. Hodder & Stoughton, London

Vollan O (1956) Den Norske klippfiskhandels historie.

Zimmerling D (1976) Die Hanse. Handelsmarkt im Zeichen der Kogge. Econ, Düsseldorf Wien

3 Mechanization of Fishing

Anon (1921) The history of trawling. Fish Trades Gazette London 19 March: 21-71, London

Anon (1969) Die Fischwirtschaft der Bundesrepublik Deutschland. Land- und Hauswirtsch. Auswertungs- und Informationsdienst, Suppl, Bonn

Collins (1887) The beamtrawl of Great Britain with notes on beamtrawling in other European countries. Bull US Fish Comm VII f

Eddie GC (1983) Engineering, economics and fisheries management. Fishing News (Books), Farnham

FAO (1985) Definition and classification of fishery vessel types. FAO Fish Tech Pap 267

Graham M (ed) (1956) Sea fisheries, their investigation in the United Kingdom. Arnold, London

Garstang W (1900) The impoverishment of the sea. J Mar Biol Assoc UK NS 6: 1-69

Hovart P (1985) Zeevisserijbeheer in vroegere Eeuwen (Sea-fishery management in earlier times). Medelg Rijksstat Zeeviss Oostende 206 (with English summary)

Iversen T (1937) Tralfiskets historie. Arsber Norges Fiskerier (Bergen) 3: 1-63

Kristjonsson H (1959, 1971) Modern fishing gear of the world, vols 1, 3. Fishing News (Books), London

Kyle HM (1929) Die Seefischerei von Großbritannien und Irland. Handbuch der Seefischerei Nordeuropas VI. Schweizerbart, Stuttgart

Le Danois E (ed) (1937) Die französische Seefischerei. Handbuch der Seefischerei Nordeuropas VII, 1. Schweizerbart, Stuttgart

March EJ (1953) Sailing trawlers. Perceival Marshall, London

Meseck G (1960) Erzeugung und Absatz von Fischmehl in der Welt. Ber Landwirtsch 38 (4): 666-70

Ministry of Agriculture, Forestry and Fisheries (ed) (1987) Fisheries statistics of Japan 1987, Tokyo

Mitson RB (1983) Fisheries sonar. Fishing News (Books), Farnham

Schärfe J (1966) Germany develops midwater herring trawl. Fishing News Int 5 (7):18 - 25

Tesch JJ, de Veen J (1933) Die niederländische Seefischerei. Handbuch der Seefischerei Nordeuropas VII, 2. Schweizerbart, Stuttgart

Timmermann G (1962) Die nordeuropäischen Seefischereifahrzeuge, ihre Entwicklung und ihre Typen. Handbuch der Seefischerei Nordeuropas XI, 4. Schweizerbart, Stuttgart

White EW (1952) British fishing boats and coastal craft. I Historical survey, II Descriptive catalogue and list of plans. Sci Mus, London

Wimpenny RS (1953) The plaice. Buckland Lecture, London

4.1 Whaling

Allen KR (1980) Conservation and management of whales. Univ Washington Press, Seattle

Andrews RC (1935) Whale hunting with gun and camera. Appleton - Century, New York London

Asgrimsson H (1989) Developments leading to the 1982 decision of the International Whaling Commission for a zero catch quota 1986 - 1990. In: Andresen A, Östreng W (eds) International resources management: the role of science and politics. Belhaven, London New York, pp 221-231

Ashley CW (1938) The Yankee whaler. Houghton Mifflin, Boston

Bennet AG (1931) Whaling in the Antarctic. London

Budker P (1958) Whales and whaling. Harrap, London

Chapman DG (1973) Management of international whaling and North Pacific fur seals: implications for fisheries management. J Fish Res Board Can 30: 2419-2426

Clark AH (1887) The whale fishery. I History and present conditions of the fishery. In: Brown Goode G (ed) The fisheries and fishing industry of the United States, Sect V: History and methods of the fisheries. US Gov Print Off Washington, pp 3-293

Deimer P (1983) Das Buch der Wale. Heyne, München

De Jong C (1978) A short history of old Dutch whaling. Univ S Afr, Pretoria

De Jong C (1983) The hunt of the Greenland whale: a short history and statistical sources. Rep Int Whal Comm Spec Issue 5: 83-106

Dow GF (1925) Whale ships and whaling. A pictoral history of whaling during three centuries, with an account of whale fishery in colonial New England. Mar Res Soc Salem Mass Publ 10

Fraser FC (1948) Whales and dolphins. In: Norman JR, Fraser FC (eds) Giant fishes, whales and dolphins. Putnam, London

Hawley F (1958-60) Whales and whaling in Japan, vol 1, pt 1. Misc Jpn 2

Hjort J, Lie J, Ruud T (1932-1937) Pelagic whaling in the Antarctic, 1 - 6. Hvalradets Skr 3, 7, 8, 9, 12, 14, 18

Hoydal K (in press) Data on the long finned pilot whale *(Globicephala melaena* Traill) in Faroe waters and an attempt to use the 274 years time series of catches to assess the state of the stock. IWC Pap SC/38/SM7

Jackson G (1978) The British whaling trade. Black, London

Jenkins JT (1921) A history of the whale fisheries from the Basque fisheries of the tenth century to the hunting of the finner whale at the present date. Witherby, London

Jenkins JT (1932) Whales and modern whaling. Witherby, London

Kühn H (1952) Die Felsbilder Europas. Kohlhammer, Stuttgart

Lindemann M (1869) Die arktische Fischerei der deutschen Seestädte 1620 - 1868. Petermanns Mitt Erg H 26

Lynge E (1936) Der Walfang. Ein Beitrag zur Weltwirtschaft der Fettstoffe. Wandlungen der Weltwirtschaft 7. Leipzig

Mackintosh NA (1965) The stocks of whales. Fishing News (Books), London

Marquette WM, Bockstoce JR (1980) Historical shore based catch of bowhead whales in the Bering, Chuckchi and Beaufort Seas. Mar Fish Rev 42 (9-10): 5-19

Matthews LH (ed) (1968) The whale. Allen & Unwin, London

Melville H (1851) Moby-Dick; or, the Whale. New York

Münzing J (1987) Der historische Walfang in Bildern. Sammlungen des Altonaer Museums in Hamburg 13. Koehler, Herford

Oesau W (1937) Schleswig-Holsteins Grönlandfahrt auf Walfischfang und Robbenschlag. Augustin, Glückstadt

Oesau W (1939) Die deutsche Südseefischerei auf Wale im 19. Jahrhundert. Augustin, Glückstadt

Peters N (1937) Über Hochseewalfang und Tierleben im Südlichen Eismeer. Der Fischmarkt. Vorrath, Cuxhaven

Peters N (1938) Der neue deutsche Walfang. Hansa, Hamburg

Risting S (1922) Av hvalfangstens historie. Cappelens, Kristiania

Risting S, Paulsen HB (1938) Der norwegische Walfang. Handbuch der Seefischerei Nordeuropas VIII, 1b. Schweizerbart, Stuttgart

Scammon CM (1874) The marine mammals of the north-western coast of North America, described and illustrated: together with an account of the American whale fishery. Carmany, San Francisco

Schubert K (1955) Der Walfang der Gegenwart. Handbuch der Seefischerei Nordeuropas 11. Schweizerbart, Stuttgart

Scoresby W (1820) An account of the Arctic regions with a history and description of the northern whale fishery, vol 2: The whale fishery. Constable, Edinburgh

Starbuck A (1878) History of the American whale fishery from its earliest inception to the year 1876, vols 1, 2. Rep Comm US Fish Fisheries 1875-76 App: 769 pp (reprinted 1964 by Argosy-Antiquarian, New York

Tesch JJ, de Veen J (1933) Die niederländische Seefischerei. Handbuch der Seefischerei Nordeuropas VII, 2. Schweizerbart, Stuttgart

Tönnesen JN (1970) Norwegian Antarctic whaling, 1905-68: an historical appraisal. Polar Rec 15 (96): 283-290

Tönnesen JM, Johnsen AO (1982) The history of modern whaling. Hurst, London

Tower WS (1916) A history of the American whale fishery. Univ Penn Ser Polit Econ Publ Law 20, Philadelphia

Tressler DK, Mc W Lemon J (1951) Marine products of commerce, 2nd edn. Reinhold, New York

Walton DWH (ed) (1987) Antarctic science. Cambridge Univ Press, Cambridge

Winterhoff E (1974) Walfang in der Antarktis. Schrift Dtsch Schiffahrtsmus Bremerhaven 4. Stalling, Oldenburg Hamburg

Zeller HL (1915) Der Walfischfang der Basken. Seerechtl Forsch 1

Zorgdrager CG (1727) Bloeijende opkomst de aloude en hedendaagsche Groenlandsche Visschery. Van Thol & Alberts, Gravenhage

4.2 Sealing

Barthelmess K (1987) Kommerzieller Robbenfang. In: Deimer P Das Buch der Robben. Rasch & Röhring, Hamburg

Beddington JR, Beverton RJH, Lavigne DM (eds) (1985) Marine mammals and fisheries. Allen & Unwin, London

Bonner WN (1982) Seals and man. Washington Sea Grant Progr Seattle

Clark AH (1887) The North Atlantic seal fishery. In: Brown Goode G (ed) The fisheries and fishing industry of the United States, sect V: The history and methods of the fisheries. US Gov Print Off Washington, pp 474-483

Coffey DJ (1977) The encyclopedia of sea mammals. Hart-Davis, Mac Gibbon, London

Everson I (1987) Life in a cold environment. In: Walton, DWH (ed) Antarctic science. Cambridge Univ Press, Cambridge

Haug T, Kroyer AB, Nilssen KT, Ugland KI, Aspholm PE (in press) Harp seal *(Phoca groenlandica)* invasions in Norwegian coastal waters: age composition and feeding habits. J Cons int Explor Mer

ICES (1987) Reports of ICES Working Group on Seals in the Baltic and of Working Group on Harp and Hooded Seals in the Greenland Sea. ICES Coop Res Rep 148, Copenhagen

Iversen T (1938) Der norwegische Robbenfang. Handbuch der Seefischerei Nordeuropas VIII, 1b. Schweizerbart, Stuttgart

Kenyon KW, Scheffer VB, Douglas DG (1954) A population study of the Alaska fur seal herd. US Dep Int Spec Sci Rep Wildlife 12

Mohr E (1952) Die Robben der europäischen Gewässer. Monographie der Wildsäugetiere XII. Schöps, Frankfurt/Main

NOAA (ed)(1977) The story of the Pribilof fur seals. US Dep Comm, NOAA, Washington

Oesau W (1955) Hamburgs Grönlandfahrt auf Walfischfang und Robbenschlag vom 17. - 19. Jahrhundert. Augustin, Glückstadt Hamburg

Ridgeway SH, Harrison RJ (eds) (1981) Handbook of marine mammals, vol 1: The walrus, sea lions, fur seals and sea otter. Academic Press, London New York

Riley F (1967) Fur seal industry of the Pribilof Islands. US Dep Commerce Fish and Wildl Serv Circ 275

Savini MJ (1974) Report on international and national legislation for the conservation of marine mammals, pt 1: International legislation. FAO Fish Circ 326, Rome

Sergeant DE (1965) Exploitation and conservation of harp and hooded seals. Polar Rec 12 (80): 541-551

Sergeant DE (1976) History and present status of populations of harp and hooded seals. Biol Conserv 10: 95-118

Stonehouse B (1972) Animals of the Antarctic. Eurobook, London

5.1 Statistical Account of World Fishing

FAO (1947-1989) Yearbook of fishery statistics, vols 1 - 66. FAO, Rome

Lebret LJ, Sauvée J (1950) Pêcheries mondiales et marché du poisson. Paris

Lindemann M (1880) Die Seefischereien, ihre Gebiete, Betrieb und Erträge in den Jahren 1869 - 1878. Petermanns Mitt Erg H 60

Lübbert H (1950) Die großen Seefischereien der Erde, ihre Fanggründe und ihre Fangprodukte. Girardet, Hamburg

5.2.1 Japan

AICAF (ed)(1984) Fisheries and fishery products in Japan. Jpn Agric Rev 9

Anon (1948a) The Japanese tuna fisheries. US Fish Wildl Serv, Fish Leaflet 297

Anon (1948b) Canned crab industry of Japan. US Fish Wildl Serv, Fish Leaflet 314

Asada Y, Hirasawa Y, Nagasaki F (1983) Fishery management in Japan. FAO Fish Tech Pap 238

Bartz F (1965) Die großen Fischereiräume der Welt, II: Asien mit Einschluß der Sowjetunion. Steiner, Wiesbaden

Borgstrom G (1964) Japan's world success in fishing. Fishing News (Books), London

Bourgeois (1951) Japanese offshore trawling. US Fish Wildl Serv, Fish Leaflet 389

Doumenge F (1962) Le developpement de la grande pêche industrielle japonaise. pêche Mar 41 (1008): 147-155

FAO (1985) Definition and classification of fishery vessel types. FAO Fish Tech Pap 267

Fujinaga M (1954-1955) Japanese fishery: brief history of Japanese fishery. Contemp Jpn 23: 711-725

Fujinami N (1965) Japan - modern development of the fishing industry. Fishing News Int 4 (4): 405-407

Japan Fish Association (ed)(1966) Japan's fishery industry. Tokyo

Japan Fish Association (ed)(1984) Fisheries of Japan 1984. Tokyo

Kafuku T, Ikenoue H (1983) Modern methods of aquaculture in Japan. Development in Aquaculture and Fisheries Science 11. Kondasha, Tokyo; Elsevier, Amsterdam, Oxford, New York

Kasahara H (1972) Japanese distant-water fisheries: a review. NMFS Fish Bull 70 (2): 227-282

Kondo K (1980) The recovery of the Japanese sardine. The biological basis of stock size fluctuations. Rapp P-v Réun Cons int Explor Mer 177: 332-354

Ministry of Agriculture and Forestry (ed) (1973,1983) The fifth (seventh) fishery census of Japan. Tokyo

Ministry of Agriculture, Forestry and Fisheries (ed) (1987) Fisheries statistics of Japan 1987. Tokyo

Motoda S, Hirano Y (1963) Review of Japanese herring investigations. Rapp P-v Réun Cons int Explor Mer 154 (35): 249-261

Nagasaki F (1973) Long-term and short-term fluctuations in the catches of coastal pelagic fisheries around Japan. J Fish Res Board Can 30: 2361-2367

Ohira Z (1957) Fishery problems between USSR and Japan. Jpn Ann Int Law 3: 109-125

Okutani T (1977) Stock assessment of cephalopod resources fished by Japan. FAO Fish Tech Pap 173

Pruter AT (1973) Development and present status of bottomfish resources in the Bering Sea. J Fish Res Board Can 30: 2373-2385

Shapiro S (1948) The Japanese tuna fisheries. US Fish Wildl Serv, Fish Leaflet 297

Shapiro S (1950) The Japanese longline fishery for tuna. Comm Fish Rev 12 (4)

Shimada BM (1951) Japanese tuna-mothership in the western equatorial Pacific Ocean. Comm Fish Rev 13 (6)

Ventilla RF (1982) The scallop industry in Japan. In: Blaxter JHS, Russell F, Yonge M (eds) Advances in marine biology 20. Academic Press, London New York, pp 309-382

Vogt R (1953) Die Bedeutung der Fischerei in der japanischen Wirtschaft. Fischwirtschaft 10: 10-14

Yamaguchi K (1947) Nippon gyogyö shi (A history of Japanese fisheries). Seikatsusha, Tokyo (Engl Transl 1964: Univ Press, Tokyo)

5.2.2 Russia - Soviet Union

Andrusov N (1888) A survey in the history of the development of the Caspian Sea and its inhabitants. Bull Imp Russ Geogr Soc 24 (in Russian)

Anon (1959) High seas fisheries of the USSR. US Bur Comm Fish, Fish Leaflet 482

Bajevsky B (1927) Fisheries of Siberia. Appendix for the fiscal year 1926. Rep US Comm Fish

Bartz F (1965) Die großen Fischereiräume der Welt, II: Asien mit Einschluß der Sowjetunion. Steiner, Wiesbaden

Borgstrom G (1961) The Atlantic fisheries of the USSR. In: Borgstrom G, Heighway AJ (eds) Atlantic Ocean fisheries. Fishing News (Books), London, pp 282-315

Brady P (1965) Focus on the Soviet fleet. Fishing News Int 4 (4): 392-400

Breitfuss (1905) Die Fischerei an der Murmanküste. Rapp p-v Réun Cons int Explor Mer 3

Breitfuss (1930) Die Erschließung des eurasiatischen hohen Nordens. Petermanns Mitt Erg H 207

Carré F (1978) Les pêches en mer Caspienne. Ann Géogr 479

Classen C (1924) Russische Fischerei und Fischindustrie. Der Fisch, vol 2. Lübeck

Douglas WF (1930/1940) Das Fischereiwesen Rußlands. Arch Hydrobiol Suppl, vol X, 1. Schweizerbart, Stuttgart, pp 1-206

Kellner P (1980) Fischereiwirtschaft und Fischproduktion der Sowjetunion. Giessener Abhandlungen zur Agrar- und Wirtschaftsforsch des europäischen Ostens 105. Duncker & Humblot, Berlin

Kravanja MA (1964) Soviet Far East fisheries expansion. Comm Fish Rev 26 (11a)

Moisseev PA (1955) Geographical distribution of fishes and other commercial species in the Okhotsk and Bering Seas. Trud Inst Okean XIV. Moscow (in Russian)

Nansen F (1914) Sibirien. Leipzig

Ovchynnyk MM (1961) Development of some marine and inland Russian fisheries, and fish utilization. In: Borgstrom G, Heighway AJ (eds) Atlantic Ocean fisheries. Fishing News (Books), London, pp 267-281

Zenkevich LA (1963) Biology of the seas of the USSR. Allen & Unwin, London

Zenkevich LA (1967) Seas of the USSR; their fauna and flora. Uchpedgiz, Moscow (in Russian)

5.2.3 China

Anon (1961) China's fishing industry is developing fast. World Fishing 10 (12): 51

Bartz F (1965) Die großen Fischereiräume der Welt, II: Asien mit Einschluß der Sowjetunion. Steiner, Wiesbaden

Bureau of Aquatic Products, Ministry of Agriculture, Animal Husbandry and Fishery (eds) (1985) China's fishery. Beijing

Chen CS, Lin SY (1935) The fish fry industry of China. Bull Chekiang Prov Fish Exp Stn Tinghai 1 (4)

Dabry de Thiersant PD (1872) La pisciculture et la pêche en Chine. Paris

FAO (1983) Freshwater aquaculture development in China. FAO Fish Tech Pap 215

FAO (1990) Aquaculture production (1985-88). FAO Fish Circ 815, Rev 2

Lin MC, Wu YC (1936) A survey of the fishery industry of Ningpo. Bull Chekiang Prov Fish Exp Stn 1 - 2

Mohan Joseph M (ed) (1990) Aquaculture in Asia. Asian Fish Soc, Indian Branch, Mangalore

Nichols JT (1943) The fresh-water fishes of China. Nat Hist Centr Asia 9

Ruddle K, Zhong G (1988) Integrated agriculture - aquaculture in South China. Cambridge Univ Press, Cambridge

Szczepanik EF (1961) The fisheries of mainland China I. Fishing News Int 1 (1): 39-42

Von dem Borne, M (1894) Fischzucht der Chinesen. Teichwirtschaft 4th edn. Thaer Bibl. Parey, Berlin, pp 134-143

Wei-ching WY (1910) The fisheries of China. Bull Bur Fish 28.1 (1908)

Zhu de-Shan (1980) A brief introduction to the fisheries of China. FAO Fish Circ 726

5.2.4 North America

Ahlstrom EH, Radovich J (1970) Management of the Pacific sardine. A century of fisheries in North America. Spec Publ Am Fish Soc 7: 183-193

Anderson WM (1958) The shrimp and the shrimp industry of the southern United States. US Bur Comm Fish, Fish Leaflet 472

Anderson WM, Lindner MJ, King JE (1949) The shrimp fishery of the southern United States. Comm Fish Rev 11 (2): 1 - 17

Bartz F (1942) Fischgründe und Fischereiwirtschaft an der Westküste Nordamerikas. Schr Geogr Inst Univ Kiel

Bartz F (1974) Die großen Fischereiräume der Welt, III: Neue Welt und Südliche Halbkugel. Steiner, Wiesbaden

Borgstrom G (1961) US Atlantic fisheries and current trends in supply and use. In: Borgstrom G, Heighway AJ (eds) Atlantic Ocean fisheries. Fishing News (Books), London, pp 225-255

Brown Goode G (1879) The natural and economic history of the American menhaden. US Comm Fish Fisheries, Rep Comm 1877

Brown Goode G (ed) (1887) The fisheries and fishing industry of the United States, sect V: The history and methods of the fisheries. US Comm Fish Fisheries. Gov Print Off, Washington

Brown Goode G, Collins JW, Earll RE, Clark AH (1884) Materials for a history of the mackerel fishery. Rep Comm Fish Fisheries 4: 93-531

Culley M (1971) The pilchard. Biology and exploitation. Pergamon, Oxford

De Witt G (ed) (1968) The future of the fishing industry of the United States. Fish NS Univ Washington 4

Gregory HE, Barnes K (1939) North Pacific fisheries with special reference to Alaska salmon. Stud Pac 3, Am Counc Inst Pac Relat, San Francisco New York Honolulu

Henry KA (1971) Atlantic menhaden (*Brevoortia tyrannus*) resource and fishery - analysis of decline. US Natl Mar Fish Serv Spec Sci Rep Fish 642

Hourston AS (1980) The decline and recovery of Canada's Pacific herring stocks. Rapp P-v Réun Cons int Explor Mer 173: 143-153

Hoy DL, Clark GM (1967) Atlantic mackerel fishery 1804 - 1965. US Fish Wildl Serv, Fish Leaflet 603

Judah CB (1933) The North American fisheries and British policy to 1713. Univ Illinois Bull 31

June FC (1961) The menhaden fishery of the United States. US Bur Comm Fish, Fish Leaflet 521

Kinloch J (1961) The fisheries of Canada. In: Borgstrom G, Heighway AJ (eds) Atlantic Ocean fisheries. Fishing News (Books), London, pp 256-259

McFarland, R (1911) A history of the New England fisheries. Appleton, New York

Murphy GI (1966) Population biology of the Pacific sardine *(Sardinops caerulea)*. Proc Cal Acad Sci 34 (1): 1-84

Power EA (1959) United States tuna fishery, 1911 - 1958. US Bur Comm Fish, Fish Leaflet 484

Rothschild BJ, Parrack ML (1981) The US Gulf of Mexico shrimp fishery. In: Worksh Sci Basis for the management of penaeid shrimp, Nov 1981, Key West, Florida

Schaaf WE (1975) Status of the Gulf and Atlantic menhaden fisheries and implications for resource management. Mar Fish Rev 37 (9): 1-9

Schaaf WE (1979) An analysis of the dynamic population response of Atlantic menhaden, *Brevoortia tyrannus*, to an intensive fishery. Rapp P-v Réun Cons int Explor Mer 177: 243-251

Templeman WF (1966) Marine resources of Newfoundland. Bull Fish Res Board Can 154: 1-172

Thompson WF, Freeman N (1930) History of the halibut fishery. Rep Int Fish Comm 5

Vaughan DS, Merriner JV, Smith JW (1988) The US menhaden fishery: current status and utilization. In: Davis N (ed) Fatty fish utilization: upgrading from feed to food. UNC Sea Grant Publ 88-04, Raleigh, NC

Walford LA (1947) Fishery resources of the United States. Public Affairs Press, Washington, DC

5.2.5 Peru and Chile

Ancieta F (1963) La pesqueria del bonito en el Perú. FAO Fish Rep 6 (3): 1607-1619

Anon (1981) Chilean fisheries resources catalogue. Inst Fomento Pesquero, Santiago

Arntz WE (1986) The two faces of El Niño 1982-83. Meeresforschung 31: 1-46

Arntz WE, Fahrbach E (1991) El Niño. Klimaexperiment der Natur. Birkhäuser, Basel

Barber RT, Chavez FP (1983) Biological consequences of El Niño. Science 222 (4629): 1203-1210

Bartz F (1974) Die großen Fischereiräume der Welt, III: Neue Welt und Südliche Halbkugel. Steiner, Wiesbaden

Boerema LK, Gulland JA (1973) Stock assessment of the Peruvian anchovy *(Engraulis ringens)* and management of the fishery. J Fish Res Board Can 30 (12): 2226-2235

Brandhorst W (1965) Die chilenische Fischerei und ihre weiteren Entwicklungsmöglichkeiten. Ber Landwirtsch 43 (1): 148-187

Edwards CR Aboriginal watercrafts on the Pacific coast of South America. Ibero-Americana 47. Univ Cal Press, Berkeley Los Angeles

Fiedler RH (1944) The Peruvian fisheries. Geogr Rev 34, 1

Furness RW (1982) Competition between fisheries and seabird communities. In: Blaxter JHS, Russell F, Yonge M (eds) Advances in marine biology 20. Academic Press, London, pp 225-307

Idyll CP (1973) The anchovy crisis. Sci Am 228 (6): 22-29

Loftas T (1972) Where have all the anchoveta gone? New Sci 55 (813): 583-586

Murphy RC (1981) The guano and the anchoveta fishery. In: Glantz MG, Thompson JD (eds) Resource management and environmental uncertainty. John Wiley & Sons, New York, pp 81-106

Nicholls N (1987) The El Niño / Southern Oscillation phenomenon. In: Glantz M, Katz R, Krenz M (eds) The social impacts associated with the 1982-83 worldwide climate anomalies. Natl Cent Atmosph Res, Boulder, Col, pp 2-10

Paulik GJ (1971) Anchovy, birds and fishermen in the Peru current. In: Murdoch WW (ed) Environment, resources, pollution and society. Sinauer, Stanford, pp 156-185

Pauly D, Tsukayama I (eds) (1987) The Peruvian anchoveta and its upwelling ecosystem: three decades of change. ICLARM Stud Rev 15: 1-351

Schwabe GH (1941) Über Mariscos und Mariscofischerei. Z Fisch Hilfswissensch 39: 313-347

Schweigger E (1959) Die Westküste Südamerikas im Bereich des Peru-Stromes. Geographische Handbücher. Keyser, Heidelberg

Serra BR (1987) Impact of the 1982-83 ENSO on the Southeastern Pacific fisheries, with an emphasis on Chilean fisheries. In: Glantz M, Katz R, Krenz M (eds) The social impacts associated with the 1982-83 worldwide climate anomalies. Natl Cent Atmosph Res, Boulder, Col, pp 24-29

Troadec J-P, Clark WG, Gulland JA (1980) A review of some pelagic fish stocks in other areas. Rapp P-v Réun Cons int Explor Mer 177: 252-277

Valdez - Zamudio F (1973) Impacto de medidas regulatorias en la industria pesquera peruana. J Fish Res Board Can 30 (12): 2242-2253

Valdivia JE (1978) The anchovetta and El Niño. Rapp P-v Réun Cons int Explor Mer 173: 196-202

5.2.6 South and Southwest Africa (Namibia)

Bartz F (1974) Die großen Fischereiräume der Welt, III: Neue Welt und Südliche Halbkugel. Steiner, Wiesbaden

Cushing DH (1968) The abundance of hake off South Africa. Fish Invest London (II) 25 (10): 1-20

Davies DH (1957) The biology of the South African pilchard. Invest Rep 32. Div Fish Union S Afr, Cape Town

Dreosti CM (1961) The development of the South African fish industry. In: Borgstrom G, Heighway AJ (eds) Atlantic Ocean fisheries. Fishing News (Books), London, pp 182-191

Gertenbach LPD (1962) Regulation of the South African west coast shoal fisheries. FAO Fish Rep 5: 427-458

Irvin & Johnson (1963) South African fish and fishing. Cape Town

Jones BW, van Eck TM (1967) The Cape hake, its biology and the fishery. S Afr Shipping News 22 (11): 90-97

Matthews JP (1962) The rock lobster of South West Africa. Invest Rep 7, Admin of SW Afr, Mar Res Lab, Windhoek

Newman GG (1977) The living marine resources of the Southeast Atlantic. FAO Fish Tech Pap 178

Newman GG, Crawford RJM (1980) Population biology and management of mixed-species pelagic stocks off South Africa. Rapp P-v Réun Cons int Explor Mer 177: 279-291

Payne AIL, Gulland JA, Brink KH (eds) (1987) The Benguela and comparable ecosystems. S Afr J Mar Sci 5, 957 pp

Scott P (1949) Otter-trawl fisheries of South Africa. Geogr Rev 39, 4

Scott P (1951) Inshore fisheries of South Africa. Econ Geogr 27 (II)

Stander GH, Le Roux PJ (1968) Notes on fluctuations of the commercial catch of the South African pilchard *(Sardinops ocellata)* 1950 - 65. Invest Rep 65, Div Fish Union S Afr

Thompson WW (1913) The sea fisheries of the Cape Colony. Miller, Cape Town

Von Brandt A (1966) Die Fischerei von Süd- und Südwestafrika. Allg Fisch Z 2/3 1966

5.2.7 Europe

Anon (1924-1990) Jahresbericht über die deutsche Fischerei (since 1964: Fischwirtschaft) Bundesmin Ernähr Landwirtsch Forsten, Bonn

Anon (1985) Los pescados azules de nuestras costas. Min Agric Pesca Aliment, Madrid

Bartz F (1964) Die großen Fischereiräume der Welt, I: Atlantisches Europa und Mittelmeer. Steiner, Wiesbaden

Bas C, Morales E, Rubió M (1955) La pesca en España. I Cataluña. Inst Invest Pesquero, Barcelona

Blanke W (1956) Die Seefischerei Nordwest-Europas. Struktur und Probleme. Kröger, Hamburg

Blegvad H (ed) (no date) Fiskeriet i Danmark, vols 1, 2. Selskabet til Udgivelse af Kulturskrifter, Copenhagen

Borgstrom G, Heighway AJ (eds) (1961) Atlantic Ocean fisheries. Fishing News (Books), London

Brochmann BS, Josefsen O (1984) Fiskerinaeringen. Tiden Norsk, Bergen

Buchmann E (1974) Die Fischereiwirtschaft Islands unter besonderer Berücksichtigung ihrer Entwicklung seit dem Ende des II. Weltkrieges. Thesis, Univ Mainz

Coull JR (1972) The fisheries of Europe. An economic geography. Bell, London

Dill WA (1990) Inland fisheries of Europe. EIFAC Tech Pap 52. FAO, Rome

FAO (1947-1966) Yearbook of fishery statistics, vols 1-66. FAO, Rome

FAO (1975-1991) Fishery country profiles. FAO, Rome

ICES (1903-1987) Bulletin statistique, Vols 1-72. ICES, Copenhagen

Jüngst P (1968) Die Grundfischversorgung Großbritanniens. Häfen, Verarbeitung und Vermarktung. Marburger Geogr Schr 35

Kühnhold WW, Kirchner WP (1990) Fischerei und Fischwirtschaft der DDR. Infn Fischw Auslds 40(1/2): 1-184

Lübbert H, Ehrenbaum E (eds) (1929-1936) Handbuch der Seefischerei Nordeuropas, vols III - VIII. Schweizerbart, Stuttgart

Schumacher A (1980) Review of North Atlantic catch statistics. Rapp P-v Réun Cons int Explor Mer 177: 8 - 22

Shapiro S (1965) The fisheries of Norway. US Fish Wildl Serv, Circ 235

Vanneste O, Hovart P (1959) La pêche maritime belge. Bruges

Wood H (1956) Fisheries of the United Kingdom. In: Graham M (ed) Sea fisheries, their investigation in the United Kingdom. Arnold, London, pp 10-79

5.3 International Law of the Sea

Anderson LG (ed) (1977) Economic impacts of extended fisheries jurisdiction. Ann Arbor Sci Publ, Michigan

Carnahan EL, King LR (1983) The law of the sea treaty and the United States. Marine advisory. Sea Grant Coll Prog Texas A & M Univ

Dannenbring FE (1952) Das geltende Fischereirecht in Bremen und seine geschichtliche Entwicklung. Bremen

FAO (1973) Limits and status of the territorial sea, exclusive fishing zones, fishery conservation zones and the continental shelf. FAO Fish Circ 127, Rev 1

Gündling L (1983) Die 200 Seemeilen-Wirtschaftszone. Springer, Berlin, Heidelberg, New York, Tokyo

Gulland JA (1979) The new ocean regime: winners and losers. Ceres 70: 19-23, Rome

Lucas KC, Loftas T (1982) FAO's EEZ Program: helping to build the fisheries of the future. Ocean Yearb 3: 38-76

Meseck G (1960) Ergebnisse und Auswirkungen der II. Internationalen Seerechtskonferenz und neue Entwicklungstendenzen in der Fischwirtschaft der Welt. Ber Landwirtsch 38 (4): 773-786

Miles E, Gamble JK (eds) (1977) Law of the sea: conference outcomes and problems of implementation. Ballinger, Cambridge, Mass

Moore G (1981) Legislation on coastal state requirements for foreign fishing. FAO Legislative Study 21

Rothschild BJ (ed) (1983) Global fisheries. Perspectives for the 1980s. Springer, Berlin Heidelberg New York

Weise E (1935) Die Rechtsverhältnisse der sächsischen Elbfischerinnungen. Arch Fischereigesch 19: 1-77

Winder G (1961) International territorial limits in relation to fishing. In: Borgstrom G, Heighway AJ (eds) Atlantic Ocean fisheries. Fishing News (Books), London, pp 49-54

Uthoff D (1990) Auswirkungen des neuen Seerechts auf die Seefischerei. Geogr Rundsch 12/1990: 672-679

6 Fisheries Science

Cushing DH (1988) The provident sea. Cambridge Univ Press, Cambridge

Cutting CL (1955) Fish saving. A history of fish processing from ancient to modern times. Hill, London

Gerlach R. (1950) Die Fische. Claassen, Hamburg

Jackson RI, Royce WF (1986) Ocean forum, an interpretive history of the International North Pacific Fisheries Commission. Fishing News (Books), Farnham

Koers AW (1973) International regulation of marine fisheries. Fishing News (Books), Farnham

Lenz W, Deacon M (eds) (1990) Oceanic sciences: their history and relation to man. Dtsch Hydrogr Z Erg-H B (22): 1-603

Mills EL (1989) Biological oceanography. An early history, 1870 - 1960. Cornell Univ Press, Ithaca

Royce WF (1987) Fishery development. Academic Press, Orlando

Saetersdal G (1989) Fish resources research and fishery management: a review of nearly a century of experience in the Northeast Atlantic and some recent global perspectives. J Cons int Explor Mer 46: 5-15

Sahrhage D (1989) The role of science in international management of fish resources. In: Andresen S, Östreng W (eds) International resources management: the role of science and politics. Belhaven, London New York, pp 207-220

Thomasson EM (1981) Study of the sea. The development of marine research under the auspices of the ICES. Fishing News (Books), Farnham

Went AEJ (1972) Seventy years agrowing. A history of the International Council for the Exploration of the Sea. Rapp P-v Réun Cons int Explor Mer 165: 1-252

7 Perspectives for the Future

Alverson DL, Paulik GJ (1973) Objectives and problems of managing aquatic living resources. J Fish Res Board Can 30: 1936-1947

Andresen S, Östreng W (eds) (1989) International resources management: the role of science and politics. Belhaven, London New York

Bilio M, Rosenthal H, Sindermann CJ (eds) (1986) Realism in aquaculture: achievements, constraints, perspectives. Eur Aquacult Soc, Bredene

Caddy JF (ed) (1983) Advances in assessment of world cephalopod resources. FAO Fish Tech Pap 231

Crutchfield J (1965) The fisheries: problems in resource management. Univ Washington Press, Seattle

Everson I (1977) The living resources of the Southern Ocean. GLO/SO/77/1. FAO, Rome

FAO (1980) ACMRR Working Party on the scientific basis of determining management measures. FAO Fish Rep 236

FAO (1985) Prevention of post-harvest food losses. FAO Training Ser 10

FAO (1987) Review of the state of world fishery resources. FAO Fish Circ 710, Rev 5

Gerlach SA (1981) Marine pollution: diagnosis and therapy. Springer, Berlin Heidelberg New York

Gjösaeter J, Kawaguchi K (1980) A review of the world resources of mesopelagic fish. FAO Fish Tech Pap 193

Gulland JA (ed) (1977) Fish population dynamics. John Wiley & Sons, London

Gulland JA (1980) Some problems of the management of shared stocks. FAO Fish Tech Pap 206

Gulland JA (1983) World resources of fisheries and their management. In: Kinne O (ed) Marine ecology. John Wiley & Sons, London New York, pp 839-1060

Harden-Jones FR (ed) (1974) Sea fisheries research. Elek, London

Johnston R (ed) (1976) Marine pollution. Academic Press, London

Kapetsky JM (1981) Some considerations for the management of coastal lagoons and estuarine fisheries. FAO Fish Tech Pap 218

Kirk R (1987) A history of marine fish culture in Europe and North America. Fishing News (Books), Farnham

Laevastu T, Favorite F (1988) Fishing and stock fluctuations. Fishing News (Books), Farnham

Larkin PA, Wilimovsky NJ (1973) Contemporary methods and future trends in fishery management and development. J Fish Res Board Can 30: 1948-1957

Lee A (1982) How sick are the oceans? International Yearbook and statesmen who's who

Okutani T (1977) Stock assessment of cephalopod resources fished by Japan. FAO Fish Tech Pap 173

Petersen CGJ (1900) What is overfishing? J Mar Biol Assoc UK 6: 587-594

Pillay TVR (ed) (1967/1968) Proc FAO World Symp Warm-water pond fish culture, Rome 18-25 May 1966. FAO Fish Rep 44, vols 1-6

Pillay TVR (1973) The role of aquaculture in fishery development and management. J Fish Res Board Can 30: 2202-2217

Pillay TVR (1990) Aquaculture: Principles and Practices. Fishing News (Books) Blackwell, Oxford

Pillay TVR, Dill WA (eds) (1979) Advances in aquaculture. Fishing News (Books), Farnham

Robinson MA (1982) Prospects for world fisheries to 2000. FAO Fish Circ 722, Rev 1

Robinson MA (1984) Trends and prospects in world fisheries. FAO Fish Circ 772

Ruivo M (ed) (1972) Marine pollution and sea life. Fishing News (Books), London

Saetersdal G (1980) A review of past management of some pelagic stocks and its effectiveness. Rapp P-v Réun Cons int Explor Mer 177: 505-512

Sahrhage D (1981) Lebende Ressourcen des Meeres. Naturwiss Rundsch 34 (4): 137-142

Sahrhage D (1989) Antarctic krill fisheries: potential resources and ecological concerns. In: Caddy JF (ed) Marine invertebrate fisheries: their assessment and management. John Wiley & Sons, New York, pp 13-33

Saila SB (1983) Importance and assessment of discards in commercial fisheries. FAO Fish Circ 765

Saville A (ed) (1980) The assessment and management of pelagic fish stocks. Rapp P-v Réun Cons int Explor Mer 177: 1-517

Suda A (1973) Development of fisheries for unconventional species. J Fish Res Board Can 30: 2121-2158

Tanaka S (1960) Studies on the dynamics and the management of fish populations. Bull Tokai Reg Fish Res Lab 28: 1-200

Tiews K (1975) Pollution of the sea and its significance for fisheries. Applied sciences and development 5. Inst Sci Coop, Tübingen

Index

Printing: Mercedesdruck, Berlin
Binding: Buchbinderei Lüderitz & Bauer, Berlin

SPRINGER NATURE

GPSR Compliance

The European Union's (EU) General Product Safety Regulation (GPSR) is a set of rules that requires consumer products to be safe and our obligations to ensure this.

If you have any concerns about our products, you can contact us on ProductSafety@springernature.com

In case Publisher is established outside the EU, the EU authorized representative is:

Springer Nature Customer Service Center GmbH
Europaplatz 3
69115 Heidelberg, Germany

FSC
www.fsc.org
MIX
Papier | Fördert
gute Waldnutzung
FSC® C083411

Kolibri 360 GmbH
Ferdinand-Jühlke-Straße 7
99095 Erfurt, Deutschland
produktsicherheit@kolibri360.de